AF806075

Global Perspectives on Health Geography

Series Editor

Valorie Crooks, Department of Geography, Simon Fraser University,
Burnaby, Canada

Global Perspectives on Health Geography showcases cutting-edge health geography research that addresses pressing, contemporary aspects of the health-place interface. The bi-directional influence between health and place has been acknowledged for centuries, and understanding traditional and contemporary aspects of this connection is at the core of the discipline of health geography. Health geographers, for example, have: shown the complex ways in which places influence and directly impact our health; documented how and why we seek specific spaces to improve our wellbeing; and revealed how policies and practices across multiple scales affect health care delivery and receipt.

The series publishes a comprehensive portfolio of monographs and edited volumes that document the latest research in this important discipline. Proposals are accepted across a broad and ever-developing swath of topics as diverse as the discipline of health geography itself, including transnational health mobilities, experiential accounts of health and wellbeing, global-local health policies and practices, mHealth, environmental health (in)equity, theoretical approaches, and emerging spatial technologies as they relate to health and health services. Volumes in this series draw forth new methods, ways of thinking, and approaches to examining spatial and place-based aspects of health and health care across scales. They also weave together connections between health geography and other health and social science disciplines, and in doing so highlight the importance of spatial thinking.

Dr. Valorie Crooks (Simon Fraser University, crooks@sfu.ca) is the Series Editor of Global Perspectives on Health Geography. An author/editor questionnaire and book proposal form can be obtained from Publishing Editor Zachary Romano (zachary.romano@springer.com).

Bikash Dutta • Manas Das • Sutapa Rath
Sutapa Das

Using GIS to Identify Opportunities for Improved Health in Urban Slums

The Applied Case of Midnapore, India

Springer

Bikash Dutta
Department of Geography
Nistarini College
Purulia, West Bengal, India

Sutapa Rath
Department of Geography
Vidyasagar University
Midnapore, West Bengal, India

Manas Das
Department of Mining Engineering
Indian Institute of Technology (Indian School of Mines) Dhanbad
Dhanbad, Jharkhand, India

Sutapa Das
Department of Economics
Vidyasagar University
Midnapore, West Bengal, India

ISSN 2522-8005 ISSN 2522-8013 (electronic)
Global Perspectives on Health Geography
ISBN 978-3-031-89616-3 ISBN 978-3-031-89617-0 (eBook)
https://doi.org/10.1007/978-3-031-89617-0

© The Editor(s) (if applicable) and The Author(s), under exclusive license to Springer Nature Switzerland AG 2025

This work is subject to copyright. All rights are solely and exclusively licensed by the Publisher, whether the whole or part of the material is concerned, specifically the rights of translation, reprinting, reuse of illustrations, recitation, broadcasting, reproduction on microfilms or in any other physical way, and transmission or information storage and retrieval, electronic adaptation, computer software, or by similar or dissimilar methodology now known or hereafter developed.
The use of general descriptive names, registered names, trademarks, service marks, etc. in this publication does not imply, even in the absence of a specific statement, that such names are exempt from the relevant protective laws and regulations and therefore free for general use.
The publisher, the authors and the editors are safe to assume that the advice and information in this book are believed to be true and accurate at the date of publication. Neither the publisher nor the authors or the editors give a warranty, expressed or implied, with respect to the material contained herein or for any errors or omissions that may have been made. The publisher remains neutral with regard to jurisdictional claims in published maps and institutional affiliations.

This Springer imprint is published by the registered company Springer Nature Switzerland AG
The registered company address is: Gewerbestrasse 11, 6330 Cham, Switzerland

If disposing of this product, please recycle the paper.

To
My beloved mother and father
Late Dulirani Dutta and Ranjit Dutta
&
Dr. Utpal Roy
Professor
Dept. of Geography
University of Calcutta, India

Preface

A coping mechanism is exclusively associated with health and well-being that prevails in every person, including healthy environments, responsive parenting, a sense of belonging, healthy activities, resilience, and treating illness. Therefore, health coping mechanisms are the practices and activities of a single day that make health better and prevent diseases, which are directly allied with physical and mental health. Despite the development of urban areas in developing countries, the urban poor suffering from malnutrition, bad hygienic behavior, disease epidemics, etc. are pushed into a vicious cycle of health risk and socioeconomic conditions. Most researchers have addressed coping with climate change and their livelihoods or coping with disaster incidents. This study conducted a cross-sectional survey of slum neighborhoods in Midnapore municipality to appraise the different problems and challenges slum dwellers face with resulting health coping mechanisms. It also investigates the availability of overall health infrastructure and the perception of slum dwellers regarding the provision, utilization, and accessibility of the healthcare system. Likewise, this study also analyzes the factors affecting the health coping mechanism of slum dwellers. The primary data were collected through a well-structured questionnaire schedule to explore the objectives. Secondary data was collected from governmental organizations and Midnapore municipality, and geospatial data of different periods was collected from the Earth Explorer website of the USGS. The data have been analyzed using statistical tools in SPSS v26 and the GIS algorithm.

The study finds that around 50% of households face moderate to high level of social deprivation and multidimensional poverty, resulting in moderate to poor quality of life. Limited hygiene practices contribute to increased health risks and vulnerability to health shocks with minimal preparedness. Due to the lack of availability and proper services from health care facility centers, dwellers mainly prefer government hospitals and quack doctors for their foregone care with moderate to high satisfaction. Meanwhile, there is a serious lack of government healthcare services. The study also finds that the health coping mechanisms of the women, child, adolescent, and working group are not at a satisfactory level due to lack of awareness and housing accommodation, a backward socioeconomic and educational status, lack of proper food and nutrition, unemployment, high healthcare expenses, lack of basic amenities and healthcare services, etc. A strong correlation exists between personal hygiene and health shock incidence, quality of life, and health security. Moreover, this study explored a strong negative correlation between health shocks and the whole year's health coping strategies (WYHCS). Though the central and state governments have implemented different programs and policies for slum upgrading, the possibilities for improving livelihood systems through individual action are very limited. Therefore, more refinement of the data structure with advanced geospatial techniques (Health GIS) may be appended to high exactitude over a spatial frame.

This book represents different geospatial and statistical modeling for a better understanding of the health coping behavior of the urban poor of Midnapore municipality at a micro-level scale. Also, this volume presents separate methodologies with respect to previous case studies, which increases the research interest in health coping behaviors. We establish and discuss how environmental and socioeconomic variables and health are related using modern technologies. Using geospatial and statistical techniques would be useful for decision-makers, practitioners, academics, and researchers for their own studies at any scale. Finally, all outcomes will benefit the readers in solving their own health-related problems with respect to this study.

Purulia, West Bengal, India — Bikash Dutta
Dhanbad, Jharkhand, India — Manas Das
Midnapore, West Bengal, India — Sutapa Rath
Midnapore, West Bengal, India — Sutapa Das

Acknowledgment

The research work *Using GIS to Identify Opportunities for Improved Health in Urban Slums*: *The Applied Case of Midnapore, India* has not been an isolated effort. Many persons have contributed to the success of our endeavor, which we could not have completed without their loyalty, vision, and friendship. First, we thank Vidyasagar University for allowing us to pursue this research work. On this occasion, we express our sincere gratitude to all the faculty members of the Department of Geography, Vidyasagar University, and research scholars in the departments who helped us to move the research forward with their valuable advice during the research work. We are also thankful to Dr. Indranil Maity, Mr. Pradip Samanta, and Mr. Swapan Sau for their immense support, without which the surveys could not have been completed successfully.

We express our heartiest gratitude to all the people, especially those living in slum neighborhoods of Midnapore town, where we have carried out field surveys over the years. They have provided information, extended their full cooperation, and helped us a lot during the fieldwork. We would like to thank Midnapore Municipality for providing us with valuable information regarding slum neighborhoods in the Midnapore municipal area and for unbelievable cooperation.

Valuable editorial advice, including systematic guidance from Dr. Valorie Crooks, the Publishing Editor of Global Perspectives on Health Geography, Springer Nature, continuously helped us to enhance the content and improve the eminence of this book. We are obliged to Dr. Zachary Romano, Project Coordinator of Global Perspectives on Health Geography, Springer Nature, who took responsibility for the project organization and oversaw the entire production process.

We oblige our parents and all family members from the bottom of our hearts. Their love, warmth, constant support, and inspiration have oriented us toward research work, providing a platform and confidence to overcome all hurdles and face challenges. Finally, we are thankful to everyone who has encouraged us to fulfill the dream of our research work in due time.

Bikash Dutta
Manas Das
Sutapa Rath
Sutapa Das

Contents

Abbreviations

AHP	Analytical hierarchy process
AIDS	Acquired immune deficiency syndrome
ANC	Antenatal care
ANM	Auxiliary nurse-midwife
ARI	Acute respiratory infection
ARSH	Adolescent reproductive and sexual health
ASHA	Accredited Social Health Activist
AWW	Anganwadi worker
BCG	Bacille Calmette-Guerin
BPL	Below poverty level
CBD	Central business district
CHC	Community health centers
COVID-19	Coronavirus 2019
DBMS	Database management system
DH	District hospitals
DPT	Diphtheria–pertussis–tetanus
EFS	Emotion-focused strategy
EMI	Equated monthly instalment
ESRI	Environmental Systems Research Institute
ETM	Enhanced thematic mapper
FGD	Focus group discussion
FTS	First track supervisor
GCS	Geographic coordinate system
GIS	Geographic information system
GPS	Global Positioning System
GSS	Geospatial site selection
HCF	Health center facilities
HCR	Headcount ratio
HCS	Health coping strategies
HHW	Honorary Health Worker
HIB	Haemophilus influenzae type b

HIV	Human immunodeficiency virus infection
HRI	Health risk index
HSCI	Health shock coping index
HSI	Health security index
HVRI	Hazards and Vulnerability Research Institute
IBM	International business machines
ICDS	Integrated Child Development Services
ICT	Information and communication technology
IDW	Inverse distance weight
JSY	Janani Suraksha Yojana
KIOSK	Kommunikasjon Integrert Offentlig Service Kontor
KMO	Kaiser–Meyer–Olkin
LST	Land surface temperature
MC	Medical colleges
MCDM	Multi-criteria decision making
MCDSS	Multi-criteria decision support system
MLE	Maximum likelihood estimation
MNLR	Multinomial logistic regression
MoHFW	Ministry of Health and Family Welfare of India
MoSPI	Ministry of Statistics and Program Implementation
MPI	Multidimensional poverty index
NDVI	Normalized difference vegetation index
NFHS	National Family Health Survey
NGO	Non-governmental organization
NHP	National Health Policy
NNR	Nearest neighbor ratio
NUHM	National Urban Health Mission
OLI	Operational Land Imager
OLS	Ordinary least squares regression
PC	Principal component
PCA	Principal component analysis
PFS	Problem-focused strategy
PHC	Primary health centers
PHCF	Primary healthcare facilities
PHI	Personal hygiene
PHS	Primary healthcare sites
PMAY	Pradhan Mantri Awas Yojana
QoLI	Quality of life index
RCH	Reproductive and Child Health
RCHC	Reproductive and Child Health Care
RDI	Recommended dietary intake
RSBY	Rashtriya Swasthya Bima Yojana
SCDAB	Slum Community Development Authority Board
SD	Standard deviation
SDH	Sub-district hospitals

SE	Standard error
SHC	Sub-healthcare centers
SHG	Self-help group
SHP	Sub-health post
SoDI	Social deprivation index
SoVI	Social vulnerability index
SPSS	Statistical Packages for Social Sciences
STIs	Sexually transmitted infections
UFWC	Urban Family Welfare Center
UHND	Urban health and nutrition day
UHRC	Urban Health Resource Center, Govt. of India
UN	United Nations
UN DESA	United Nations Department of Economic and Social Affairs
UN-Habitat	United Nations Human Settlement program
UNICEF	United Nations Children's Fund
USAID	United States Agency for International Development
USGS	United States Geological Survey
VBA	Microsoft Visual Basic Software
WBCPB	West Bengal Pollution Control Board
WGS	World Geodetic System
WHO	World Health Organization
WYHCI	Whole Year Health Coping

Chapter 1
Health Coping Mechanism: A Global Challenge

1.1 Introduction

Coping is part of health and well-being that prevails in every person, and it includes healthy environments, responsive parenting, a sense of belonging, healthy activities, resilience, and the treatment of illness. Folkman and Moskowitz (2004) describe "coping as the thoughts and behaviors mobilized to manage internal and external stressful situations." Venner (1988) defines "coping as subconscious or unconscious adaptive responses that aim to reduce or tolerate stress." Coping means a conscious effort to solve personal and interpersonal problems to minimize or tolerate environmental stresses. In psychology, coping mechanisms are commonly called coping strategies or coping skills. Coping generally refers to adaptive (constructive) strategies that reduce stress. Generally, it is well known that high- and middle-income group people in urban areas are adopting some daily strategies (e.g., diet control, healthy food intake, physical exercise, health insurance, etc.) to improve their health status. Income, education, and health consciousness are more prevalent in the high- and middle-income groups than in the urban counterpart, i.e., the urban poor. Due to lack of income, education, health consciousness, etc., urban poor people are ignoring their health problems. Furthermore, basic amenities and services are key factors that control health coping mechanisms (Adam, 2013). Although health structures and services prevail in urban areas, poor people intentionally or unintentionally adopt healthy coping strategies, either local folk (common adaptive methods) or low-cost, long-term methods.

Worldwide urbanization has been faster than before, with overpopulation within limited space accompanied by sustainability challenges. Urbanization is changing and restructuring the urban area's physical and socioeconomic landscape. The modification of morphology in the urban area creates environmental problems (D'Souza, 2003) due to the faster growth rate of the urban population. At present, more than 56.20% of people live in towns and cities worldwide, and it is estimated that by

© The Author(s), under exclusive license to Springer Nature Switzerland AG 2025

B. Dutta et al., *Using GIS to Identify Opportunities for Improved Health in Urban Slums*, Global Perspectives on Health Geography,
https://doi.org/10.1007/978-3-031-89617-0_1

2050, they will be 68.4% (UN DESA, 2018). The growth rate of the world's urban population is 1.90% from 2015 to 2020, which will increase to 1.73% between 2020 and 2025 (UN DESA, 2018). However, Africa (43.5%) and Asia (51.1%) are the least urbanized, far from developed countries, and even below the average world urbanized population (56.2%) (UN DESA, 2018). However, it is estimated that by 2050, the urban population in Asia and Africa will be 58.9% and 66.2%, respectively (UN DESA, 2018). The estimated faster urban population growth is in Asia and Africa rather than urban development (Sánchez-Martín et al., 2019). Thus, the level and pattern of urbanization differ between and within the regions. The future scenario is that seven urban dwellers out of ten and 50% of the world's urban population will live in Africa and Asia (UN DESA, 2018). At the same time, the faster population growth rate is also shown in meta- and megacities, which have more than ten million populations, as well as small towns and cities worldwide (UN DESA, 2018). In the developing world, millions of cities and emerging cities are increasing by absorbing more population, acting as a significant center of urban growth rather than development. The larger absorption of the population in the cities and towns in developing countries is consequently faced with a lack of resources, housing inadequacy, clean water supply, sanitation, waste collections, and jeopardizing healthcare services (Sánchez-Martín et al., 2019). However, continuous urbanization is assumed to bring economic gain, which significantly impacts the health outcomes of urban dwellers. Contrary to this assumption, people living in the urban areas of developing countries face inequalities in basic amenities and healthcare services, which are detrimental to personal hygiene, quality of life, and environmental standards.

Thus, we are witnessing an exponentially expanding city population with a lack of accessibility to basic needs of life in addition to deprivation and poverty. The inequalities are proliferating, evident by the emergence of agonized neighborhoods, the so-called slums in developing countries. WHO defines "a slum household as a group of individuals living under the same roof lacking one or more of the following conditions: access to improved water, access to improved sanitation, sufficient living area, the durability of housing and security of tenure" (UN-Habitat, 2003, page 10). According to the Census of India (2011), "all clusters of 20–25 or more households, having no roofs or non-concrete roofs, and not having any facility of drinking water, toilets or drainage will be considered as slums" (Census of India, dated: 20.01.2010, pages 2–3). Hence, a slum manifests deprivation and poverty concerning decaying infrastructures. In slum neighborhoods, there is a vast gap between demand and supply in the context of basic amenities and healthcare services. Slum dwellers are highly exposed to health risks in their surrounding environment.

By 2050, out of the six billion population, three billion will live in slums. By 2030, the United Nations estimated that one in four people will live in a slum or informal settlement. Agenda 21 in the Conference on Environment and Development by the United Nations (1992) highlighted poverty and accessibility to essential social services and provided practical mechanisms for poor people to fight against poverty. Similarly, the World Conference on Human Rights (2019) exhibits how the urban environment significantly impacts health risks to the urban poor. World conferences have emphasized the issues related to the quality of life by ensuring the improvement of healthcare services, reproductive health, and overall health and well-being. Although disease prevalence has shifted from communicable to non-communicable, problems persist in the slums of the urban areas, and slum dwellers are affected by the brunt of diseases. These conditions, especially for slum dwellers, are due to inequalities among the resources and services in the urban area (UN-Habitat, 2008). In this context, socioeconomic status, environmental factors, and accessibility are significant factors in evaluating the overall situation of the urban poor.

Many studies show that health facilities in the urban areas of developing countries are generally better than their rural counterparts, even though the urban poor face higher health risks than the average national resident and sometimes the rural poor (Smith & Mark, 2011). The reason behind this disparity is the deficiency of basic capabilities of poor households, which includes poor living conditions, hazardous work, food insecurity, illiteracy, unhealthy sanitary and dietary practices, high fertility, vulnerability to catastrophic illness, and poor coping mechanisms. Consequently, the urban poor are highly exposed to environmental health hazards due to poor living conditions. Indeed, people experiencing poverty who are forced to live in the hazardous nature of the location in the urban area due to low rent and lesser chance of eviction comparatively face higher health risks (Arefin et al., 2010). The relationship between poverty and ill health is multifaceted and bidirectional in that poverty can affect poor health; at the same time, ill health can exacerbate poverty. Chronic poverty also increases the health risks of the poor with exposure to different environmental hazards and lack of access to essential services; in contrast, ill health tends to intensify poverty due to low educational levels, loss of income owing to sickness, and increased spending on healthcare (Fig. 1.1) (Amdework, 2008). In this environment, healthcare is complex, driven by poverty, environmental problems and challenges, and formal healthcare infrastructures (Naydenova et al., 2017). Therefore, living in such conditions lacking basic amenities has been hazardous and despondent to such communities' health, safety, and well-being (Hardoy et al., 1997).

India has been experiencing rapid urbanization from 2001 to 2011 and onward. During the same period, the first absolute increase in urban population exceeded the rural population, and an additional 2532 census towns were also increased (Census of India, 2011). In India, it is estimated that by 2050, the absolute number of rural populations will decrease with a negative growth rate (−1.06), while the absolute number of urban populations will increase with a positive change of growth rate (+1.54) (UN DESA, 2018). Rapid urbanization would overburden the urban infrastructure, amplifying demand for basic amenities and services such as housing, water supply, sanitation, sewerage, solid waste management, etc. Poor slum dwellers are forced to live in unhygienic conditions due to the inadequacies of basic amenities and services, which enhance the vulnerability and resulting health shocks incidence. World Bank (2021) reported that 0.9 physicians were available per 1000 people in India in 2018, below the WHO standard of 1:1000 ratios. This condition is due to the ill-equipped primary healthcare system at the regional and state levels.

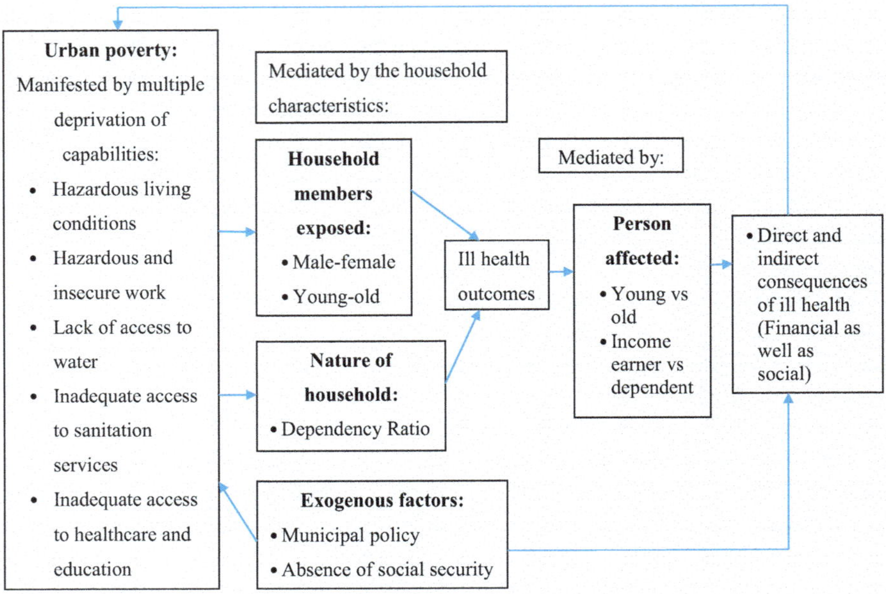

Fig. 1.1 Linkage between Ill-health and chronic poverty (after E. Amdework, 2008, adopted from Grant, 2005)

Although the government of India has decided to meet the requirement as per the WHO standard by 2031, there is still a difference in healthcare facilities between slum and non-slum dwellers. Hence, to make an India with slum-free urban development, the provision of housing and healthcare has become the foremost essential elements.

The government of India has taken different initiatives, programs, and policies to tackle the deterioration of healthcare and well-being in urban areas, such as the Pradhan Mantri Awas Yojana (PMAY), National Urban Health Mission (NUHM), etc. Thus, it is necessary to assess the complex relationship between urban development and health well-being and community participation in a comprehensive way, as both are important in a nation's development agenda worldwide. Rapid urbanization profoundly impacts the environment, affecting the urban poor, mainly due to inadequate basic necessities and well-being services. This led the urban poor into the vicious cycle of poverty and created poor neighborhoods, which in turn affected their health, well-being, and health shock incidence. Furthermore, poor neighborhoods are again unable to make rational decisions to overcome the poor situation by accessing the basic needs of their daily livelihood and healthcare. In this regard, it is essential to examine the health coping mechanism of the urban slum dwellers in the emerging city.

The health of the urban poor is worse than that of the rural population. The features of each city and town may vary by local context; general urban health and social challenges include overcrowding; indoor air pollution; rising levels of risk factors like tobacco use, unhealthy diet, physical inactivity, and the harmful use of alcohol; road traffic injuries; inadequate infrastructure and transport facilities; poor solid waste management systems; and insufficient access to health facilities in slum areas. Most of these cities also face various health challenges, such as communicable diseases, non-communicable diseases, maternal and child health problems, environmental hazards, and risks of remerging and emerging diseases. As per the UHRC (Urban Health Resource Centre, Govt. of India), in urban slums of India, one in every ten children do not breathe to see their fifth birthday, and only 58% of slum children do not receive all the recommended vaccinations. Over half (56%) of delivery takes place at home in slums, which may cause serious risk to both mother and newborn. Moreover, inadequate sanitation in slums may also contribute to the high burden of diseases; for example, two-thirds of slum households do not have access to toilets, and almost 40% do not have piped water supply at home. In India, healthcare facilities focus more on rural areas than urban areas, as urban areas have an organizational structure with government hospitals and private healthcare centers. From this perspective, there is a considerable deficiency of primary government healthcare centers, particularly for low-income people in urban areas. Additionally, the rapid growth of the urban population has overburdened the existing healthcare system; it does not maintain its carrying capacity, and the cost and accessibility of private healthcare facilities are out of reach for most of the poor urban dwellers (Lahon, 2017). In those circumstances, our study seeks to determine health coping strategies that urban poor dwellers have been adapting to meet their health improvement and well-being.

1.2 Concept and History of Coping Mechanism

The coping mechanism is a psychological term commonly known as coping strategies or coping skills. The concept of coping mechanisms or strategies invariably relates to the circumstances of survival and threat. There is no standard coping mechanism that has been identified due to its changing nature across the world, i.e., coping mechanisms vary from culture to culture, place to place, even in the community, social group, household, gender, age, economic status, season, and time in history (WHO, 1999). People's coping mechanisms depend on experience deeply influenced by sociocultural factors, irrespective of their level of deprivation, caste, religion, and social status. Coping is the capacity to respond and recover from a stressful situation, whether natural or man-made (WHO, 1999; Zamble & Gekoski, 1994; Papathanasiou et al., 2015; Folkman & Lazarus, 1980; Fontana, 1996; Costa & Pinto, 2017; Lim et al., 2013). Coping strategies or mechanisms are remedial activities adopted by people whose existence and living are compromised or vulnerable. WHO (1999) defines "coping mechanisms" as a range of activities that represent the commonality on one side, i.e., community self-help initiatives, but, on the other side, represent the misery and low standard of life, i.e., migration, begging, child labor, violence, prostitution, even selling organs, etc. The concept of coping mechanisms is intrinsically associated with poor people because poor people cope with stressful situations due to low living standards, while rich people manage stressful situations.

Historically, the concept of coping mechanisms was developed from the conception of S. Freud's (1926) "classic" defense mechanisms, which were based on a state-oriented macro-analytic approach. The term was first used by psychologists Abraham Maslow and Bela Mittelmann in their book *Principles of Abnormal Psychology* (1941). However, Lazarus and Folkman (1984) have expanded the concept in their most influential work on "Transactional Theories of Stress and Coping." Lazarus and Folkman (1984) define a coping mechanism as the "cognitive and behavioral efforts to manage specific external and internal demands that are appraised as exceeding the person's resources." Lazarus and Folkman (1984) mainly focused on two types of coping strategies: firstly, problem-focused coping represents the efforts used to change stressful situations through problem-solving, decision-making, and direct action and, secondly, emotion-focused coping exhibits strategies to control cognitive distress in the actual situation (Folkman & Lazarus, 1985). Moreover, Carver et al. (1989) designed three types of coping mechanisms. The first is problem-focused coping, which encompasses active coping, planning, restraint coping, seeking social support for instrumental reasons, and suppression of competing activities. The second is emotion-focused coping involves strategies like positive reframing, acceptance, humor, religious belief, and emotional social support. The third one is dysfunctional coping, which includes focusing on and venting emotions, denial, behavioral disengagement, mental disengagement, and alcohol/drug use. Lazarus developed two approaches: (a) trait-oriented versus state-oriented and (b) micro-analytic versus macro-analytic approaches. The trait-oriented (or dispositional) strategy is the early identification of individuals' stressful situations with a lack of coping resources aimed at primary prevention. At the same time, state-oriented strategy represents actual coping with more general objectives focusing on individual coping efficiency based on certain coping efforts or variables of adaptation outcomes, e.g., health status aims to improve coping efficacy. In contrast, micro-analytic approaches focus on specific coping strategies to a greater extent, while macro-analytic approaches are more fundamental coping strategies with a higher level of abstraction (Krohne, 2002).

1.3 Review of the Past Work

1.3.1 Urbanization and Slum

Studies published in national and international journals and periodicals conducted by researchers at different periods have been reviewed. They are relevant to healthcare accessibility and health coping mechanisms of slum dwellers. Studies on the health coping mechanisms of the slum dwellers are limited, and most of the studies are focused only on living and health problems, disease prevalence, etc. among the slum dwellers rather than focusing on what they are doing during their stressful situations, the so-called coping strategies. Therefore, different studies have neglected the health coping mechanism of slum dwellers. Lack of daily income, food, housing, and access to healthcare and other essential services, poverty, and lack of awareness have an immense impact on the health-coping behavior of the urban poor.

One of the most remarkable features of the twenty-first century has been the spectacular growth of the urban population in the world. Developing countries, particularly, have experienced rapid urbanization and the mushrooming of huge metropolises, cities, and towns. By 2050, the urban population in Asia and Africa will be 58.9% and 66.2%, respectively (UN DESA, 2018). India has been experiencing mighty growth in its urban population. According to the Census of India (2011), India holds approximately 31% urban population to the total population. It is estimated that, by 2030, the urban population will be 40.1% and, by 2050, it will reach 52.8% (UN DESA, 2018). India grew its population from 2001 to 2011 with an average annual growth rate of 1.76%, but the urban population grew by 3.18%. Slum populations rose by 2.51% (to the total slum population), but the urban population's growth has slowed in the megacities considerably during the last decade (2001–2011). These statistics indicate that the population is primarily concentrated in a million cities, cities, and small towns rather than metropolitan and megacities; for example, 38.1% of the slum households to the total slum households are living in 46 million cities, and the remaining 61.9% in other cities probably in a small town. In India, 17% of the population lived in slums, 30% of the urban population in 2001 (Census of India, 2001), which has doubled within 10 years from 43 million in 2001 to 93 million in 2011, with a growth rate of 5% (Census of India, 2011). The Census of India (2011) data also shows that the population of four metropolitan cities was much higher: 41.3% in Mumbai, 29.6% in Kolkata, 28% in Chennai, and about 15% in Delhi. *The Times of India* (20 August 2013) reported that by 2017, the slum population had reached 104 million, around 9% of the projected national population of 1.28 billion. About 35.2% of the population living in slums (to the total urban population) in India in 2018 was reported by the World Bank (World Bank, 2021). Although UN-Habitat (2020) reported that the slum population of India has substantially reduced from 35.4% in 2016 to 34.8% in 2018, the absolute number increased from 155.434 million in 2016 to 160.33 million in 2018 (UN DESA, 2018).

Thus, it is clear that the population in urban areas is increasing steadily worldwide; most of them live in urban slums, informal surroundings, or footpath shelters. By 2025, the world population is expected to reach 8.1 billion, with most of this growth in developing countries (UN, 2013). Nowhere else will this growing urban population be concentrated in existing severely populated informal settlements of cities known as slums. Slums epitomize a significant transformation of urban areas' physical and socioeconomic environment and human health (Riley et al., 2007). Nearly one-third of the urban population of India lives in slums, which are characterized by overcrowding, poor hygiene and sanitation, and the absence of proper civic services (Kantharia, 2010). A few studies have suggested that significant factors contributing to the decision-making of slum communities include accessibility of healthcare from a public and private sector with distance, family decision-making structure, perceived quality of care, etc.

1.3.2 Milieu of Urbanization and Slum in India

Gupta et al. (2008) highlighted that the proportion of urban poverty to the total poverty in India has gradually increased; for instance, the urban population increased from 15% in the early 1970s to about 25% in the mid-1990s and 34.4% in 2004–2005. According to the Tendulkar method (Planning Commission of India, 2013), in the years 2011–2012, the urban population below the poverty line of West Bengal was 14.66%. The Census of India (2011) revealed that, in West Bengal, the total slum reported towns is 122, in which the population of notified slums reported 48,918, recognized slums were 37,03,852, and identified slums reported 26,65,824. The total slum population in 2011 was 6,418,594 compared to the 4,115,980 slum population in the 2001 census, i.e., the slum population increased by 2,302,614 within 10 years (about 6% per annum to the total urban population) and about 1% to the total slum population of India (Census of India, 2011).

Traditionally, slums refer to the socioeconomic and environmentally deprived housing areas in the parts of the town and city. The definition of "slum" probably derives from the Irish phrase lomé (pron. S'lum ae), meaning "it is a bleak or destitute place." In their legislation, the government of India (1988) defined slums as "environmentally and structurally deficient areas." In the 1981 Census of India, the Slum Area Act 1956 defines slums as "areas where buildings are unfit for human habitation for reasons such as dilapidation, overcrowding, faulty arrangement of streets, and lack of ventilation, light, sanitation facilities, etc." The Census of India in 2001 adopts an obvious definition for designating an urban settlement as a "slum"—"a densely packed area with a population of around 300 (60–70 households) living in a congested, dense, and unhygienic environment." The housing infrastructure is mostly kutcha or semi-pucca without proper sanitary or drinking water facilities. Populations leave their ancestral homes in rural areas and migrate to rented areas in towns and cities with visions of better education, employment, and quality of life. However, the present set of slum conditions poses a nightmare for them. Slums, in most circumstances, are unauthorized colonies developed along with hazardous sites with a lack of formally established basic facilities like water, electricity, sanitation, health services, etc. (Ray, 2017). According to the Ford Foundation, "a slum is a residential area in which the housing is so deteriorated and substandard or as unwholesome as to be a menace to the health, safety, morality or welfare of the occupants" (Rathor, 2003). According to the Census of India (2011), a slum is "all clusters of 20–25 or more households, having no roofs or non-concrete roofs, and not having any facility of drinking water, toilets or drainage will be considered slums."

Urban poor live in slums, which are characterized as overcrowded, often polluted, and lacking basic civic amenities. Most of the urban poor are engaged in the informal sector with the continuous threat of eviction, removal, and almost lack of social security. Demographic trends of India show that the overall population growth rate was stabilized at 2% from 1991 to 2001, where the urban population growth rate was 3% and the slum growth rate doubled to 5–6% in the same period (Agarwal & Taneja, 2005). The decadal growth rate from 2001 to 2011, the slum population in all towns grew by 25.1%, with slum households comprising approximately 37.1% of the total (Census of India, 2011). This notable growth of urban slums in India reflects the urbanization of poverty and the reorganization of urban areas, which gives form to budding "risk spaces" (Fitzpatrick & LaGory, 2010). Indeed, "urban slum areas habitually display low levels of education, lack of economic opportunities, poor health outcomes and low life expectancy rates relative to non-slum urban populations" (UN-Habitat, 2006). The outlines of slum conditions from the *UN-Habitat World Cities Report 2020* are as follows:

- Currently, 1 billion populations are living in slums and informal settlements, and 1.6 billion or 20% of the world's population live in inadequate, crowded, and unsafe housing.
- The growth of slums, particularly in developing regions, is due to a lack of urban agenda such as rapid urbanization; ineffective planning; lack of affordable housing for the urban poor; dysfunctional urban, land, and housing policies; a scarcity of housing finance; poverty; and low incomes.
- Populations of informal settlements and slums engaged in informal sectors have been more economically vulnerable.
- Populations of informal settlements and slums have been more economically vulnerable to climate change and natural disasters than the other settlements.
- Slums are characterized by the most stable faces of poverty, inequality, exclusion, and deprivation.
- More than one billion people live in slums and informal settlements, of which 50% live in East, Southeast, Central, and South Asia and 23% in Sub-Saharan Africa.
- Empirical analysis estimated that a 1% growth in urban population will increase the prevalence of slums in Africa and Asia by 2.3% and 5.3%, respectively (Arimah, 2011).
- Three hundred million slum children are disadvantaged and challenged due to a detrimental built environment, lack of affordable housing, high costs of living, and inadequate distribution and access to services.
- Sixty percent of slum households live in densely poor, serviced, and hazardous locations, for example, Dharavi slum in Mumbai of India (population density of 270,000 km^2) and Cox's Bazar in Bangladesh (with a population of about three million).
- In developing countries, 75% of the urban poor lack handwashing facilities with soap and water. In 2017, globally, households with 3 billion populations had no handwashing facilities, while 1.6 billion lacked handwashing facilities with soap and water, and 1.4 billion people still do not have handwashing facilities.

- Slum dwellers must cope with inadequate access to potable water, poor sanitation, overcrowding, poor-quality housing in hazardous locations, insecure tenure and risk of eviction, food insecurity, malnutrition, poor health, unemployment, and stigmatization. These circumstances make them highly vulnerable to COVID-19 and other pandemics. Therefore, slum dwellers are at the most risk for needed actions such as economic, environmental, and social infrastructure for this particular population.
- Urban areas in India have less than 4% of government primary healthcare facilities.
- On average, there is only 1 health facility for every 1,50,000 people in India's largest cities.

The Census of India (2011) **revealed that:**

- Twelve percent of the urban slum population lived in class I cities, which is higher in the larger cities.
- Seventeen percent of the urban population does not have a "pucca" house, while 32% of urban households live in a single room, and more than 3% do not have a private room.
- Roughly 19% of households have no latrine facilities within their premises in the urban area. In comparison, around 10% have no improved drinking water or sanitation facilities, and about 25% of households do not have bathing places within their premises.

Pandey and Sengupta, in their book *State of India's Environment, 2019 (CSE)*, revealed that:

- In India, one in every six urban people is living in slums unfit for human habitation. For example, among three urban households in Andhra Pradesh, one lives in a slum.
- In Odisha, among ten slum households, nine are living without a drainage connection (90.6%), and 64.9% of slum households do not have access to treated tap water.
- Sixty-five percent of the Indian towns are characterized by slum areas.
- Slum dwellers are the most deprived section in urban society. Six among ten slum dwellers reside near unsanitary drains, and about four of every ten slum households use untreated water.
- Thirty-five percent of slum households in India still do not have access to treated tap water. Sixty percent of slum households in Tamil Nadu, Andhra Pradesh, Madhya Pradesh, Karnataka, and West Bengal still use untreated water for drinking.
- In India, slums in the urban area will grow exponentially in the next 10 years.
- Overall, in India, 63% of slum households live without a proper drainage connection or are connected to open drains, of which 44% have open drains and about 19% do not have any drainage connection. For example, 61% of slum households in Maharashtra and four other states do not have drainage connections.

- About one million people live in Dharavi slum (the world's largest slum in Asia) in Maharashtra within a 2 km^2 area.
- West Bengal has the third largest slum population, about 64 lakhs (it had 122 slum-reported towns out of 129 statutory towns) after Maharashtra and Andhra Pradesh.

The rural-to-urban migration stream has been the most influential factor in urban growth in developing countries since rural push by poverty and urban pull by city light (Pernia, 1994; Mehta, 1995). Rural migrants generally find the informal job sector in urban areas, and subsequently, they fail to enter the high-wage or formal employment sector. This leads to a shift of rural poverty into urban areas (Karn et al., 2003) and the development of poor urban neighborhoods in a backward space called "slums." The high density and poor living arrangements may help to increase open defecation, garbage dumping, and the spread of infectious diseases. Therefore, it should be necessary to investigate intra-urban differentials in health (Montgomery, 2009) while keeping in mind that slums are not homogeneous entities (Dolui et al., 2014) but rather complex and dynamic (Nolan, 2015). Lifestyle behavior includes social relations, consumption, entertainment, physical exercise, dress, etc. Many studies conclude that children and youth in slums are more affected by the lifestyle and dietary patterns of the family (Sundari, 2003). Social connections (Resnick et al., 1997a, b), family structure, norms, culture, attitudes (Ford & Kelly, 2005; Svetaz et al., 2000) and education, knowledge of caring, physical exercise, family culture, food habits, substance use, etc. are the most important factors that protect against poor health outcomes and serve as a domain of lifestyle behavior and health awareness. Smoking, alcohol use, and violent behaviors depend on the behavior of the family members. Slum dwellers in India are highly vulnerable due to the use of substances like gutka, Khaini, pan, smoking, and alcohol, which are abundantly available and affect their health and income. For example, many studies show that smoking, tobacco use, drinking alcohol, or engaging in violence are more likely behaviors among the youth whose parents are engaged in the same behaviors. Turin et al. studied urban slum children in Delhi. They found that the lack of macronutrients and micronutrients in school children's diets was the leading cause of undernourishment.

The study also found that several unhealthy practices, such as untimely dietary practices (fast-food consumption), little physical activity, drinking alcohol, smoking, and the prevalence of obesity, are common among youth. Similar studies by Mackeown and Faber (2002) and Ross and Chaloupka (2004) show the effect of smoking and the use of alcohol hurts the health status of youth. According to NFHS-5 (2019–2020), 44.7% and 8% of men and women (above 15 years), respectively, use any kind of tobacco, while 18.9% and 0.8% of men and women, respectively, consume alcohol in the urban area of West Bengal. There is no solid urban-rural differential in the case of alcohol consumption and smoking. Tanaka et al. (2002) reveal that health is directly linked with physical and mental health, and habits such as exercise, regular sleeping, and eating are essential for maintaining quality of life. A healthy lifestyle and behaviors can help to achieve sustainable health and enhance well-being.

1.3.3 Healthcare Problems, Provision, and Distribution in Urban Areas

In India, the population is rising more in small towns than in metropolitan cities (Sharma, 2020; Alabi, 2011; Danjuma, 2015; Lawal & Anyiam, 2019). For the rapid population explosion in the small town, there is a rational need to study the spatial association and distribution patterns of healthcare systems and facilities and their spatial allocation (Alabi, 2011; Danjuma, 2015; Ajala et al., 2017). The healthcare system in India is undoubtedly facing challenges in the healthcare provision and distribution across space (Moisi et al., 2010; Danjuma, 2015) due to swift population growth in existing towns and cities (Onda et al., 2019; Lawal & Anyiam, 2019), distance to the nearest healthcare center (Lawal & Anyiam, 2019; Parry et al., 2018; Yang, 2006; Hu et al., 2013; Wang & Luo, 2005; Shannon et al., 1969; Muller et al., 1998), lack of education and awareness, and the scarce financial support (Danjuma, 2015). For instance, growing death rates (Mansour, 2016; Shannon et al., 1969; Muller et al., 1998) of child and pregnant women are greatly influenced by household distance farther away from the healthcare center. However, the government has not taken the initiative to plan (Yang, 2006) and provision healthcare infrastructures over the years (Danjuma, 2015; Savigny & Wijeyaratne, 1995; Onokerhoraye, 1999) in India. To achieve the third Sustainable Development Goals (SDGs) (Lawal & Anyiam, 2019; Wang et al., 2018), healthcare facilities are essential indicators, especially in developing countries like India, for socioeconomic development as well as social well-being (Mansour, 2016; WHO, 2001; Alabi, 2011; Danjuma, 2015). However, the rapid growth of the population within the small urban area demands affordable healthcare facilities (Lawal & Anyiam, 2019; Alabi, 2011; Danjuma, 2015) and cannot maintain its carrying capacity; meanwhile, unplanned urban growth results in inequitable distribution (Alabi, 2011; Danjuma, 2015; Brockerhoff, 2000; Lawal & Anyiam, 2019) of healthcare facility centers unable to deliver services to its catchment area (Murad, 2008).

The proper allocation and accessibility to healthcare centers and services to everyone can protect against unavoidable disability, illness, and diseases (Ajala et al., 2017; Perry & Gesler, 2000; WHO, 1981; Moisi et al., 2010; Alma Ata Conference, 1978). Recently, continuous population growth and demand for improved healthcare facilities challenge urban planners (Murad, 2008) regarding how to provide standard healthcare infrastructure and services (Lawal & Anyiam, 2019; Murad, 2008; McGrail & Humphreys, 2014; Mansour, 2016; Perry & Gesler, 2000). Besides, urban healthcare centers act as block headquarters (Roy, 2016) or district hospital points providing services to their rural counterparts (Mansour, 2016; Christaller, 1966; Sedenu et al., 2016). Thus, urban dwellers are deprived of proper treatment or neglected to maintain a massive flux of population from the rural-urban fringe (Mansour, 2016; Dinda et al., 2018; Danjuma, 2015). Therefore, the urban area should have a distinguishable healthcare infrastructure and facilities (Tangcharoensathien et al., 2015; Lawal & Anyiam, 2019). In this context, geospatial analysis (Guiqin et al., 2009; Gorsevski et al., 2012; Danjuma, 2015) with Geographic Information System (GIS) algorithm (Zarkesh, 2005; Rahimi et al.,

2017; Chen, 2014; Ahmed et al., 2016; Wang et al., 2018; Malczewski, 2006; Murseli, 2014; Youzi et al., 2017; Yang, 2006; Dom et al., 2012; Mansour, 2016; Murad, 2008) to be helpful for allocation of new primary healthcare centers and enhance planning (Murad, 2008; Shukla et al., 2017; Sandipan et al., 2013) to the public health administrators (Cheng & Chu, 2007) and municipal authorities, primarily, for cost-benefit analysis (Pyrialakou et al., 2016) and municipal planning (Nikolakaki, 2004).

WHO (1989) defined "primary healthcare" (PHC) as essential healthcare with the objectives of promoting health, preventing and curing diseases, and rehabilitation, i.e., everyday living after an illness and disability (Mansour, 2016; Akinsola, 1993) and which is accessible and available to all people at a minimum cost by the homeland for their development at every stage (Akinsola, 1993). In India, healthcare service systems is characterized by a three-tier hierarchical referral system, i.e., primary, secondary, and tertiary (Chokshi et al., 2016; Uzochukwu et al., 2016; Fiedler, 1981; Kroeger, 1983) public healthcare service facilities, which are abstracted as basic healthcare service systems, i.e., sub-health centers (SHC) and primary health centers (PHC) are monitored by community health centers (CHCs) and sub-district hospitals (SDH) are supervised by district hospitals (DH) and medical colleges (MC) (Ministry of Health and Family Welfare of India, 2011; Roy, 2016). The service quality of the primary healthcare system, in turn, reduces the stress on the sub-district hospitals and district hospitals (Roy, 2016; Muller et al., 1998). In hilly, tribal, or difficult areas, the primary healthcare center covers a population of 20,000, while 30,000 are in easily accessible areas with 6 surveillance beds. Primary healthcare centers are the point of a basic level of services in the healthcare system that act as referral units for six sub-centers and transfer critical cases to community healthcare (30 bedded hospitals) and hospitals at sub-district and district units (Department of Health and Family Welfare, India, 2013; Roy, 2016). Therefore, primary healthcare centers are the most convenient way to prevent disease prevalence (Guagliardo, 2004) at the primary stage in a large area. The government of India has initiated the National Urban Health Mission (NUHM) introduced on 1 May 2013, covering all state capitals, district headquarters, and other cities or towns with a population of 50,000 and above (as per census 2011) to provide requisite primary healthcare services and several schemes to the urban poor (Perry & Gesler, 2000) specifically for the slum dwellers and cut their expenses for treatment under the Ministries of Urban Development, Housing, and Urban Poverty Alleviation, Human Resource Development, and Women and Child Development (Department of Health & Family Welfare, India, 2013).

"Prevention is better than cure" is usually an accepted reality by all citizens that prevents hygiene and food levels, which can reduce vulnerability to diseases and maximize well-being. Updated healthcare information can increase health awareness and thereby reduce health risks. In developing countries like India, where healthcare facilities and services are comparatively underprivileged in quality to some extent, public awareness tools should be effective healthcare measures (Nishtar, 2010). Recently, in digital India, mass media and, more extensively, social media through smartphones are the most influential tools in broadcasting healthcare-related information and accessibility to all groups of citizens. In this context, health awareness through healthcare facilities and services is the most effective tool for better handling the health of the urban poor. Insufficient cost of medical services and lack of capacity to work due to ill health have manifold direct and indirect economic effects on poor households. Furthermore, ill health has severe social aftermaths that go from one generation to the next generation and contribute to the further poverty of even now poor households.

The World Bank estimated that behind all diseases in developing countries, the local environment contributes one-fifth of the causal factors. Similarly, WHO (2008) has suggested that unsafe water, sanitation, indoor smoke from solid fuels, and hygiene are the leading factors in mortality risk among developing countries. Many empirical studies have found that healthcare centers and facilities are disproportionately distributed (Eyles, 1996; Oyerinde, 2006; Malqvist et al., 2010; Yang, 2006) as well as determined the social well-being by measuring accessibility to healthcare sites (Muller et al., 1998; Frech & Mobley, 2000; Jones et al., 2009; Parry et al., 2018; McGrail & Humphreys, 2014; Ajala et al., 2017; Perry & Gesler, 2000; Khan, 1992), available modes of transportation (Lawal & Anyiam, 2019; Wang et al., 2018; Kim et al., 2018), and time consuming to healthcare services (Gale et al., 2002; Moisi et al., 2010; Okwaraji et al., 2012; Malqvist et al., 2010). Kim et al. (2018) emphasize social equity in healthcare services, i.e., every person has an equal chance of accessibility (Ajala et al., 2017; Langford & Higgs, 2006). Ruishan et al. (2013) highlighted the potential spatial accessibility of healthcare services by circumventing economic and geographical impediments (McGrail & Humphreys, 2014; Lawal & Anyiam, 2019; Kim et al., 2018). Mansour (2016) and Akinsola (1993) found that with the changing socioeconomic conditions and government health policies, the provision of healthcare services varies across countries, even at the village level. Boulos et al. (2001) and Assad stated that, due to the large population with low GDP and inadequate distribution of hospitals, primary health centers and equipment in India need improved healthcare services. Continuous rural-to-urban migration leads to the rapid growth of small cities (Jain & Subbaiah, 2007; Malqvist et al., 2010), which demand healthcare facility centers that are most looming and significant (Yang, 2006) in India. Self-treatment and no treatment are common in developing countries; both are vulnerable to long-term health problems.

1.3.4 Slum and Health Coping Mechanism

The resource endowment and capacity for managing poor households are related to coping mechanisms (Rakodi, 1995). Poor households work in both formal and informal sectors, providing housing and making social networks to avoid perpetuating the poverty trap (Roberts, 1994). Likewise, slum dwellers are heterogeneous, consisting of diverse people with different interests, means, and backgrounds. Thus, slum dwellers willingly or unwillingly face stresses, pain, or changes experienced daily. Sometimes, they try to solve or cope with existing problems or adopt effective or harmful strategies. Viken et al. (2015) undertook a study on maternal health coping strategies of migrant women in Norway, which pointed out that health knowledge directly affects maternal health coping strategies, as well as migrated women taking fewer coping strategies for their maternity care (Small et al., 2014). Baqutayan (2015) stated that there is a close link between environmental stress and coping mechanisms. Effective coping strategies can reduce mental hesitation, positively managing the situation (Uchino, 2004; Lim et al., 2013; Henderson et al., 2013). Adam (2013) mentioned five types of capital that poor households used in their coping mechanism: human, social and political, physical or production, financial, and natural capital. Chetia opined that the coping strategy of the urban poor does not depend on what the household's assets have but rather depends on the capacity to manage those assets and transform them into income generation.

Urban poor naturally adopt five strategies: income raising strategy, expenditure minimizing strategy, strategies taken during the eviction, social and economic infrastructure options, and social network. Dahal (2017) found that changes in food habits, cooperation, reduced essential expenditure, and use of saving and borrowing are significant health coping strategies of poor people. Sometimes, traditional knowledge and known healthcare practices are significant health coping strategies (Haque et al., 2013) among poor households. Also, engaging in a small-scale street shop, low-paid labor, wage labor, buying on credits, vendor shop and loans, substance abuse, social networks, working longer hours, and seeking employment in the private sector are the other coping strategies of people with low incomes (Adam, 2013; DeLongis & Holtzman, 2005; Amendah et al., 2014). They are generally taken to poor medical facilities, and for their urgent treatment, they mainly depend on local medical shops (Rahman & Nahar, 2010; Malaviya & Bhagat, 2013) mainly to avoid expensive consultancy fees of clinics or doctors as well as only go to hospitals when they are poorly sick. Due to weak and informal risk management strategies, low-income people live poorly (Heltberg & Lund, 2009; Arefin et al., 2010; Dahal, 2017). Most slum dwellers are unaware of how to access and utilize urban healthcare facilities (Rasheed et al., 2016; Arefin et al., 2010). Using savings, borrowing money from informal lenders with high interest rates, selling non-productive and productive assets, and spending business capital are coping strategies during the severe illness of the urban poor.

Proper allocation and availability of basic amenities and services, socioeconomic condition, religion, free education, health shocks, children's school fees, price of essential commodities, ownership of a house, family environment, and income levels are controlling the health coping mechanism of the urban poor (Adam, 2013; Arefin et al., 2010; Rabenu et al., 2017; Farley et al., 2005; Al-Dubai et al., 2011; Amendah et al., 2014; Roy, 2016). Literacy impacts health coping mechanisms; for example, a literate person has more knowledge about their health than a non-literate person. Therefore, slum dwellers are suffering more from communicable diseases than their urban counterparts due to a lack of proper information and awareness (Lahon, 2017; Arefin et al., 2010). The coping mechanism is also influenced by gender and age differences due to genetic factors and social conditions (Washburn-Ormachea et al., 2004; Davis et al., 1999; Amendah et al., 2014). Due to caring for family members, women have taken more survival strategies than men (Al-Dubai et al., 2011; Henderson et al., 2013). However, middle-aged men are shown more coping strategies than younger adults (Aldwin & Levenson, 2001). Sheuya (2008) opined that slum people have improved their health and well-being by setting out strategies rather than depending on the government. The use of smoking or drinking alcohol as a coping effort has long-term adverse effects that create additional health shocks, which again reduces the strain of coping strategies. Family can be protected from stressful situations if they fully cope with their resources (McCubbin et al., 1980; Judge, 1998). Positive strategies help to increase the relationship among the family members and neighbors, improving the quality of life. It can also help the urban poor to provide steady and reliable sources of income and effective livelihood planning (Amendah et al., 2014).

1.3.5 Study Site and Literature Gap

The Midnapore town has been developed with its population since 1991 in different directions. Dinda et al. (2018) assessed that the built-up area in 1991 was 15.29 km^2, which increased in 2001 to 20.10 km^2 with a growth rate of 31.48%, although it was high between 2001 and 2017, about 35.39%. The settlement areas also developed outside the municipal boundary and gradually expanded, a sign of reliance on healthcare services from the municipal healthcare centers. Constant rural-to-urban migration brought the development of slum and squatter settlements to dwell in most of the unoccupied land illegally and unplanned urban development, which reduces the proper allocation of healthcare facilities. Dolui et al. (2014) estimated that the built-up area increased by 72.35%, while dense and degraded vegetation cover area significantly decreased by 83.63% and 30.04%, respectively, from 1991 to 2011. This study also shows that surface radiant temperature increased by 6.52 °C due to urban development. The reason behind the growth of the urban area is the gradual increase of migrated population in the town. According to the Census of India (2011), Midnapore, among the urban areas of Paschim Medinipur district, was the landmark of spectacular population growth between 1991 and 2011 (Table 1.1). Paul (2018) found that the land use pattern in the periphery of the town has changed due to the lack of basic amenities. Mandal et al. (2016) stated in their studies that the spacing plan in the Midnapore urban center has not been adequately maintained. The average spacing between the buildings is less than 1 m, spreading diseases such as cholera, viral flu, and dengue. Due to a lack of space, the hazard and disaster rescue operation could not have been conducted properly. Roy (2016) studied the neighborhood level in Midnapore. He found that the health status of the urban poor significantly depends on healthcare facilities, health-related behavior, health perception, and awareness rather than socio-demographic and political factors. This study also pointed out that poor neighborhoods are characterized by high health, healthcare, socioeconomic, and demographic homogeneity. Mukherji (2001) opined that most city dwellers are not interested in actively participating in the decision-making process associated with their living conditions.

Table 1.1 Share of urban population to district urban population, Paschim Medinipur, 1991 and 2011

Sub-division/ municipality	Area (sq. km) in 2011	Population (number) in 2011	Density of population (per sq. km)	% population to district population in 2011	Share of district's urban population in 2011	Share of district's urban population in 1991	Difference
Jhargram sub-div.	3037.6	1,136,548	374	19.22	–	–	–
Jhargram (M)	17.04	61,712	3622	1.04	10.92%	9.53%	1.39%
Sadar sub-div.	2441.5	1,435,321	588	24.27			0.00%
Medinipur (M)	**14.78**	**169,264**	**11,452**	**2.86**	**29.96%**	**28.30%**	**1.66%**
Kharagpur sub-div.	2913.1	2,293,909	787	38.79	–	–	0.00%
Kharagpur (M)	90.65	207,604	2290	3.51	36.75%	40.12%	−3.37%
Ghatal sub-div.	953.09	1,047,679	1099	17.72	–	–	0.00%
Chandrakona (M)	16.58	23,629	1425	0.4	4.18%	3.77%	0.41%
Khirpai (M)	11.65	16,384	1406	0.28	2.90%	2.78%	0.12%
Ramjibanpur (M)	10.36	19,611	1893	0.33	3.47%	3.38%	0.09%
Kharar (M)	10.36	12,118	1170	0.21	2.15%	2.28%	−0.13%
Ghatal (M)	10.36	54,591	5269	0.92	9.66%	9.83%	−0.17%
District total	9368	564,913	631	100	100%	100%	–

Source: Bureau of Applied Economics and Statistics, Department of Programme Implementation, Government of West Bengal, 2014

During the questionnaire survey in the study area, most of the slum dwellers raised the problems of health, essential services, and healthcare facilities, which are summarized through the word clouds using N-Vivo as (Fig. 1.2):

Similarly, after careful review, most of the literature related to the study represented health coping strategies of the urban slum people with environmental stress, which are represented through the word clouds using N-Vivo as (Fig. 1.3):

The explosive growth of the urban population with the absolute number of slum population in India faces significant health problems. There is still a gap between the real need and what is provided. Slum dwellers are facing uneven health challenges such as reproductive and maternal health, child and adolescent health behaviors, alcohol and tobacco use, and non-communicable diseases

Fig. 1.2 Word clouds to identify the factors related to healthcare issues in Midnapore municipality from informants' view using N-Vivo (Source: Author)

Fig. 1.3 Word clouds to identify popular literatures in health and health coping research in slum area using N-Vivo (Source: Author)

related to poor diet and sedentary lifestyles, along with a continuing high prevalence of infectious diseases such as dengue, cold, fever, gastro/diarrhea and stomach problem, skin diseases, diabetes, asthma/dyspnea, etc. The existing gaps in sanitation and personal hygiene education, child and reproductive health and HIV/AIDS education, use of sanitary napkins, and the building of household latrines in the slum areas must be handled by the municipal authorities, NGOs, and other government officials. Therefore, the present study attempts to bring out the healthcare system available for slum dwellers and their health coping mechanisms in the study area.

1.4 Geographical Perspectives of Health Coping Mechanism

Health is intrinsically linked with geography because health varies across geographical places, i.e., spatial locations. The socioeconomic, natural, and built environment within geographical boundaries affects our health and well-being, and healthcare centers, facilities, and services are also affected by the geographical conditions of the places. The geographic context of places plays a vital role in environmental risk and vulnerability, shaping health policy and strategies (Dummer, 2008). In this context, health geography as a subdiscipline of human geography emerged in the 1950s, and it deals with the interaction between people and their living environment, encompassing place, space, and society (Dummer, 2008). Health geography emphasizes the spatial patterns of healthcare centers and the planning and provision of health services within social, cultural, and political frameworks. The research in geography on health has also incorporated qualitative and quantitative data and methodologies since the 1980s, using statistical tools to explore the complex problems of community health and healthcare systems. The qualitative studies focus on the utilization and accessibility of healthcare facilities for the communities and the perception of communities on providing healthcare facilities within the particular sociocultural landscape. In contrast, quantitative studies focus on the spatial modeling of diseases, healthcare centers, and facilities.

Geographers are interested in the time-space framework linked with migration, climate, environmental change, health risks, health outcomes, and health-related practices within physical and social environments. Over time, health geography has evolved to integrate social, environmental, and epidemiological dimensions of physical and mental well-being. It examines health inequalities, service utilization, risk factors, and spatial analysis of healthcare infrastructure across rural and urban areas (Dummer, 2008). Since the 1980s, geographical research has been shifted increasingly toward the assessment of accessibility, restructuring, and provision of healthcare rather than the distribution of healthcare resources. There has been significant variation in the healthcare system in developed and developing countries since the 1980s. Developed countries emphasize the spatial equality of healthcare services, health policy, and intervention, while developing countries focus more on developing hospitals in central city precincts. However, in developing countries, small towns have increased populations since the twenty-first century due to continuous migration from rural areas to the low-cost, unprivileged areas of the town. Continuous population and resulting slum growth promoted unhygienic living conditions and demand for healthcare facilities. At the same time, healthcare services have been increasing in the major urban centers as well as government initiatives and community healthcare services. Nonetheless, health themes are the prime research area in urban landscapes. Rapid urbanization and resulting environmental contaminations lead to the community's vulnerability and exposure to health risks. Despite the government healthcare facilities, people must adopt strategies to overcome the day-to-day health risks and improve their health and well-being. Geographers are now interested in researching community health coping strategies for different age groups such as childhood, adolescent, parental, and aged. At the same time, research on health coping mechanisms has been extended to different service sectors such as nursing, the private sector, farmers, business people, etc. Therefore, trends of geographical research recently increased toward community health perception, the nature and causes of the utilization of existing healthcare services, coping strategies within living circumstances, etc. So, it is noteworthy that the future geography of health research would be more on community health, health perception and awareness, and self-initiatives toward well-being and healthy living.

1.5 Scope of the Present Study

In the present day, the urban slum is a sensitive issue of human rights and also a dark chapter of any country. Slum conditions slow down people's standards, mental intellect, spiritual, emotional-moral, social growth, and physical health. Slum development is not only the result of the underdevelopment of human beings but also the economic problem behind it that leads to decent work prospects, health consequences, and poverty traps. Moreover, slum dwellers are mainly influenced by the living environment around them.

According to the Census of India (2011), out of the total urban population, 31.14% of the slum population lives in the Midnapore urban area (class I), and the decadal growth rate (2001–2011) is 35.54%. The overall growth of the urban population is higher (1.66%) between 2001 and 2011 than the other urban areas in Paschim Medinipur district within the same administrative boundary. In a spatial context, the accommodation of a vast slum population within municipal boundaries, irrespective of the unchanged area, one can easily understand the pressure and problems over slum dwellers in their daily livelihood.

Different government reports reveal the unhygienic conditions of the urban poor. Most health indicators fail to prove that knowledge and use of modern health facilities are lower in urban areas than in rural areas, sometimes the national average. However, most modern health facilities have been established in the urban area. It has also been found that fertility and mortality rates are declining in urban areas, which keeps in mind that some underlying factors or traditions or practices command the health and well-being of the urban poor.

Slums of Midnapore municipality, under non-notified conditions, have no precise information and data, and no health-related data are available (Table 1.2). In this dark situation, the study attempted to collect health behavior, perception of the healthcare system and accessibility of health services, health practices and coping information of the urban poor, health mapping, and an overall health data bank. Moreover, this micro-spatial study would be helpful to the urban authorities, health personnel, district planning commission, and authority under disaster management in the development of healthcare services and policymakers in taking suitable

Table 1.2 Overall health scenario and indicators in Midnapore municipality

Demographic and health indicators	Number or percentage	Sources
City population	169,264	Census of India (2011)
Total no. of slums (non-notified)	156	Midnapore Municipality (2022)
Total slum population (2011)	56,096	Census of India (2011)
Slum households	12,061	Census of India (2011)
Average family size	4.65	Census of India (2011)
Percentage of slum population to the total urban population of Midnapore City in 2011	33.14%	Census of India (2011)
Decadal growth rate (2001–2011)	35.54%	Census of India (2011)
No. of hospitals	02	Midnapore Municipality (2022)
No. of nursing home	15	Midnapore Municipality (2022)
Urban community health service center/SHP (sub-health post)	8	Midnapore municipality (2022)
ICDS center	102	Midnapore Municipality (2022)
FTS (first-track supervisor)	8	Midnapore municipality (2022)
HHW (honorary health worker)	34	Midnapore municipality (2022)
Health camps in f.y. 2018–2019 (till July 2018)	55	Midnapore municipality (2022)
ANM (auxiliary nurse midwife)	3	Midnapore municipality (2022)
The average distance of the nearest AWC	0.5–1 km	Midnapore municipality (2022)
Location and distance of nearest primary health center/UHP/UFWC	1–1.5 km	Midnapore municipality (2022)
Location and distance of nearest primary school	1–2 km	Midnapore municipality (2022)
Quality of housing (kutcha/pucca/mixed)	Mixed	Midnapore municipality (2022)
Close open drainage system	3	Census of India (2011)
Community private latrine	17	Census of India (2011)

Source: Midnapore municipality

policies and programs, resource allocation, and health risk management strategies in the light of potential impacts, recovery, and health shocks incident among the slum neighborhoods in the broadest perspective of well-being and welfare in Midnapore City as well as provide a broader scope for replication in other places in West Bengal and beyond.

1.6 Methodology and Design of the Study

Research in different disciplines is a painstaking and methodical inquiry of the phenomenon under study, which is an objective of progressing knowledge. In social research, a systematic method has been employed to explore the new facts and verify the old ones. Generally, all kinds of health studies are carried out with the help of quantitative and qualitative methodology. Thus, the methodology is the process of systematically solving the research problem through the design of the whole study. In health geography, quantitative and qualitative methods have been used during data collection and analysis, and with the integration of GIS, spatial distribution, interpretation, and understanding of ill health, disease, and disability in the context of place can be visualized for describing the multidimensional features.

In the present study, the primary data were collected through a well-structured questionnaire schedule, and secondary data were collected from different governmental organizations and Midnapore municipality, as well as geospatial data of different periods has been collected from Earth Explorer-USGS open website (https://earthexplorer.usgs.gov). Both data have been analyzed with statistical tools and a GIS algorithm. Some derivation, mathematical formulas, models, and geospatial techniques have been employed to visualize neighborhood spatial variation through cartographic techniques such as graphs, charts, plots, and GIS mapping. The analyzed data is then used to draw interpretations and to discover the results and conclusions. The whole design of the study is presented in Fig. 1.4, and the methodological approaches (Fig. 1.5) are given below.

1.6.1 Data Collection

Many significant methods are used for collecting qualitative and quantitative data. The present study employed three primary methods for the collection of both the data as follows:

- Interviews.
- Participant observation.
- Focus groups.

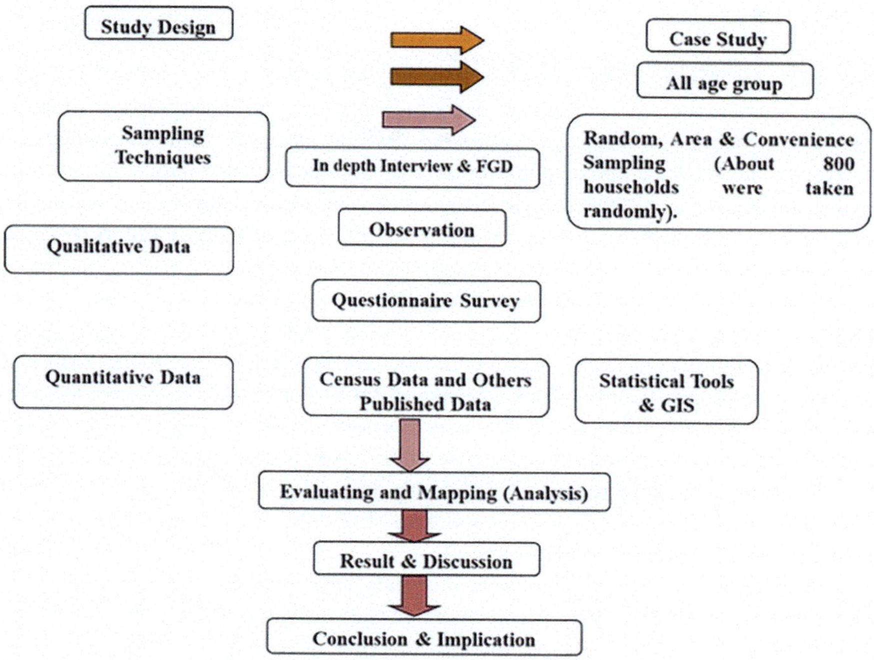

Fig. 1.4 Methodological flowchart of the study (Source: Author)

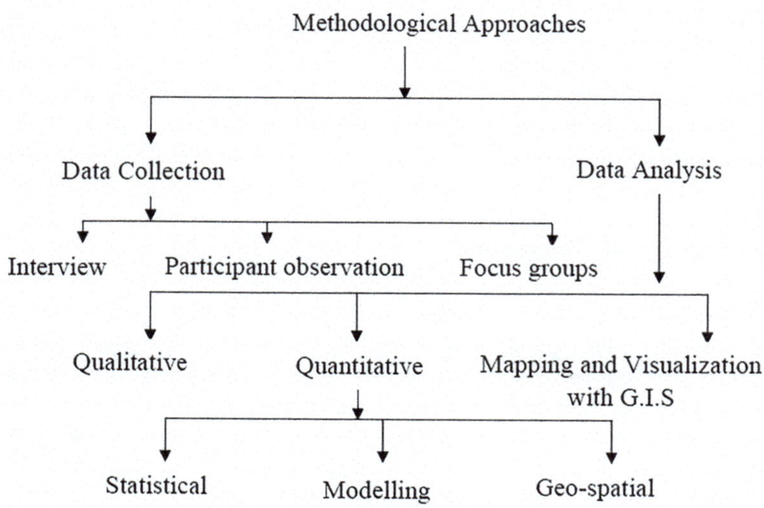

Fig. 1.5 Methodological approaches of the study (Source: Author)

1.6.1.1 Interviews

The interview plays a vital role in collecting information from individuals and groups, which is the most significant part of research in health geography. An interview is a process of asking questions and deriving answers from individuals or groups. In the health context, quantitative and qualitative information about individuals or groups is required to describe and explain their experiences of health, well-being, and illness. Therefore, in-depth conversation helps researchers to know people's values, beliefs, and feelings during their livelihood.

The study used the interview as a data collection tool through a structured questionnaire. The pre-tested questionnaire includes qualitative and quantitative questions such as living conditions, health behaviors and awareness, disease patterns and treatment processes, health infrastructure and provision of the healthcare system, utilization and accessibility of the healthcare system, and health coping strategies of poor neighborhoods in the study area. The interview was done door to door, and the valuable responses were recorded in the schedule. The study also incorporates a voice recorder, which records the respondents' life experiences and valuable comments as a qualitative data collection tool.

A detailed survey was conducted in the study area to assess conventionally different aspects of health coping mechanisms in the study period from 20 October to 13 December 2022 based on spontaneous respondents using key informant interviews (Solangi et al., 2019). Overall, data from different indicators were collected from 156 notified slum neighborhoods through a well-structured questionnaire survey. Simple random sampling methods were employed to collect the data from roughly 800 households, almost 10% of the entire slum households in the study area. In general, five to ten households, based on population composition in each slum household, were randomly selected for the survey. Eight hundred respondents, such as household heads or other senior members, have been targeted for the interview and actively participated in the abovementioned interviews, taking their eager consent. During the interview, the average age of the respondents was 48.02 years.

1.6.1.2 Participant Observation

Many researchers argue that participant observation is the best way to collect information by actively participating in and observing human activity. Participant observation is a method in which the investigator is personally involved during observation. This method motivates and stimulates investigators to get insight into the group's behavior. So far as the study is concerned, this method has been used to study the infrastructure of the primary healthcare centers and their locations in the municipal wards, dwelling house structure and assets, road conditions, or groups of individual behaviors such as alcoholics. The study was also integrated by taking photographs to understand the actual situation and for future reference.

1.6.1.3 Focus Group Discussion

A focus group discussion (FGD) is a qualitative data collection tool that gathers deep information from the community about thoughts and opinions on a topic under study. It is a predetermined semi-structured interview in which people of a similar community have the same background or experiences and involves discussing a specific topic of interest. FGD has been used in the study for two purposes: first, to collect the qualitative data and understand the community perceptions, attitudes, beliefs, opinions, or ideas about the study and, second, to affirm the reliability of the collected data through the questionnaire and collect the additional information regarding the study. FGD is used as a qualitative data collection method with a semi-structured interview to elicit information about the community's perception of health and well-being and self-initiatives to promote their future health. Health coping mechanisms are multidimensional and encompass all facets of human livelihoods within their living milieu. Different theories have measured various aspects of human health and its security based on different concepts of deprivation, quality of life, accessibility of resources, and well-being. Subjective and objective indicators have been designated based on empirical observation and focus group discussion to develop various indices in the study context. The principal indicators include a household's socioeconomic status, built environment, social relationships to their breathing environment, availability of basic amenities, healthcare facilities, accessibility and services, and physical and psychological health.

Twenty focus group discussions were conducted among the slum (e.g., Natun Pally, Mahatabpur Sasan, Talpukur Colony, Sudam pukur PARH, Mandal Mahalla, Hasan pukur Muslim bastee, Adibasi Para, Porabanglo bastee, etc.) in the study area to collect conventionally qualitative data on different aspects of health coping mechanisms in the study period from 20 October to 13 December 2022 based on spontaneous responses from the slum dwellers of different age, sex, and religion. Each focus group consisted of at least 12 to 15 slum dwellers; sometimes, more than 15 slum dwellers eagerly attended this group discussion. They spontaneously express their problems, perceptions, and opinions about their health coping mechanisms. The FGD incorporates voice recording for a smooth interpretation of the community's perceptions and future references.

1.6.2 Data Analysis

1.6.2.1 Quantitative Methods

Quantitative methods are the process of numerical data analysis, encompassing objective, statistical, or mathematical measurements of data collected through questionnaires, surveys, or computational tools to generalize and explain a particular phenomenon (Babbie, 2010). Quantitative methods can analyze data at different measurement scales (nominal, ordinal, interval, and ratio) in a systematic, standardized approach without losing original characteristics. Generally, quantitative methods are used to explore the differences between groups and relationships among the variables and test the hypothesis.

Quantitative methods of data analysis typically involve three tasks. The first task is the statistical method, which explores the data and finds spatial patterns and probable associations with other variables. The second one is modeling, where we emphasize a recognized statistical model of our data with the testing hypothesis. In health geography, modeling represents the actual situation in a graphical form or other forms with a desirable significance level. The third one is the geospatial analysis, which gathers, displays, and manipulates satellite imagery, GPS, aerial photography, and historical and geographical data with geographic coordinates integrated into geographic models. In practice, geographers often simultaneously use all three quantitative methods using modern statistical (e.g., Excel, SPSS, etc.), remote sensing (e.g., ERDAS Imagine), and GIS (e.g., ArcGIS) software for the inferential analysis, extracting the data and fitting the models and finally mapping the results.

1.6.2.2 Spatial Data (Statistical) Analysis

Spatial data analysis is how raw data is transformed into convenient information. This type of analysis entails techniques designed to find the relationship and variances or test hypotheses and models based on spatial data. Statistical methods are more useful when facing the problem of mapping health data for a set of areal units. Geographers have worked with statistical measures to devise spatial autocorrelation, test hypotheses, and validate models (Cliff & Haggett, 1988) in characterizing health data, which is now more extensive. A set of statistical tools has been used for the present study, discussed below.

1.6.2.3 Principal Component Analysis (PCA)

The principal component analysis is a data reduction methodology (inductive approach) that recognizes significant and more meaningful components from a larger dataset depending on the observed variance (Abdi & Williams, 2010; Hoseinzadeh et al., 2016). PCA summarizes many variables using the first few principal components that are rank-ordered by variance, uncorrelated, and low in number while ignoring the rest. It also preserved as much statistics as possible. The principal component analysis (PCA) is the exploratory data analysis process for building predictive models. PCA was used in this study to extract major components through a comprehensive explanation of the data by grouping variables into components. Components are extracted through the dimension reduction tool of

SPSS v26 using Kaiser normalization and varimax rotation methods with eigenvalues higher than 1.0 and names to specify dormant variables. A dataset of variables is needed to build the indices. To calculate the value of indices for every spatial unit (156 slum areas) of the study area by using extracted principal components as (Eq. 1.1)

$$I_{\text{slum}} = \Sigma \text{PC}_i \quad (1.1)$$

where I_{slum}, the rescaled variance, explains the weight of the index score, which is rescaled to 100% for each factor, assigning weights based on the explained variance and aggregates of the variables (Reams et al., 2012), and PC_i is the principal component.

1.6.2.4 Pearson Correlation

Pearson linear correlation is used to find the correlation among the variables. Generally, it measures the linear correlation between variables developed by Karl Pearson. Also, it is derived by the ratio between the covariance of two variables with the product of their standard deviations, the so-called normalized estimation of the covariance. The value always remains between −1 and + 1. The Pearson correlation matrix is more useful when a large number of variables are essential for the explanation of a particular phenomenon.

1.6.2.5 Ordinary Least Square Regression (OLS)

Ordinary least square regression (OLS) is known as a linear regression algorithm that measures the unknown parameters based on linear regression (simple or multiple) models with the conditions of linearity; no endogeneity, normality, and homoscedasticity; no autocorrelation; and no multicollinearity. The coefficient minimizes the sum of squared residuals (SSR) and is estimated based on the principle of least squares. Thus, with the linear function, OLS estimated the relationships among the observed dependent variables and predicted variables, which are derived as (Eq. 1.2)

$$Y = B_0 + \Sigma_j = 1 \ldots p\beta_j x_j + \varepsilon \quad (1.2)$$

where Y is the dependent variable, β_0 is the model's intercept, X_j corresponds to the jth explanatory variable of the model (j = 1 to p), and ε is the random error with expectation 0 and variance σ^2. Besides, for the ith observation, the predicted dependent variable (Y) is determined by the equation (Eq. 1.3)

$$Yi = B_0 + \Sigma_j = 1 \ldots p\beta_j x_{ij} \quad (1.3)$$

The objective of the OLS method is to reduce the sum of square differences between the observed and predicted values. This reduction leads to the estimation of the parameters of the model (Eq. 1.4)

$$\left[\beta = \left(x'Dx\right)^{-1} x'Dy\sigma^2 = 1 / \left(W - p*\right)\Sigma_i = 1\ldots W_i\left(y_i - y_j\right)\right] \tag{1.4}$$

where β is the vector of the estimators of the β_i parameters, X is the matrix of the explanatory variables preceded by a vector of 1 s, y is the vector of the n observed values of the dependent variable, p^* is the number of explanatory variables to which added one if the intercept is not fixed, w_i is the weight of the ith observation, and W is the sum of the w_i weights. D is a matrix with the w_i weights on its diagonal. Similarly, the vector of the predicted values can be estimated as (Eq. 1.5)

$$Y = X\left(X'DX\right)^{-1} X'Dy \tag{1.5}$$

OLS is applicable when bulky explanatory variables are available. Thus, OLS estimates the unknown parameters to increase the number of observations and explain the effects of explanatory variables.

1.6.2.6 Linear Regression

Linear regression is a useful predictive model of the observed datasets that explain the relationship between a dependent variable (response) and an independent variable (explanatory or predictor). In linear regression, the relationships are based on linear predictor functions, and the unknown parameters are estimated from the observed datasets. This model can be simple or multiple based on the number of independent variables. The linear regression model is used to predict the response variable and explain the variation in the response variable attributed to variation in the predictive variables. The linear regression model based on the "least squares" method is derived as (Eq. 1.6)

$$Y_i = \beta_0 + \beta_1 X_i + \varepsilon_i \tag{1.6}$$

where β_0 is the intercept, β_1 is the coefficient, and X_i and Y_i are the explanatory and response variables, respectively. $\beta_0 + \beta_1 X_i$ is the linear component and ε_i is the random error or random error component.

1.6.2.7 Multinomial Logistic Regression (MNLR)

When predicting a nominal dependent variable based on multiple continuous or categorical independent variables, we generally use a multinomial logistic regression model called "multinomial regression." It is also used to explain the relationship of one nominal dependent variable with one or more independent variables. Multinomial regression is a more critical predictive analysis than the other regression analysis because it does not assume normality, linearity, outlier, or homoscedasticity. However, careful consideration should be given to factors such as sample size, linear relationship between dependent and independent variables, and mutually exclusive and exhaustive categories of dependent variables measured at the nominal level before using multinomial regression. This model also predicts the probabilities of possible outcomes for each dependent variable with the given set of independent variables. Therefore, the probability of the possible outcomes is derived as (Eq. 1.7)

$$P(y_i = j) = \frac{e^{X_i \beta_j}}{\sum_{j=0}^{j} e^{X_i \beta_j}} \tag{1.7}$$

where i represents the sample, j represents the dependent variables, X_i represents the independent variables, and β_j is the parameter to be estimated.

The model was estimated using the maximum likelihood estimation (MLE) technique. The value of the logistic coefficient is near zero, indicating that the predictor has less influence. Furthermore, Exp(B) is the odds ratio allied with each predictor. The values of Exp(B) >1.0 and <1.0 denote greater and fewer influences on the predictor, while Exp(B) of 1.0 indicates that it does not affect the predictor.

1.6.2.8 Deductive Approach: Standardization (Z-Score, Ratio of Value, and min-Max Rescaling)

For the development of an index with limited variables in the context of social or physical aspects based on the knowledge of prior theories, we need to adopt a deductive approach (Yoon, 2012). Generally, researchers have used different kinds of deductive approaches. However, in the present study, only three approaches were used for analytical purposes: z-score transformation, maximum value transformation, and min-max rescaling. These methods have been used to develop the final index due to their ability to transform the data unit to a small-scale range.

Z-Score Transformation

The z-score is used to devise a composite index score (Yoon, 2012; Zahran et al., 2008). It is an ideal average deviation with variation and the standard normal distribution similar to the standard deviation, whose value is equal to zero. Z-score is estimated with the following derivation (Eq. 1.8):

$$Z = \frac{(\text{Score} - \text{mean})}{\text{Standard deviation}} \tag{1.8}$$

Maximum Value Transformation (Ratio of Value)

The maximum value transformation is used to devise an index score, defined as the proportion of the value of the variables (X_i) to the maximum value (X_{max}) among the variables. The value of derived rescaling is ranging from zero to one. The higher value (near to one) represents the higher index score and vice versa. The derivation of the index is as follows (Eq. 1.9):

$$R_i = \frac{X_i}{X_{max}} \tag{1.9}$$

where R_i is the normalized value of a variable of X_i.

Min-Max Rescaling Transformation

One of the most standardized methods is the min-max rescaling, in which each variable is rescaled into a range of scores between zeros and one. The rescaling score is obtained by the proportion of minimum value (X_{min}) in the dataset, subtracted from each variable (X_i), divided by the range of variable value, i.e., maximum value (X_{max}) subtracted from the minimum value (X_{min}) among the variables. The min-max rescaling value toward zero indicates the most unpleasant and toward one indicates the more applicable for developing an index. The mathematical formula is derived by (Cutter et al., 2010) (Eq. 1.10)

$$V_i = \frac{X_i - X_{min}}{X_{max} - X_{min}} \tag{1.10}$$

where V_i is the normalized value of a variable of X_i.

Modeling

Modeling involves developing a model representing some aspects of a phenomenon in the real world. Scientific modeling signifies the logical selection and identification of empirical objects, environmental processes, and relevant phenomena into different models to quickly comprehend, explain, quantify, and visualize common knowledge. Thus, modeling is simply the process of transforming or augmenting human thoughts into models in a logical way.

Different models are usually used to show the spatial distribution of geographical phenomena. For example, if we want to show the disease pattern, we need an empirical model to validate the situation. Statistical techniques are used to validate the empirical models. The regression models, principal component analysis, standardization methods, and other relevant statistical methods have been used to validate the different empirical models and explain the reasons behind the facts in the study.

Many empirical models have been developed for understanding the actual situation in the study area, such as social deprivation (SoDI), multidimensional poverty (MPI), quality of life (QoLI), social vulnerability (SoVI), personal hygiene (PHI), and health risk (HRI), health security (HSI), whole year health coping (WYHCI), and health shock coping (HSCI). These models have been devised in the study in a single figure through the different indicators into a single index for each slum neighborhood that represents the overall health status of the community. Thus, the models can serve as a foundation for distributing resources, assessing existing and proposed programs, and planning interventions. Various indicators have been used to directly or indirectly contribute to the models.

- *Morbidity indicators*: numerous indicators of health status, disability, and acute and chronic diseases.
- *Outcome indicators*: poverty, death rates, maternal and infant deaths, life expectancy, disease prevalence, health shocks, health security, etc.
- *Utilization indicators*: healthcare facilities, hospital admission rate, physician visit rate, etc.
- *Resource availability indicators*: the ratio of healthcare centers to the population, the ratio of physicians to the population, basic amenities, and other measures of resources.
- *Functional status indicators*: employment status, school days lost, deprivation, coping strategies, etc.

Assume that all indicators should be assigned equal weights and scores on a scale appropriate for each geographical administrative unit. This index for every empirical model could be used to compare from one community to another over time.

Geospatial Analysis

Geospatial analysis is gathering, manipulating, exploring, modeling, making predictions, and displaying and interpreting georeferenced data from geographical data, satellite imagery, GPS, and aerial photography within the computer-based algorithm. Geospatial analysis extracted the pertinent data from irrelevant data and conceptualized and visualized the hidden facts from the disorderly geographically organized data such as proximity, contiguity, affiliation, etc. Thus, geospatial analysis finds the complex relationship between people in places and their surrounding living environment through modeling, visualizations, and predictions. Moreover, geospatial analysis can incorporate and process massive geospatial and geometric data in a space-time continuum with huge geographical locations to visualize the changing pattern over time. For the present study, different geospatial techniques have been adopted to visualize the variation in the spatial distribution pattern of basic amenities, healthcare facilities, and other needs among the slum neighborhoods.

1.6.2.9 Cluster-Outlier Analysis

Anselin Local Moran's I is a spatial statistic tool in ArcGIS v10.4 environment to classify identified spatial clusters according to the type of association of a dataset based on the method of inverse distance weight (IDW) (Sánchez-Martín et al., 2019; Jana & Sar, 2016; Aksha et al., 2019) using a fixed distance band of 928 meters. This analysis identifies whether spatial features are in a statistically significant spatial pattern or inconsistency. The result was in four typological gradations with a particular statistical significance level as (1) high-high (HH) and (2) low-low (LL), a feature with similarly high and low scores surrounded by features with high and low scores, respectively, and indicating positive spatial autocorrelation; (3) high-low (HL) representing a feature with dissimilarly high score surrounded by features with a low score; and (4) low-high (LH), a feature with a low score surrounded by features with high scores, denoting negative spatial autocorrelation. The mathematical formula for the Local Moran's I statistic to express spatial association (ESRI) is derived by (Eqs. 1.11 and 1.12)

$$I_i = \frac{x_i - \overline{X}}{S_i^2} \sum_{j=1, j \neq 1}^{n} w_{i,j} \left(x_j - \overline{X} \right) \tag{1.11}$$

where x_i is an attribute value for the feature i, $\overline{X}$1$\overline{X}$ is the mean of the corresponding attribute value, and $x_{i,j}$is the spatial weight between feature i and j.and

$$S_i^2 = \frac{\sum_{j=1, j \neq 1}^{n} \left(x - \overline{X} \right)}{n - 1} \tag{1.12}$$

with n equating to the total number of features and the z_{I_i} is computed as (Eqs. 1.13–1.15)

$$z_{I_i} = \frac{I_i - E[I_i]}{\sqrt{V[I_i]}} \tag{1.13}$$

where

$$E[I_i] = -\frac{\Sigma_{j=1, j \neq 1}^{n}}{n - 1} \tag{1.14}$$

$$V[I_i] = E\left[I_i^2\right] - E[I_i]^2 \tag{1.15}$$

The z-score and p-value were predicted to evaluate the statistical significance level. A very high positive or negative z-score represents similarity in a score of surrounding to a spatial unit. A low positive or negative z-score denotes spatial outlier consistency. A less statistically significant score does not make any associations.

1.6.2.10 Hot Spot Analysis

Hot spot analysis was deployed to identify groupings within a spatial dataset using **Getis-Ord Gi*** statistic as a geostatistical tool in the ArcGIS environment. In this study, a fixed distance band has been used with a threshold distance value of 928.23 m to identify cold spots and hot spots of the conditions of spatial units (Sánchez-Martín et al., 2019; Jana & Sar, 2016). The mathematical formula for hot spot analysis (ESRI) is derived by (Eqs. 1.16–1.18)

$$G_i = \frac{\sum_{j=1}^{n} w_{i,j} w_j - \bar{x} \sum_{j=1}^{n} w_{i,j}}{S\sqrt{\frac{\left[n\sum_{j=1}^{n} w_{i,j}^2 - \left(\sum_{i,j}^{n} w_{i,j}\right)^2\right]}{n-1}}} \quad (1.16)$$

where x_j is the attribute value of j , $w_{i,j}$ is the particular weight between the feature i and j, and n is equal to the total number of features.

and

$$\bar{X} = \frac{\sum_{j=1}^{n} x_j}{n} \quad (1.17)$$

$$S = \sqrt{\frac{\sum_{j=1}^{n} x_j^2}{n} - (\bar{X})^2} \quad (1.18)$$

G_i^* statistic is denoted by a z-score. The score of the nearest entity is an essential factor in establishing statistically significant spatial grouping. Statistically significant hot spots and cold spots are characterized by high and low scores, respectively, in the surroundings of local entities (Sánchez-Martín et al., 2019). The spots indicate that the decent condition of spatial units reached a high score and vice versa. The positive and negative z-scores were used to identify the high (hot spot) or low (cold spot) score spatial clustering. In contrast, a z-score in a zero neighborhood indicates no significant spatial clustering.

1.6.2.11 Analytical Hierarchy Process (AHP) Model

In the late 1970s, Thomas Saaty has developed the Analytical Hierarchy Process (AHP) in the framework of the multi-criteria decision-making (MCDM) system as a pairwise comparison technique where various weights and scores are assigned, which represents the relative position of pair's criteria (Saaty, 1994) based on his scale (Saaty, 1980).

In medical geography, the application of AHP is rare and has not been copiously explored (Soltani & Marandi, 2010). Siddiqui et al. (1996) first combined the GIS and AHP models to solve the site selection problems (Ahmed et al., 2016). They apply the AHP model in decision-making to solve complex problems concerning subjective and objective aspects. The AHP generates a score or weight for the evaluation criterion through the decision-maker's pairwise comparisons by which bias in the decision-making processes can be reduced and make strong reliability of the evaluations by decision-makers (Wang et al., 2018; Ahmed et al., 2016; Hao & Dai, 2013; Korpela & Tuominen, 1996). With the help of GIS, SPSS, and Microsoft Excel, the evaluation process is accomplished through a pairwise comparison matrix based on selecting different criteria. Author-induced scores and weights are combined by the measurement scale of Saaty (Abdullahi & Pradhan, 2012). For spatial mapping, environmental and socioeconomic factors have been incorporated with experts' knowledge, which is normalized and converted to Saaty's scale. Finally, the model has developed suitability classes in the GIS algorithm, which are classified from 1 (very poor) to 4 (most suitable) as per the needs of the objectives.

1.6.2.12 Location-Allocation Model

Location allocation is mathematical programming as an optimization model, which is designed to locate optimum sites for the new service provisions as well as identify the coverage and non-coverage population demand points across a city or region in the context of the spatial distribution of the resources with the aim of proper allocation of services. Facility, demand, distance cutoff, and location are the major components of this model. The basic principles of this model are multiple services providing locations and places of demand where services can be allocated. The significant areas of applying the location-allocation model are increasing service accessibility, minimizing cost, and planning interventions (Church & Medrano, 2018). Location-allocation analysis is an essential tool in the ArcGIS Network Analyst extension. The present study has used this model to find new healthcare sites and maximize accessibility to healthcare centers.

1.6.2.13 Speed Model Builder

Speed model builder is a visual programming language in ArcGIS Pro that is a predictive model that recognizes the spatial accessibility of facilities across a city or region. Road conditions, especially distance from the road and slope of the terrain, are the prerequisite conditions in this model (Table 1.3), determining the vehicle speed and spatial accessibility to the facilities. This model can also find the shortest path to reach the accessible service points. Moreover, it can use a speed model builder as a map tool in a geoprocessing system. The study used this model to find spatial accessibility to public healthcare facility centers.

Table 1.3 Outlook and conditions of speed model builder (Source: Das et al., 2023)

Conditions	Speed (km/h)	Distance from road (m)	Slope (degree)
On-road, high speed—low slope	30	≤20	≤9
On-road, high slope—moderate speed	20	≤20	>9
Off-road, low slope—low, moderate speed	15	≥20	≤9
Off-road, high slope—low speed	10	≥20	>9

1.6.2.14 Qualitative Methods

Already, we have understood the application part of quantitative approaches, which represent the quantitative value of the spatial point patterns and give more weight to data analysis. This approach focuses little on people's feelings, experiences, beliefs, and attitudes. We need to understand the facts of individual health coping mechanisms rather than quantification on a map. Therefore, qualitative methods have been given more emphasis on data collection. Hence, we must understand people's experiences by participating in their daily livelihoods and spending a long time with them. The primary data collection methods are interviews, participant observation, questionnaires, and focus group discussions. Many methods analyze qualitative data, such as enumerative methods, content analysis, rating scale, investigative methods, discourse analysis, iterative methods, coding methods, grounded theory, etc. Different computer software packages have been used to analyze the data and visualization on maps (MacKian, 2000).

Recently, geographers have emphasized the mixed method, where qualitative and quantitative data are collected and analyzed simultaneously, the so-called triangulation. In the context of health geography, both methods are needed to understand the people's overall health scenario and planning interventions.

1.6.3 Visualization and Mapping

We will wish to represent the spatial and non-spatial data through a map to show the spatial pattern of a particular phenomenon, known as visualization. In health geography, geographic information systems (GIS) have been used widely for visual evidence of the correlation between social and environmental factors. GIS is a computer-based programming system that can edit, manipulate, integrate, visualize, and analyze spatial and non-spatial data and visualize the analyzed data derived from other programming systems (e.g., SPSS, Excel, etc.) through geospatial techniques. Moreover, GIS can decide based on the input database and modeling for an output, which may contribute to the planning processes. Geospatial analysis through GIS in the context of health status has been used in the study for measuring accessibility to the services, risk zone, disease prevalence, connectivity, spatial modeling for the healthcare facilities, finding the shortest distance and travel time, and decision support system. Nowadays, these visualizations have been designed to be available to everyone through different programming applications for better healthcare information.

The Global Positioning System (GPS) is another tool for collecting locational data about healthcare facilities, disease progression, or particular areas. It plots these data on a map to represent the distributional pattern of such phenomenon across the study area. In the deficient database region, GPS can help to collect the database locally. Satellite imagery can be used to detect temporal information for a particular region. With the help of remote sensing techniques, we can detect temporal changes over the region. Cartographic representation of maps showing different types of cartograms and maps is used to find the spatial distribution pattern in the study area, such as choropleth maps, isodemographic maps, health status maps, etc.

1.6.3.1 Design of the Study

The present study is divided into seven chapters. The first chapter traces the general introduction and methodological design of the study. The second chapter provides a detailed description of the socioeconomic entity of slum neighborhoods in the study area through different indices using inductive and deductive approaches. The third chapter describes lifestyle behavior and health awareness among the slum neighborhoods in the study area. This chapter analyzes personal hygiene, health risks, and health security among slum dwellers. Empirical data was collected from the slum neighborhoods and analyzed using principal component analysis and constructing a health index. The fourth chapter explains the infrastructure and provision of healthcare services by assessing the healthcare system and facilities, using disease prevention equipment, and identifying suitable sites for healthcare centers using statistical and geospatial analysis. The fifth chapter examines the utilization and accessibility of healthcare services in slum neighborhoods. This chapter also studies the spatial accessibility to healthcare centers using geospatial analysis and emphasizing women and child healthcare services. The sixth chapter deals with the nature of health coping mechanisms among slum dwellers in different age groups. This chapter is also concerned with building coping indices using empirical data through statistical analysis. The seventh chapter is the last chapter of the study, which deals with the study's significant findings, conclusions drawn from overall studies, and primary suggestions for the overall development of livelihood status among the urban poor in the study area.

1.7 Selection of the Study Area

Midnapore municipality is an essential administrative headquarter since the undivided Medinipur district is located in the central part of Paschim Medinipur district in West Bengal, established in 1865. Medinipur was established after the name of

Medinikar (from 1200 to 1431 A.D.), the famous author of "Medinikosh." It has had a long, magnificent urban history since the sixteenth century. Initially, Midnapore was under the kingship of Jaleswar, and the city was popularly known as a city of lavishness. After Medinikar, Midnapore was under the regime of Muslim rule, and Nawabs like Alibardi, Siraj-ud-Daulah, Mir Jafar, Mir Qasim, and others stayed here for a long time. Subsequently, Midnapore town was recognized as the district headquarter on the political map of Bengal based on geopolitical, commercial, and cultural dominance, and it gained urban status on 22 September 1783 under British rule. Thus, Midnapore has been well known as an old town that serves as the solitary most influential urban center of a vast hinterland with poor development.

Now, the Midnapore municipal area consists of 25 wards (in 2013) with an area of 2000 hectares (20 km^2), and it has been recognized as a class I city with a population of 169,127 (Census of India, 2011). It extends from 87°17′ 18.57″ to 87°20′ 30.12″ east longitude and 22° 23′ 44.56″ to 22° 26′ 34.91″ north latitude (Fig. 1.6). It is bounded by the banks of the Kangsabati River (differently known as Kasai and Cossye) in the south. It is well-connected with roads (NH-60) in the east; railway

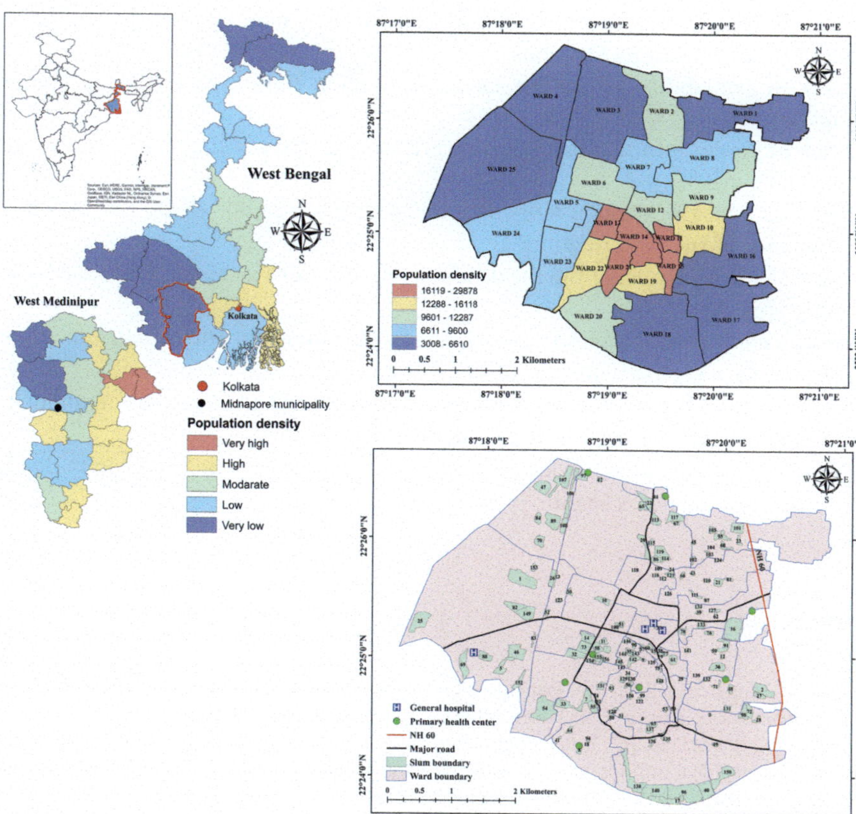

Fig. 1.6 Location map of the study area

lines to the state capital, Kolkata; forest and agricultural land to the west; and agricultural land in the north. Recent traffic signals, good infrastructure facilities, multistoried buildings and shopping stores, small-scale industries, educational centers, healthcare centers, etc. are well organized. Therefore, Midnapore has been multiplying with the population, which indicates that it is a developed city due to rapid rural-to-urban migration and dependency on its hinterlands by providing administrative, educational, and healthcare facilities. However, beyond the curtain, there lies the unhygienic environment of the slum with no proper sanitation, no sufficient space for dwelling, and no sign of least comfort of day-to-day life, which leads to poor living conditions, chronic public health, as well as child health problems.

1.8 Limitation of the Present Study

There are some constraints during the survey as well as in different steps under study, which are given below:

- Lack of information regarding doctor-patient ratio, number of beds, and other facilities in the district hospitals.
- Lack of availability of health-related information from private nursing homes and hospitals.
- Lack of proper information on healthcare services for working male groups and aged persons.
- There is a lack of data on non-communicable diseases due to the respondents' avoidance of such information.
- Lack of secondary data from the different organizations and authorities about the study area.
- More samples are needed to validate the results, but due to affordability and time constraints, a standard number of samples were collected for analysis.
- Lack of information regarding women's healthcare and problems due to unintentional responses among the women in the study area.
- More trained women interviewers have been required to collect data on in-depth problems of women's health.

References

Abdi, H., & Williams, L. J. (2010). Principal component analysis. *Wiley Interdisciplinary Reviews: Computational Statistics, 2*, 433–459. https://doi.org/10.1002/wics.101

Abdullahi, S., & Pradhan, B. (2012). Spatial modeling of site suitability assessment for hospitals using geographical information system-based multicriteria approach at Qazvin city, Iran. July 2015. https://doi.org/10.1080/10106049.2012.752531.

Adam, A. W. (2013). *Perceptions of slum dwellers and municipal officials on factors impacting the provision of basic slum Services in Accra, Ghana*. International Institute of Social Studies.

Agarwal, S., & Taneja, S. (2005). All slums are not equal: Child health condition among the urban poor. *Indian Pediatrics, 12*.

Ahmed, A. H., Mahmoud, H., & Aly, A. M. M. (2016). Site suitability evaluation for sustainable distribution of hospital using spatial information technologies and AHP : A case study of upper Egypt, Aswan City. https://doi.org/10.4236/jgis.2016.85048

Ajala, O. A., Sanni, L., & Adeyinka, S. A. (2017). Accessibility to health care facilities: A panacea for sustainable rural development in Osun State Southwestern, Nigeria Accessibility to Health Care Facilities: A Panacea for Sustainable Rural Development in Osun State Southwestern, Nigeria. 9274. https://doi.org/10.1080/09709274.2005.11905819

Akinsola, H. A. (1993). Community Health and social medicine in medical and nursing practice. 3 AM communications, Ibadan, Nigeria.

Aksha, S. K., Juran, L., Resler, L. M., & Zhang, Y. (2019). An analysis of social vulnerability to natural hazards in Nepal using a modified social vulnerability index. *International Journal of Disaster Risk Science, 10*(1), 103–116. https://doi.org/10.1007/s13753-018-0192-7

Alabi, M. O. (2011). Towards a sustainable distribution of health centers using GIS. *American Journals of Tropical Medicine and Public Health.*

Al-Dubai, S. A., Al-Naggar, R. A., Alshagga, M. A., & Rampal, K. G. (2011). Stress and coping strategies of students in a medical faculty in Malaysia. Malays. *Journal of Medical Sciences, 18*(3), 57–64. PMID: 22135602; PMCID: PMC3216229.

Aldwin, C. M., & Levenson, M. R. (2001). Stress, coping, and health at midlife: A developmental perspective. In M. E. Lachman (Ed.), *Handbook of midlife development* (pp. 188–214). John Wiley & Sons, Inc.

Amdework, E. (2008). Urban poverty, environment and III-Health a sociological analysis of exposure to environmental health hazards among residents of Akaki-Kaliti Sub-City. *Journal of Ethiopian Studies, 41*(1/2). http://www.jstor.org/stable/41967611

Amendah, D. D., Buigut, S., & Mohamed, S. (2014). Coping strategies among urban poor: Evidence from Nairobi, Kenya. *PLoS One, 9*, 1. https://doi.org/10.1371/journal.pone.0083428

Arefin, K., et al. (2010). Health awareness and knowledge of the slum dwellers: A case study in Bangladesh. *Asian Profile, 38*(4), 427–436.

Arimah, B. C. (2011). Slums as expressions of social exclusion: Explaining the prevalence of slums in African countries. In proceedings of the international conference on social cohesion and development, Paris, 20–21 January 2011 (1–16).

Babbie, E. (2010). *The practice of social research* (12th ed.). Wadsworth.

Baqutayan, S. (2015). Stress and coping mechanisms: A historical overview. *Mediterranean Journal of Social Sciences*. https://doi.org/10.5901/mjss.2015.v6n2s1p479

Boulos, M. N. K., Roudsari, A. V., & Carson, E. R. (2001). Health geomatics: An enabling suite of technologies in health and healthcare, *219*, 195–219. https://doi.org/10.1006/jbin.2001.1015

Brockerhoff, M. P. (2000). An urbanizing world. *Population Reference Bureau (PRB), 55*, 1–48.

Carver, C., Scheier, M., & Weintraub, J. (1989). Assessing coping strategies: A theoretically based approach. *Journal of personality and social psychology., 56*, 267–283. https://doi.org/10.1037/0022-3514.56.2.267

Census of India.: http://censusindia.gov.in/; (2001 and 2011).

Census of India 1961. (1961, 2011).

Chen, J. (2014). GIS-based multi-criteria analysis for land use suitability assessment in the City of Regina. *Environmental Systems Research, 3*, 13. https://doi.org/10.1186/2193-2697-3-13

Cheng, P. W. H., & Chu, T. M. (2007). Health care insurance in Taiwan. *Asian Studies Review, 33*, 303–317.

Chokshi, M., et al. (2016). Health system in India. *Journal of Perinatology, 36*, S9–S12. https://doi.org/10.1038/jp.2016.184. www.nature.com/jp

Christaller, W. (1966). *Central places in Southern Germany*. Prentice-Hall.

Church, R., & Medrano, F. A. (2018). Location-allocation modeling and GIS. The Geographic Information Science & Technology Body of Knowledge (3rd Quarter 2018 Edition). John P. Wilson (Ed). https://doi.org/10.22224/gistbok/2018.3.4

Cliff, A. D., & Haggett, P. (1988). Methods for the measurement of epidemic velocity from time-series data. *International Journal of Epidemiology, 11*, 82–89.

Costa, B., & Pinto, I. C. (2017). Stress, burnout and coping in health professionals: A literature review. *Journal of Psychology and Brain Studies., 1*(4), 1–8.

Cutter, S. L., Burton, C. G., & Emrich, C. T. (2010). Disaster resilience indicators for benchmarking baseline conditions. *Journal of Homeland Security and Emergency Management, 7*, 1–22. https://doi.org/10.2202/1547-7355.1732

D'Souza, S. V. (2003). Urbanization as a perspective for social and historical analysis in developing societies. In R. S. Sandhu (Ed.), *Urbanization in India sociological contributions* (pp. 49–57).

Dahal, K. (2017). Livelihood and coping strategies among urban poor people in post-conflict period: Case of the Kathmandu, Nepal. *The Geographical Journal of Nepal, 10*, 73–88.

Danjuma, I. (2015). Locational analysis of healthcare facilities in Keffi Local Government Area of Nasarawa State, Nigeria. Ph.D. Thesis. http://kubanni.abu.edu.ng/jspui/bitstream

Das, et al. (2023). Spatial accessibility modeling to healthcare facilities in the case of health shocks of Midnapore municipality, India. *GeoJournal, 88*, 3749–3772. https://doi.org/10.1007/s10708-023-10838-1

Davis, M. C., Matthews, K. A., & Twamley, E. W. (1999). Is life more difficult on Mars or Venus? A meta-analytic review of sex differences in major and minor life events. *Annals of Behavioral Medicine, 21*(1), 83–97. https://doi.org/10.1007/BF02895038

DeLongis, A., & Holtzman, S. (2005). Coping in context: The role of stress, social support, and personality in coping. *Journal of Personality, 73*(6), 1633–1656. https://doi.org/10.1111/j.1467-6494.2005.00361.x

Dinda, S., Chaterjee, N. D., Das, K., & Ghosh, S. (2018). *Integration of GIS and statistical approach in the mapping of urban sprawl and predicting future growth in Midnapore town*. Springer.

Dolui, G., Soumen, D., & Sujan, S. (2014). An application of Remote Sensing and GIS to Analyze Urban Expansion and Land use the Land cover change of Midnapore Municipality, WB, India. *The International Research Journal of Earth Science, 2*(5), 8–20. ISSN 2321–2527.

Dom, N. C., et al. (2012). Manifestation of GIS tools for spatial pattern distribution analysis of dengue fever epidemic in the city of Subang Jaya, Malaysia. *EnvironmentAsia, 5*(2), 82–92.

Dummer, T. J. B. (2008). Health geography: Supporting public health policy and planning. *CMAJ Canadian Medical Association Journal, 178*(9), 1177–1180. https://doi.org/10.1503/cmaj.071783

Eyles, J. D. (1996). Poverty as inequality: A case study. *Progress in Human Geography*, 55–64.

Farley, T., Galves, A., Dickinson, L. M., & Perez, M. J. (2005). Stress, coping, and health: A comparison of Mexican immigrants, Mexican-Americans, and non-Hispanic whites. *Journal of Immigrant Health, 7*(3), 213–220. https://doi.org/10.1007/s10903-005-3678-5

Fiedler, J. L. (1981). A review of the literature on access and utilization of medical care with special emphasis on rural primary care. Social Science & Medicine. Part C. *Medical Economics, 15*(3), 129–142. https://doi.org/10.1016/0160-7995(81)90028-9

Fitzpatrick, K., & LaGory, M. (2010). *Unhealthy cities: Poverty, race, and place in America* (1st ed.). Routledge. https://doi.org/10.4324/9780203843765

Folkman, S., & Lazarus, R. S. (1980). An analysis of coping in a middle-aged community sample. *Journal of Health and Social Behavior, 21*, 219–231.

Folkman, S., & Lazarus, R. S. (1985). If it changes it must be a process: Study of emotion and coping during three stages of a college examination. *Journal of Personality and Social Psychology, 48*(1), 150–170. https://doi.org/10.1037/0022-3514.48.1.150

Folkman, S., & Moskowitz, J. T. (2004). Coping: Pitfalls and promise. *Annual Review of Psychology, 55*, 745–774.

Ford, M. E., & Kelly, P. A. (2005). Conceptualizing and categorizing race and ethnicity in health services research. *Health Services Research, 40*(5 II), 1658–1675. https://doi.org/10.1111/j.1475-6773.2005.00449.x

Fontana, D. (1996). Psychology for educators. Savvalas Editions.
Frech, H. E., & Mobley, L. R. (2000). Efficiency, growth, and concentration: An empirical analysis of hospital markets. *Economic Inquiry, 38*(3), 369–384. https://doi.org/10.1111/j.1465-7295.2000.tb00024.x
Freud S. (1926). Inhibitions, symptoms and anxiety. Standard edition;20:87–187
Gale, S., Lovett, A., Haynes, R., & Sabel, C. E. (2002). Car travel time and accessibility by bus to general practitioner services: A study using patient registers and GIS. *Social Science & Medicine, 55*(1), 97–111. https://doi.org/10.1016/S0277-9536(01)00216-3
Gorsevski, P. V., Twumasi, Y. A., & Turki, J. M. (2012). Integrating multi-criteria evaluation techniques with geographic information systems for landfill site selection: A case study using ordered weighted average. *Environmental Management, 49*(2), 287–296. https://doi.org/10.1007/s00267-011-9781-4
Grant, U. (2005). Health and poverty linkages: Perspectives of the chronically poor (chronic poverty research Centre working paper). *Chronic Poverty Research Centre*. https://assets.publishing.service.gov.uk/media/57a08c97e5274a27b20012d1/other-grant-health.pdf
Guagliardo, M. F. (2004). Spatial accessibility of primary care: Concepts, methods and challenges. *International Journal of Health Geographics, 3*(1), 1–13. https://doi.org/10.1186/1476-072X-3-1
Guiqin, W., et al. (2009). Landfill site selection using spatial information technologies and AHP: A case study in Beijing, China. *Journal of Environmental Management*.
Gupta, et al. (2008). Reproductive and child health inequities in Chandigarh Union Territory of India. *Urban Health, 85*(2), 291–299.
Hao, Q. L., Dai, R. (2013). Dalian LNG terminal port site selection based on the fuzzy evaluation. *39*, 103–106, 111.
Haque, M. A., Budi, A., Azam Malik, A., et al. (2013). Health coping strategies of the people vulnerable to climate change in a resource-poor rural setting in Bangladesh. *BMC Public Health, 13*, 565. https://doi.org/10.1186/1471-2458-13-565
Hardoy, et al. (1997). *Environmental problems in third world Cities*. Earth Scan Publications.
Heltberg, R., & Lund, N. (2009). Shocks, coping, and outcomes for Pakistan's poor: Health risks predominate. *The Journal of Development Studies, 45*(6), 889–910. https://doi.org/10.1080/00220380902802214
Henderson, C., Evans-Lacko, S., & Thornicroft, G. (2013). Mental illness stigma, help seeking, and public health programs. *American Journal of Public Health, 103*(5), 777–780. https://doi.org/10.2105/AJPH.2012.301056. Epub 2013 Mar 14. PMID: 23488489; PMCID: PMC3698814.
Hoseinzadeh, E., Wei, C., Safari, M., & Godini, H. (2016). Evaluation of rainwater quality using factor analysis: Case study of Khorramabad in western Iran. *Desalination and Water Treatment, 57*(53), 25345–25357. https://doi.org/10.1080/19443994.2016.1154895
Hu, R., Dong, S., Zhao, Y., Hu, H., & Li, Z. (2013). Assessing potential spatial accessibility of health services in rural China: A case study of Donghai County. *International journal for equity in health., 12*, 35. https://doi.org/10.1186/1475-9276-12-35
Jain, K., & Subbaiah, Y. V. (2007). Site suitability analysis for urban development using GIS. *Journal of Applied Sciences, 7*(18), 2576–2583. https://doi.org/10.3923/jas.2007.2576.2583
Jana, M., & Sar, N. (2016). Modeling of hotspot detection using cluster outlier analysis and Getis-Ord Gi* statistic of educational development in upper-primary level, India. *Modeling Earth Systems and Environment, 2*(2), 1–10. https://doi.org/10.1007/s40808-016-0122-x
Jones, S. G., Ashby, A. J., Momin, S. R., & Naidoo, A. (2009). Spatial implications associated with using Euclidean distance measurements and geographic centroid imputation in health care research. 316–327. https://doi.org/10.1111/j.1475-6773.2009.01044.x
Judge, S. L. (1998). Parental coping strategies and strengths in families of young children with disabilities. *Family Relations, 47*, 263–268. https://doi.org/10.2307/584976

Kantharia, S. L. (2010). Urban health issues in India- need of the day, editorial. *National Journal of Community Medicine, 1*, 1.

Karn, S. K., et al. (2003). Living environment and health of urban poor: A study in Mumbai. *Economic and Political Weekly, 38*(34), 23–29. 3575-3577+3579-3586. http://www.jstor.org/stable/4413939

Khan, A. A. (1992). An integrated approach to measuring potential spatial access to health care services. *Socioeconomic Planning Sciences, 26*(4), 275–287. https://doi.org/10.1016/0038-0121(92)90004-O

Kim, Y., Byon, Y., & Yeo, H. (2018). Enhancing healthcare accessibility measurements using GIS : A case study in Seoul, Korea. 1–19.

Knowledge Steez. (2019). 3rd World conference on human rights and intellectual property rights 2019. https://knowledgesteez.wordpress.com/2019/03/08/3rd-world-conference-on-human-rights-and-intellectual-property-rights-2019/

Korpela, J., & Tuominen, M. (1996). A decision aid in warehouse site selection. *International Journal of Production Economics, 45*, 169–180.

Kroeger, A. (1983). Anthropological and socio-medical health care research in developing countries. *Social Science & Medicine, 17*(3), 147–161. https://doi.org/10.1016/0277-9536(83)90248-4

Krohne. (2002). *Stress and coping theories*. Johannes Gutenberg-Universität Mainz Germany. https://userpage.fu-berlin.de

Lahon, S. (2017). Educational status and level of health awareness of the children of urban slums with special reference to Guwahati city—A study. *International Journal of Applied Research, 3*(3), 680–686.

Langford, M., & Higgs, G. (2006). Measuring potential access to primary healthcare services: The influence of alternative spatial representations of. *Population, 58*(March 2005), 294–306.

Lawal, O., & Anyiam, F. E. (2019). Geo-spatial information science modelling geographic accessibility to primary health care facilities : Combining open data and geospatial analysis modelling geographic accessibility to primary Health care facilities : Combining open data and geospatial analysis. *Geo-spatial Information Science, 22*(3), 174–184. https://doi.org/10.1080/10095020.2019.1645508

Lazarus, R. S., & Folkman, S. (1984). *Stress, appraisal, and coping*. Springer.

Lim, Y. M., Tam, C. L., & Lee, T. H. (2013). Perceived stress, coping strategy and general health: A study on accounting students in Malaysia. *Journal of Arts, Science and Commerce, IV*(1), 88–95.

MacKeown, J. M., & Faber, M. (2002). Urbanisation and cariogenic food habits among 4-24-month-old black South African children in rural and urban areas. *Public Health Nutrition, 5*(6), 719–726. https://doi.org/10.1079/PHN2002358

MacKian, S. (2000). Contours of coping: Mapping the subject world of long-term illness. *Health and Place, 6*(2), 95–104. https://doi.org/10.1016/S1353-8292(00)00006-X

Malaviya, P., & Bhagat, N. K. (2013). Urban poverty and health risk factors: A case study of slum dwellers of Jammu (J & K) India. *Health and Population: Perspectives and Issues, 36*(1), 16–29.

Malczewski, J. (2006). GIS-based multi-criteria decision analysis: A survey of the literature. *International Journal of Geographical Information Science, 20*(7), 703–726.

Malqvist, M., Sohel, N., Do, T. T., et al. (2010). Distance decay in delivery care utilisation associated with neonatal mortality. A case referent study in northern Vietnam. *BMC Public Health, 10*, 762. https://doi.org/10.1186/1471-2458-10-762

Mandal, K., Soumendu, C., & Chatterjee, N. D. (2016). Morphological analysis of a historical urban landscape: The case of midnapore, an early urban centre of Eastern India. *The International Research Journal of Earth Science, 4*(12), 28–37. E-ISSN 2321–2527.

Mansour, S. (2016). Geo-spatial information science spatial analysis of public health facilities in Riyadh governorate, Saudi Arabia : A GIS-based study to assess geographic variations of service provision and accessibility. *Geo-spatial Information Science, 5020*, 1–13. https://doi.org/10.1080/10095020.2016.1151205

Sánchez-Martín, J.-M., Rengifo-Gallego, J.-I., & Blas-Morato, R. (2019). Hot spot analysis versus cluster and outlier analysis: An enquiry into the grouping of rural accommodation in Extremadura (Spain). *Journal of Rural and Community Development, 14*(1), 101–115.

McCubbin, H. I., Joy, C. B., Cauble, A. E., Comeau, J. K., Patterson, J. M., & Needle, R. H. (1980). Family stress and coping: A decade review. *Journal of Marriage and the Family, 42*(4), 855. https://doi.org/10.2307/351829

McGrail, M. R., & Humphreys, J. S. (2014). Measuring spatial accessibility to primary Health care services: Utilising dynamic catchment sizes. *Applied Geography, 54*, 182–188. https://doi.org/10.1016/j.apgeog.2014.08.005

Mehta, D. (1995). Theory of slum revisited: A labour market perspective of housing problems of the urban poor. In P. Roy & S. D. Gupta (Eds.), *Urbanisation and slums*. Har Anand Publications.

Midnapore Municipality. (2022). Ward Services. https://midnaporemunicipality.com/

Ministry of Health and Family Welfare, Government of India. (2011). Annual report 2010–11. https://nhm.gov.in/index1.php?lang=1&level=3&sublinkid=957&lid=405

Ministry of Health and Family Welfare, Government of India. (2013). Annual report 2012–13. https://nhm.gov.in/index1.php?lang=1&level=3&sublinkid=957&lid=405

Moisi, J. C., et al. (2010). Geographic access to care is not a determinant of child mortality in a rural Kenyan setting with high health facility density. *BMC Public Health*. http://www.biomedcentral.com/1471-2458/10/14

Montgomery, R. M. (2009). Urban poverty and health in developing countries. Population Bulletin. Washington, DC.

Mukherji, A. (2001). Towards a sustainable city: The problem of environment and urban poverty in Medinipur. http://www.anthrobase.com/txt/M/Mukherji_A_01.htm.%0A%0A, https://doi.org/10.1016/j.envint.2005.10.003.

Muller, I., Smith, T., Mellor, S., & Genton, B. (1998). The effect of distance from home on attendance at a small rural health center in Papua New Guinea. i, 878–884.

Murad, A. A. (2008). Defining health catchment areas in Jeddah city, Saudi Arabia : an example demonstrating the utility of geographical information systems. *2*(2), 151–160.

Murseli, R., & Isufi, F. (2014). Land suitability location analysis for housing development using GIS. UBT International Conference, 18, 187–198. https://doi.org/10.33107/ubt-ic.2014.18

Naydenova, E., Raghu, A., Ernst, J., Sahariah, S. A., Gandhi, M., & Murphy, G. (2017). Healthcare choices in Mumbai slums: A cross-sectional study. *Wellcome Open Research, 2*, 115. https://doi.org/10.12688/wellcomeopenres.13127.2. PMID: 30027122; PMCID: PMC6039940.

Nikolakaki, P. A. (2004). GIS site-selection process for habitat creation: Estimating connectivity of habitat patches. *Landscape and Urban Planning, 68*, 77–94.

Nishtar, S. (2010). *Choked pipes: Reforming Pakistan's mixed health system*. Oxford University Press.

Nolan, L. B. (2015). Slum definitions in urban India: Implications for the measurement of Health inequalities. *Population and Development Review, 41*(1), 59–84. http://www.jstor.org/stable/24639396

Okwaraji, Y. B., Mulholland, K., Schellenberg, J. R. M. A., Andarge, G., Admassu, M., & Edmond, K. M. (2012). The association between travel time to health facilities and childhood vaccine coverage in rural Ethiopia. A community based cross sectional study. *BMC Public Health, 12*(1), 1–9. https://doi.org/10.1186/1471-2458-12-476

Onda, K., Sinha, P., Gaughan, A. E., et al. (2019). Missing millions: Undercounting urbanization in India. *Population and Environment, 41*, 126–150. https://doi.org/10.1007/s11111-019-00329-2

Onokerhoraye, A. G. (1999). Access and utilization of modern health care facilities in the petroleum-producing region of Nigeria: The case study of Bayelsa State. http://www.hspl.harvard.edu/research/taken./file/rep162.pdfj.

Oyerinde, O. O. (2006). Inequality in the distribution of social amenities: A case study of Ilorin Metropolis (Doctoral dissertation, B. Sc. Dissertation, Department of Geography, University, of Ilorin).

Papathanasiou, I. V., Tsoras, K., Neroliatsiou, A., & Roupa, A. (2015). Stress: Concepts, theoretical models and nursing interventions. *American Journal of Nursing Science, 4*(2–1), 45–50. https://doi.org/10.11648/j.ajns.s.2015040201.19

Parry, J. A., Ganaie, S. A., & Bhat, M. S. (2018). GIS-based land suitability analysis using AHP model for urban services planning in Srinagar and Jammu urban centers of J & K, India. *Journal of Urban Management, 7*(2), 46–56. https://doi.org/10.1016/j.jum.2018.05.002

Paul, P. (2018). Land use character in urban periphery: A case study at Midnapore Municipality, Vol. 8 (3), December 2018, www.lbp.world, ISSN: 2249-894X.

Pernia, E. M. (1994). *Urban poverty in Asia: A survey of critical issues*. Oxford University Press.

Perry, B., & Gesler, W. (2000). Physical access to primary Health Care in Andean Bolivia. *Social Science & Medicine, 50*, 1177–1188. https://doi.org/10.1016/S0277-9536(99)00364-0

Planning Commission of India. (2013). Twelfth five-year plan (2012–2017): Faster, more inclusive and sustainable growth. *NITI Aayog*. https://www.niti.gov.in/sites/default/files/2023-08/12fyp_vol1.pdf

Pyrialakou, V. D., Gkritza, K., & Fricker, J. D. (2016). Accessibility, mobility, and realized travel behavior: Assessing transport disadvantage from a policy perspective. *Journal of Transport Geography, 51*, 252–269. https://doi.org/10.1016/j.jtrangeo.2016.02.001

Rabenu, E., Yaniv, E., & Elizur, D. (2017). The relationship between psychological capital, coping with stress, Well-being, and performance. *Current Psychology: A Journal for Diverse Perspectives on Diverse Psychological Issues, 36*(4), 875–887. https://doi.org/10.1007/s12144-016-9477-4

Rahimi, F., Goli, A., & Rezaee, R. (2017). Hospital location-allocation in Shiraz using Geographical Information System (GIS). *18*(8). https://doi.org/10.5812/semj.57572.Research.

Rahman, & Nahar. (2010). Factors influencing health and healthcare delivery system for the urban poor in Chittagong city, Bangladesh. *Asian Journal of Management Research, 4*, 2.

Rakodi, C. (1995). Poverty lines or household strategies? *Habitat International, 19*(4), 407–426.

Rasheed, S., Siddique, A. K., Sharmin, T., Hasan, A. M. R., & Hanifi, S. M. A. (2016). Salt intake and health risk in climate change vulnerable coastal Bangladesh: What role do beliefs and practices play? 1–15. https://doi.org/10.1371/journal.pone.0152783.

Rathor, A. (2003). *Slum dwellers: Curse on development* (1st ed., ISBN: 81-7625-383-9). Sarup & Sons publication.

Ray, B. (2017). Quality of life in selected slums of Kolkata: A step forward in the era of pseudo-urbanization. *Local Environment, 22*(3), 365–387. https://doi.org/10.1080/13549839.2016.1205571

Reams, M. A., Lam, N. S. N., & Baker, A. (2012). Measuring capacity for resilience among coastal counties of the U. S. Northern Gulf of Mexico Region. *American Journal of Climate Change, 01*(04), 194–204. https://doi.org/10.4236/ajcc.2012.14016

Resnick, H. S., Acierno, R., & Kilpatrick, D. G. (1997a). Health impact of interpersonal violence 2: Medical and mental health outcomes. *Behavioral Medicine, 23*(2), 65–78. https://doi.org/10.1080/08964289709596730

Resnick, M. D., Bearman, P. S., Blum, R. W., et al. (1997b). Protecting adolescents from harm: Findings from the National Longitudinal Study on adolescent Health. *Journal of the American Medical Association, 278*(10), 823–832. https://doi.org/10.1001/jama.1997.03550100049038

Riley, L. W., Ko, A. I., Unger, A., et al. (2007). Slum health: Diseases of neglected populations. *BMC International Health and Human Rights, 7*, 2.

Roberts, B. (1994). Informal economy and family strategies. *Journal of Urban and Regional Research, 18*(1), 6–23.

Ross, H., & Chaloupka, F. (2004). The effect of public policies and prices on youth smoking. *Southern Economic Journal, 70*, 796–815. https://doi.org/10.2307/4135273

Roy, U. (2016). Causative factors of social inequality and its impact on community health: A neighbourhood level study in Midnapore Municipal Area, West Bengal, India. The International Archives of the Photogrammetry, Remote Sensing and Spatial Information Sciences, XLI-B8 201. https://doi.org/10.5194/isprs-archives-XLI-B8-1417-2016

Ruishan, et al. (2013). Assessing potential spatial accessibility of health services in rural China: A case study of Donghai county. *International Journal for Equity in Health, 12*(35).

Saaty, T. (1980). *The analytical hierarchy process* (pp. 20–25). McGraw-Hill.

Saaty, T. (1994). Highlights and critical points in the theory and application of the analytic hierarchy process. *European Journal of Operational Research, 74*(3), 426–447.

Sandipan, D., Anirban, B., & Sagar, M. (2013). Study on urban land suitability assessment using remote sensing and GIS: A case study of Khairagarh, in Chhattisgarh. *International Journal of Computer Applications, 74*, 10.

Savigny, D. D., & Wijeyaratne, P. (1995). GIS for health and the environment: proceedings of an international workshop held in Colombo, Sri Lanka, 5–10 Sept. 1994.

Sedenu et al. (2016). Spatial analysis of the distribution pattern of primary healthcare facilities in spatial analysis of the distribution pattern of primary healthcare facilities in Ile—Ife Metropolis using geographic information system. June. https://doi.org/10.9734/JSRR/2016/22339.

Shannon, G. W., Bashshur, R. L., & Metzner, C. (1969). The concept of distance as a factor in accessibility and utilization of health care. *Med Care Review, 26*, 143–161.

Sharma, N. C. (2020). Why are smaller cities in India witnessing faster growth? Economic & Political Weekly, 55(10), 47–53.

Sheuya, S. A. (2008). Improving the health and lives of people living in slums. *1136*, 298–306. https://doi.org/10.1196/annals.1425.003.

Shukla, A., Kumar, V., & Jain, K. (2017). Site suitability evaluation for urban development using remote sensing, GIS, and analytic hierarchy process (AHP). *Proceedings of International Conference on Computer Vision and Image Processing*. https://doi.org/10.1007/978-981-10-2107-7_34

Siddiqui, M. Z., Everett, J. W., & Vieux, B. E. (1996). Landfill siting using geographic information systems: A demonstration. *Journal of Environmental Engineering, 122*(6), 515–523.

Small, R., Roth, C., Raval, M., et al. (2014). Immigrant and nonimmigrant women's experiences of maternity care: A systematic and comparative review of studies in five countries. *BMC Pregnancy and Childbirth, 14*, 152.

Smith, A. P., & Mark, G. (2011). Occupational stress, job characteristics, coping, and the mental health of nurses. *British Journal of Health Psychology*. https://doi.org/10.1111/j.2044-8287.2011.02051.x

Solangi, G. S., Siyal, A. A., Babar, M. M., & Siyal, P. (2019). Groundwater quality evaluation using the water quality index (WQI), the synthetic pollution index (SPI), and geospatial tools: A case study of Sujawal district, Pakistan. *Human and Ecological Risk Assessment: An International Journal, 26*(6), 1529–1549. https://doi.org/10.1080/10807039.2019.1588099

Soltani, A., & Marandi, E. (2010). Hospital site selection using a two-stage fuzzy multi-criteria decision-making process. *Journal of Urban and Environmental Engineering, 5*(1), 32–43.

Sundari, S. (2003). Quality of life of migrant households in urban slums. In Martin J. Bunch, V. Madha Suresh and T. Vasantha Kumaran, eds., Proceedings of the Third International Conference on Environment and Health, Chennai, India, 15–17 December 2003. Chennai: Department of geography, University of Madras and Faculty of Environmental Studies, York University, 537–552.

Svetaz, M. V., Ireland, M., & Blum, R. (2000). Adolescents with learning disabilities: Risk and protective factors associated with emotional Well-being: Findings from the National Longitudinal Study of adolescent Health. *Journal of Adolescent Health, 27*(5), 340–348. https://doi.org/10.1016/S1054-139X(00)00170-1

Tanaka, H., Taira, K., Arakawa, M., Masuda, A., Yamamoto, Y., Komoda, Y., Kadegaru, H., & Shirakawa, S. (2002). An examination of sleep health, lifestyle and mental health in junior high school students. *Psychiatry and Clinical Neurosciences, 56*, 235–236. https://doi.org/10.1046/j.1440-1819.2002.00997.x

Tangcharoensathien, V., Mills, A., & Palu, T. (2015). Accelerating health equity: The key role of universal Health coverage in the sustainable development goals. *BMC Medicine, 13*, 101. https://doi.org/10.1186/s12916-015-0342-3

Uchino, B. N. (2004). Social support and physical health: Understanding the health consequences of relationships. Yale University Press. https://doi.org/10.12987/yale/9780300102185.001.0001.
UN-Habitat. (2003). Adequate housing and slum upgrading: Module-1.
UN-Habitat. (2020). World Cities report 2020: The value of sustainable urbanization. United Nations Human Settlements Programme.
United Nations. (1992). Agenda 21: Programme of action for sustainable development. *United Nations Conference on Environment and Development*. https://sustainabledevelopment.un.org/content/documents/Agenda21.pdf
United Nations, Department of Economic and Social Affairs. (2013). Report on the world social situation 2013: Inequality matters. https://www.un.org/en/development/desa/publications/world-social-situation-2013.html
United Nations Department of Economic and Social Affairs. (2018). The sustainable development goals report 2018. *United Nations*. https://www.un-ilibrary.org/content/books/9789213633175
United Nations Human Settlements Programme (UN-Habitat). (2006). The state of the world's cities 2006/2007: The millennium development goals and urban poverty reduction. Earthscan.
United Nations Human Settlements Programme (UN-Habitat). (2008). State of the world's cities 2008/2009: Harmonious cities. Earthscan. https://unhabitat.org/state-of-the-worlds-cities-20082009-harmonious-cities-2
Uzochukwu, B., Onwujekwe, O., Mbachu, C., Okwuosa, C., Etiaba, E., Nyström, M. E., & L. G. (2016). The challenge of bridging the gap between researchers and policy makers: Experiences of a Health Policy Research Group in engaging policy makers to support evidence-informed policy making in Nigeria. *Globalization and Health, 12*(1), 67. https://doi.org/10.1186/s12992-016-0209-1
Venner, M. (1988). Adjustment, coping and defense mechanisms--deciding factors in the therapeutic process. *Z Gesamte Inn Med, 43*(2), 40–43.
Viken, B., Lyberg, A., & Severinsson, E. (2015). Maternal health coping strategies of migrant women in Norway. Nursing research and practice, 2015.
Wang, F., & Luo, W. (2005). Assessing spatial and nonspatial factors for healthcare access: towards an integrated approach to defining health professional shortage areas. *11*, 131–146. https://doi.org/10.1016/j.healthplace.2004.02.003.
Wang, L., Fan, H., & Wang, Y. (2018). Site selection of retail shops based on spatial accessibility and hybrid BP Neural Network. https://doi.org/10.3390/ijgi7060202.
Washburn-Ormachea, J. M., Hillman, K. L., McVaugh, D. T., McAlister, L. J., & Furlong, D. B. (2004). The interplay of gender, gender-role orientation, and coping with peer and academic stressors in adolescents. *Journal of Youth and Adolescence, 33*(1), 31–40.
World Bank (2021). https://tradingeconomics.com/india/population-living-in-slums-percent-of-urban-population-wb-data.html
World Health Organization. (1981). *Global strategy for health for all by the year*, 2000. https://apps.who.int/iris/handle/10665/38893
World Health Organization. (1999). The world health report 1999: Making a difference. https://apps.who.int/iris/handle/10665/42167.
World Health Organization. (2001). The world health report 2001: Mental health: New understanding, new hope. https://iris.who.int/handle/10665/42390.
World Health Organization & United Nations Children's Fund. (1978). *September 12*. Declaration of Alma-Ata. https://www.who.int/publications/almaata_declaration_en.pdf
World Health Organization. (1989). *Strengthening the performance of community health workers in primary health care: Report of a WHO Study Group [meeting held in Geneva from December 2 to 9, 1987]*. (WHO Technical Report Series, *No. 780*. https://iris.who.int/handle/10665/39568
World Health Organization. (2008). The world health report 2008: Primary health care now more than ever. https://iris.who.int/handle/10665/43949.
Yang, D. H., et al. (2006). Comparing GIS-based methods of measuring spatial accessibility to health services. *Journal Medical Systems, 30*(1), 23–32. https://doi.org/10.1007/s10916-006-7400-5

Yoon, D. K. (2012). Assessment of social vulnerability to natural disasters : A comparative study. 823–843. https://doi.org/10.1007/s11069-012-0189-2.

Youzi, H., Nemati, G., & Emamgholi, S. (2017). The optimized location of hospital using an integrated approach GIS and analytic hierarchy process: A case study of Kohdasht City. *International Journal of Economics, 7*(1), 1–6. https://doi.org/10.4172/2162-6359.1000500

Zahran, S., Brody, S. D., Peacock, W. G., Vedlitz, A., & Grover, H. (2008). Social vulnerability and the natural and built environment: A model of flood casualties in Texas. *Disasters, 32*(4), 537–560. https://doi.org/10.1111/j.1467-7717.2008.01054.x. Epub 2008 Apr 24.

Zamble, E., & Gekoski, W. L. (1994). Coping. In V. S. Ramachandron (Ed.), *Encyclopedia of human behaviour* (Vol. 3). Academic Press.

Zarkesh M. K. (2005). Decision support system for floodwater spreading site selection in Iran. ITC.

Chapter 2
Socioeconomic Entities of Slum Neighborhoods: A Geo-Environmental Study in Midnapore Municipality

2.1 Introduction

Socioeconomic entities are significant factors in evaluating the status of households and the society they live in. In the diversified earth living space, socially defined groups are characterized by distinct cultures, economies, physical setups, social entities, psychology, and health behavior. Therefore, socioeconomic characteristics are different in the socially defined group, even at the neighborhood level, in the context of the social matrix. An urban area is a mixture of the economy in which one hand represents pleasure with a high economy or income. In contrast, the other hand represents misery, i.e., poverty, homelessness, etc. Therefore, disparity among the neighborhoods is a common phenomenon in the urban areas of developing countries. The disparity in basic amenities, socioeconomic and psychological behaviors, health, and healthcare accessibilities among the neighborhoods are distinct features in their physical and sociocultural setup, varying from space to place. So, measures of heterogeneity persist in the small space; that's why small slum neighborhoods in the urban administrative setup are taken as a unit of study. Socially deprived groups are suffering from health-related issues and are socially vulnerable as well as exposed to high health risks, which are noteworthy for planning processes or interventions.

Inequalities in health and healthcare facilities are associated with the different variables concerning the demographic, socioeconomic, and health-related attributes and healthcare facilities. This phenomenon varies at neighborhood levels within the homogeneous characteristics. Neighborhoods are now stable administrative boundaries taken as a unit of study with statistical analysis, mapping, modeling, etc. For the present study, all the slum neighborhoods in the Midnapore municipality were selected to understand the variation in socioeconomic entities of the socially poor households and the disparities in their daily livelihoods. Eight hundred families were surveyed for the study using a random sampling method. Four to five

© The Author(s), under exclusive license to Springer Nature Switzerland AG 2025

B. Dutta et al., *Using GIS to Identify Opportunities for Improved Health in Urban Slums*, Global Perspectives on Health Geography,
https://doi.org/10.1007/978-3-031-89617-0_2

households based on population composition were randomly selected from each slum neighborhood, and data was collected through a well-structured questionnaire by interview method. Different variables have been chosen based on the previous literature, such as Cutter et al. (2003), Kirby et al. (2019), Aksha et al. (2019), Roy (2016), etc., for exploring different socioeconomic entities of the study area. Some variables were added to the analysis due to the distinct socioeconomic characteristics of the study area. Socioeconomic, demographic, health, and healthcare-related quantitative and qualitative data were collected and processed into SPSS and Microsoft Excel as variables using convenient methods and tools. The selected variables were used in this chapter to explore socioeconomic status and model-building aspects, such as deprivation, poverty, quality of life, and social vulnerability.

2.2 Spatial Distribution and Pattern of Slum Neighborhoods

The rapid urbanization in Midnapore town resulted in the growing up of slum settlements all over the city. The stream of rural-to-urban migration in the deficient essential facility regions of the town with cross-cultural ethnic groups (Mandal et al., 2016) is the major factor for gradually expanding slum areas. The slums are concentrated in the city's central, northeast, and northwest parts. Nearly 156 notified slums across 25 municipal wards (Fig. 2.1 and Table 2.1) have been legally recognized by the municipal authority and provide essential services such as tap water facilities; waste material removal, electricity, and healthcare services; etc.

The nearest neighbor analysis in ArcGIS shows the spatial distribution and pattern of slum pockets or neighborhoods at <0.10 significance level. The distribution pattern of slum pockets is found to be more random than cluster in nature (Fig. 2.1), with a Z-score of 1.77. This means the slum neighborhoods are randomly distributed with clusters across the town. The study also found that slums are along the drain (the so-called nala), railway lines, peripheral sites across the river, and sometimes well-connected roads.

2.3 Physio-Climatic Features

Topographically, Midnapore town comprises three formations: lateritic upland toward the west, older alluvium to the north, and newer alluvium to the south and southeast. The study site is characterized by plain topography along the Kasai River with a slope <1° to <9° (Fig. 2.2) and elevation from the mean sea level range between 23 and 60 m (Fig. 2.3). The region is characterized as tropical monsoon humid and moist sub-humid climate ("Am" according to Koppen's and C2, BA'w according to Thornthwaite's climatic classification scheme) with an average annual

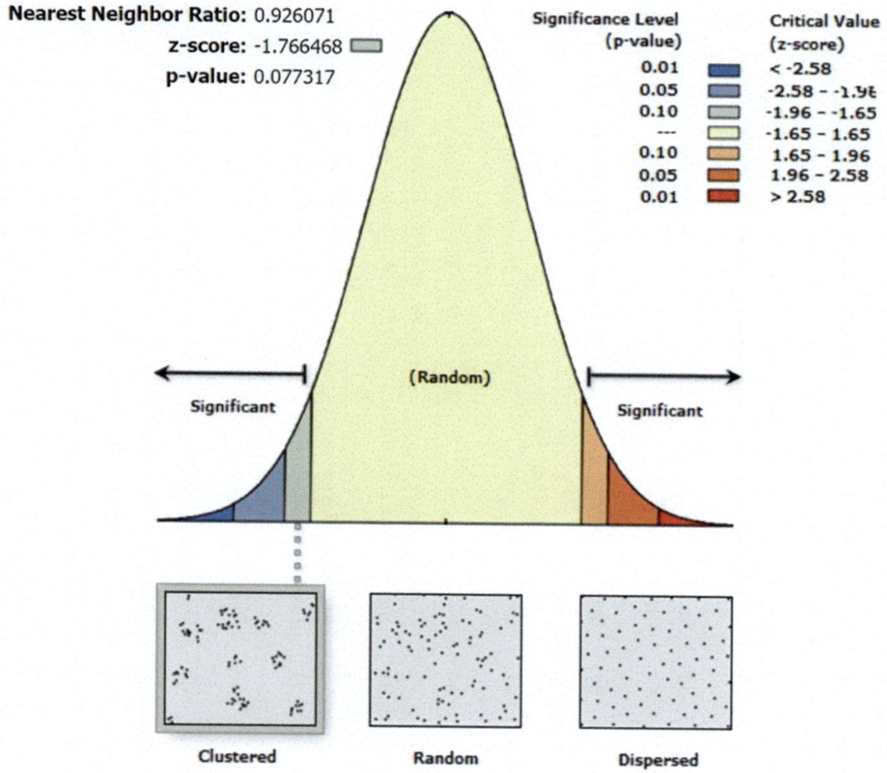

Fig. 2.1 Spatial distribution and pattern of slum neighbourhoods (Source: Author)

rainfall of 150 cm and a mean temperature of 27 °C (highest >45 °C, lowest 13 °C); the average relative humidity ranges between 64% (January) and 81% (August), and during monsoon period, humidity increases to 92–94% (Climate and Weather Averages in Midnapore, West Bengal, India), which is ideal for the growth of thick forests and disease prevalence (Mukherji, 2001).

The important climatic characteristics are the extreme heat and high relative humidity in summer, heavy rainfall in the monsoon, and chilly winter (Fig. 2.4). In this climatic condition, subtropical wet and dry deciduous forests, mainly sal and eucalyptus vegetation, have been developed on the northwest side of the town. However, due to the growing population and settlement expansion, the town's vegetation coverage gradually decreases. Figure 2.5 shows the change in vegetation coverage with the expansion of settlement areas (Figs. 2.6, 2.7, and 2.8 and Table 2.2).

Table 2.1 List of slum neighborhoods

1	*Kabardanga*	41	*Vivekananda Pally Basti*	81	*Natun Pally*	121	*Backside of Mohananda Sebashram*
2	*Hosnabad Bustee*	42	*Uday Pally Bastee*	82	*Pakija Basti (Rangamati)*	122	*Patnabazar Dompara*
3	*Brajakishorpally Bastee*	43	*Napit Para Bastee*	83	*Gate Bazar*	123	*Mitra Compound Kamar Para*
4	*Khairapara*	44	*Chalk Para Bustee*	84	*Kustha Colony Bastee*	124	*Benepukur Bastee*
5	*Khejur Tala Bastee*	45	*Bara Astana Siris Bagan Bustee*	85	*Behara Para Basti*	125	*Palmal Bastee*
6	*Mukunda Sagar Adbasi Bastee*	46	*Jharmadanga Bastee*	86	*Beltala Bastee*	126	*Hatharmath Bastee*
7	*Kamarpara*	47	*Iswarpur Bastee*	87	*Hanspukur Adibasi Para*	127	*Sarada Nagar*
8	*Sitalamandir Bauripara Bastee*	48	*Dudhar Pukur Parh Bustee*	88	*Natun Palli*	128	*Hanri Para at Mansasi School*
9	*Kargil Bustee*	49	*Minimarket Bastee*	89	*Promod Nagar*	129	*Zahangir Colony*
10	*Porabanglo*	50	*Muchipara (Sangat Bazar)*	90	*Chandan Bustee*	130	*Taltola Bastee*
11	*Pan Para*	51	*Rabindranagar Bhuniapara Bastee*	91	*Sudam Pukur Parh Bastee*	131	*Rajibnagar Bustee*
12	*Chatarpara Bastee*	52	*Ghusipara Bastee*	92	*Bandipara Bastee*	132	*Ber Ballavpur Bustee*
13	*Station Road Bhunia Para Bastee*	53	*Boxibazar Muslim Para Bastee*	93	*Patnabazar Bauripara*	133	*Piripukur Parh Bastee*
14	*Circuit House Kamar Para Bastee*	54	*Judgecourt Harijan Pally*	94	*Najargunj Mahamadan Para*	134	*Rajar Pukur*
15	*Beharapara Bastee*	55	*Idga Mahalla Bastee*	95	*Torapara Suraj Sen Adibasi Pally*	135	*Mahatabpur Shainji Baba Asram Bustee*

16	*Laldighi Bastee*	56	*Ghosh Ashram Kumar Para Bastee*	96	*Palbari Bibekananda Pally*	136	*Mahatabpur Sasan Ghat Colony*
17	*Baluchara Bastee*	57	*Majdur Nagar Bastee*	97	*Ashok Nagar Nepalipara*	137	*Mahatabpur Gandhi Bastee*
18	*Amtola Khas Basti*	58	*Momin Mahalla Bastee*	98	*Mandal Mahalla Adibasi Para*	138	*Mahatabpur Chasi Para Colony*
19	*Beharapara Bastee*	59	*Pynedanga Tentul Tola Bastee*	99	*Boxibazar Tantipara*	139	*Jelepara Bastee*
20	*Jali Pukur Adibasi Para*	60	*Harisan Dighi Bastee*	100	*Mallick Pukur Bagan Para*	140	*Palbari Ganga Mandir Jelepara Colony*
21	*Piripukur Parh Bastee*	61	*Aligunj Bastee*	101	*Sukumar Sengupata Pally Bustee*	141	*Hatisala Goli Midnapore(W)*
22	*Adibasipara Basti*	62	*Padma Pukurpar Bastee*	102	*Bara Astana Nutan Bustee*	142	*Talpukur Bastee*
23	*Torapara Bustee*	63	*Gandhi Pally Bastee*	103	*Bhuniapore*	143	*Talpukur Colony*
24	*Goalapara Bastee*	64	*Nazarganj Harijon Bastee*	104	*Ladey Para*	144	*Astana Pukur Bastee*
25	*Gabnala Bastee*	65	*Churibala Bastee*	105	*Torapara Adibasi Para*	145	*Halugali Bastee*
26	*Bhuniapara Station*	66	*Darjipara Bastee*	106	*Uttar Tantigeria*	146	*Bihari Hari Mandir Lane*
27	*Hosnabad Bustee (II)*	67	*Bauri Para Bastee*	107	*Rammohan Nagar*	147	*Ibrahimsaheb Bustee*
28	*Patharghata Bastee*	68	*Benepukur Adibasi Bustee*	108	*Nepalipara*	148	*Chirimarsai Bastee*
29	*Dhunaru Bastee*	69	*Indira Pally Bastee*	109	*Muslim Bustee in Front of Sepoybazar Mosq*	149	*Hamalpara (Rangamati)*
30	*Sekhpura Bastee*	70	*Madhyapara*	110	*Kamarara Tatul Bagan*	150	*Sialdanga Gandhighat Bastee*
31	*Patnabazar Tantipara Bastee*	71	*Dograpukur Bustee*	111	*Hanspukur Muslim Para Bastee*	151	*Jelepara Bastee*
32	*Dewan Nagar Bastee*	72	*Mandal Mahalla Bastee*	112	*Backside of Commerce College*	152	*Nibedita Pally*

(continued)

Table 2.1 (continued)

33	*Akarsha Nagar Bastee*	73	*Dewan Basti*	113	*Girza Bastee*	153	*Daspara*
34	*Mirzamahalla Jahangir Bastee*	74	*Goalapara Bastee*	114	*Sepoy Bazar Bastee*	154	*Mufti Mahalla*
35	*Talmal Bustee*	75	*Baipukur Harijon Bastee*	115	*Kasai Para Bastee*	155	*Pathan Mahalla*
36	*Absanpara Bastee*	76	*Nabina Bag Bastee*	116	*Backside of Muslim Burial Ground & Fire brigade*	156	*Shah Mahalla*
37	*Girja Bastee*	77	*Fakirdanga Bastee*	117	*Muchipara Bastee*		
38	*Singpara Bastee*	78	*Munshi Pukur Bastee*	118	*Subhasanagar Garge Gali*		
39	*Baburchipara Bastee*	79	*Adibasipara Bastee*	119	*Boroastana Bastee*		
40	*Pal Bari Bastee*	80	*Harijon Para Bastee*	120	*Patnabazar Alokbisoi Club Bustee*		

Source: Midnapore municipality (2022)

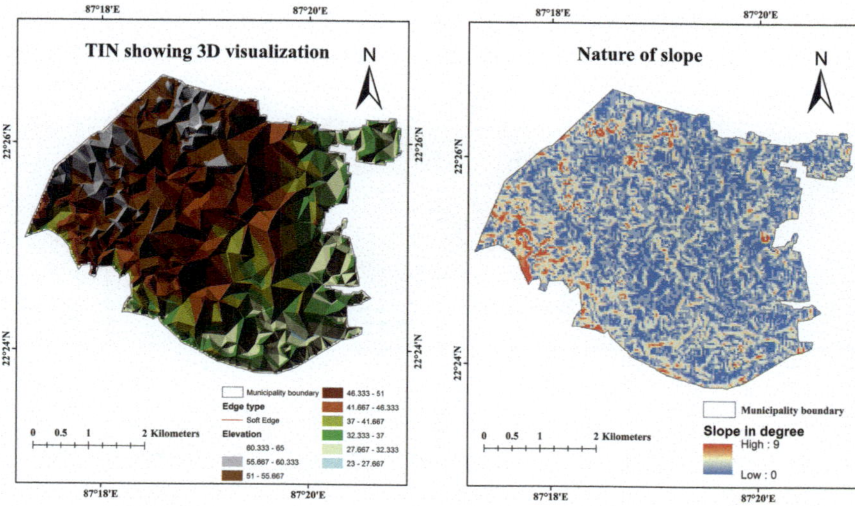

Fig. 2.2 TIN and slope map of Midnapore Municipality (Source: WWW.usgs.gov)

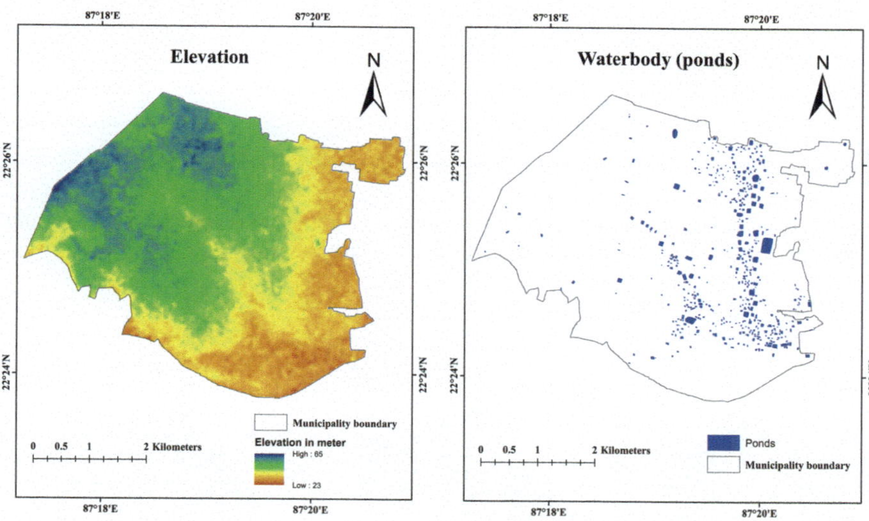

Fig. 2.3 Elevation and water bodies in the study area (Source: https://earthexplorer.usgs.gov/; Data type: ASTER DEM)

2.4 Socioeconomic and Demographic Profile

The Midnapore town is a district town that functions as an administrative, judicial, and healthcare center. Beyond this, it is an essential educational and business center around the rural area. Therefore, the population influx in the town is increasing day by day. According to the Census of India (2011), the population of Midnapore town

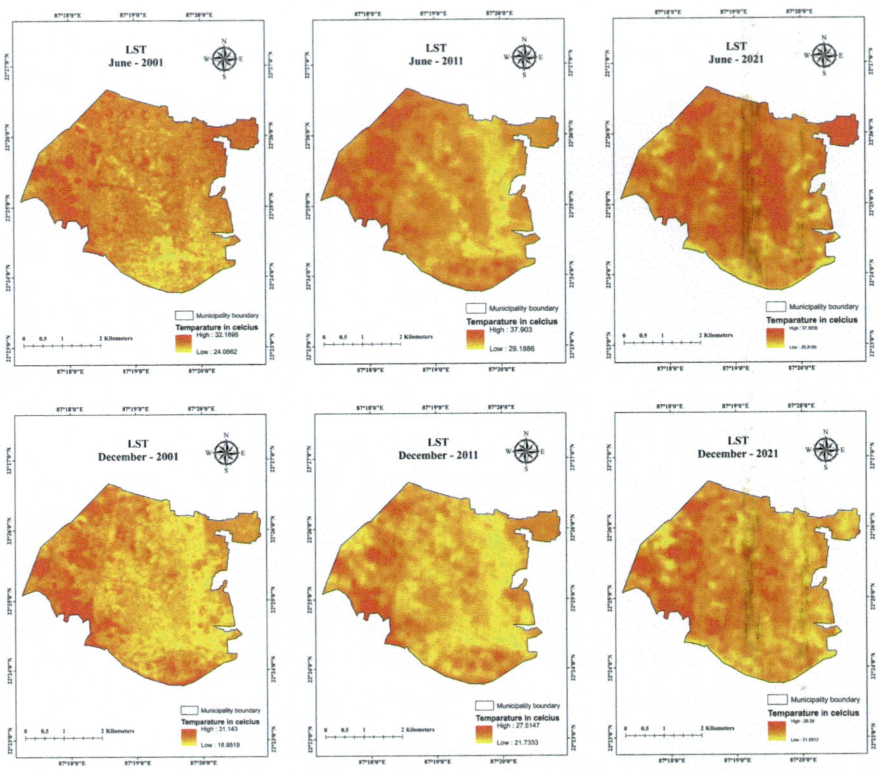

Fig. 2.4 Land surface temperature (LST) in the month of June and December from 2001 to 2021 (Data type: Landsat 4–5, 7ETM+ and 8 OLI, December, June)

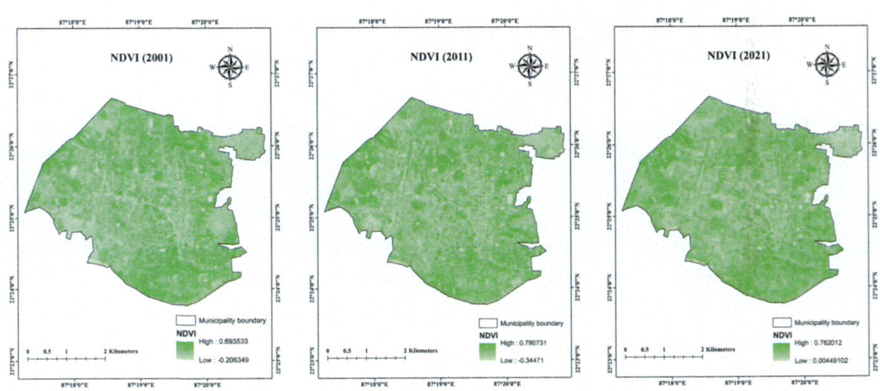

Fig. 2.5 NDVI showing the vegetation cover from 2001 to 2021

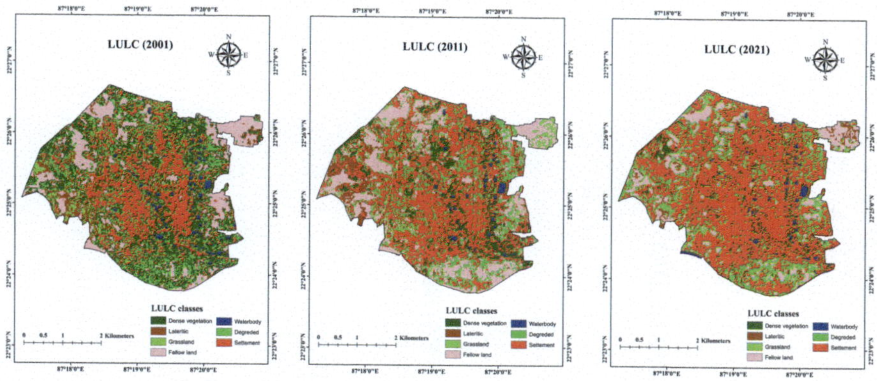

Fig. 2.6 Land use and Landcover map from 2001 to 2021

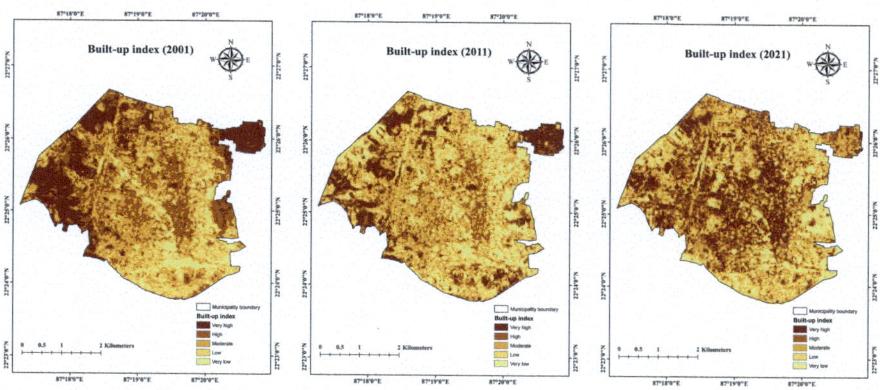

Fig. 2.7 Built-up Index from 2001 to 2021

was 1,69,264, with which male and female populations were 84,977 and 84,287, respectively; the sex ratio and the child sex ratio were 992 and 985, respectively, and literate persons are 88.99%, while male literates are 92.52% and female literates 85.44% as well as the city had 1,39,827 Hindu population and 27,238 Muslim population. The settlement density varied from 3006 to 29,000 per km^2. Besides, the slum population was 50,943 (Midnapore municipality, 2012), contributing 30.09% to the town's total population. Thus, Midnapore is the second largest city after Kharagpur in the Paschim Medinipur district.

Due to medical facilities, basic amenities, job opportunities, and other services, the population has increased to a saturation level. The estimated population composition of the town in 2021 by the Census of India is revealed in Tables 2.3 and 2.4.

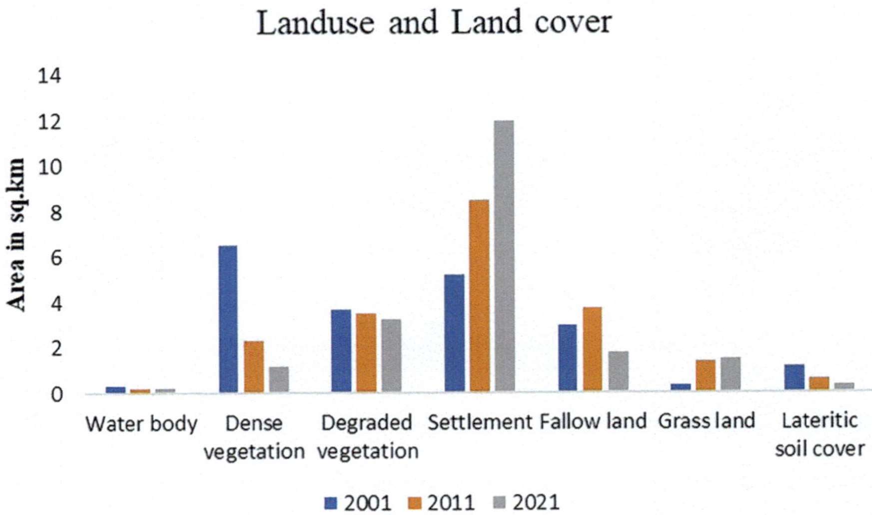

Fig. 2.8 Land use and Landcover change from 2001 to 2021 (Data type: Landsat 4–5,7ETM+ and 8 OLI, December)

Table 2.2 Land use and land cover change (km²) from 2001 to 2021

Land use and land cover (area in km²)	2001	2011	2021
Waterbody	0.35	0.23	0.25
Dense vegetation	**6.54**	**2.31**	**1.18**
Degraded vegetation	3.67	3.52	3.28
Settlement	**5.24**	**8.50**	**11.95**
Fallow land	3.00	3.74	1.79
Grassland	0.35	1.37	1.52
Lateritic soil cover	1.15	0.64	0.33
Total area in km²	20.3	20.3	20.3

Source: Author(s)

Table 2.3 Estimated population composition of Midnapore town in 2021

Description	As per the 2011 census			Estimated population in 2021
	Total	Male	Female	
City population	1,69,264	84,977	84,287	1,87,104
Literates	1,37,133	71,547	65,586	1,51,587
Children (0–6)	15,172	7644	7528	16,771
Average literacy (%)	88.99	92.52	85.44	
Sex ratio	992			
Child sex ratio	985			

Source: https://www.indiacensus.net/city/medinipur

Table 2.4 Religion-wise estimated population composition of Midnapore town in 2021

Description	Population in 2011	Percentage	Estimated population in 2021
Hindu	139,827	82.66	154,565
Muslims	27,238	16.10	30,109
Christian	1558	0.92	1772
Sikh	51	0.03	56
Buddhist	165	0.10	182
Jain	23	0.01	25
Not stated	305	0.18	337

Source: https://www.indiacensus.net/city/medinipur

2.5 Socioeconomic Profile at a Glance

The overall socioeconomic condition of slum neighborhoods is at stake in India. According to the survey in the study area, the characteristics of the socioeconomic status of the slum areas are as follows:

- The proportion of slum households to urban households in West Bengal is 21.9%, whereas 31.14% (Census of India, 2011) of the slum population live in Midnapore municipality distributed in 156 slum neighborhoods across 25 wards under Midnapore municipality with multiple deprivations such as basic amenities, sanitation, room density, land tenure, unemployment, underemployment, malnutrition, morbidity, mortality, higher incidence of crime, and social unrest.
- The slum population is continuously increasing.
- Twenty-six percent of households with roof material are asbestos sheets.
- Thirteen percent of households with wall material are mud.
- Forty-five percent of households have one room for dwelling.
- Twenty-six percent of households are still living in a rented house.
- Eleven percent of drinking water sources are away from the premises.
- Twenty-eight percent of drinking water sources are hand pumps and tube wells/boreholes.
- Twenty-nine percent of households do not have a kitchen.
- Thirty-four percent of households do not have a latrine facility.
- Thirty-three percent of households still use firewood, crop residue, cow dung cake, coal, lignite, and charcoal for cooking fuels.
- About 53.2% of households are availing banking services.
- Nineteen percent of households do not have a bathroom facility within their premises.
- Nineteen percent of households do not have drainage connectivity for wastewater outlets.

Table 2.5 also shows the household's demographic and socioeconomic characteristics according to the survey. Figure 2.9 represents the age group-wise distribution of the sample population, which means the demographic dividend, and Fig. 2.10 shows the housing structure in the study area.

Table 2.5 Household characteristics

Socioeconomic characteristics	Average
Monthly income (rupees) per family	12,263.65
Household size (persons)	4.90
The mean age of household head (years)	48.02
Literacy rate	78.28%
Literate in a household (persons)	3.54
Earning members in a household (persons)	1.43
Age dependency ratio	11.48%
Child dependency ratio	13.14%
Health shock experiences (yearly)	2.45

Source: Field survey (2022)

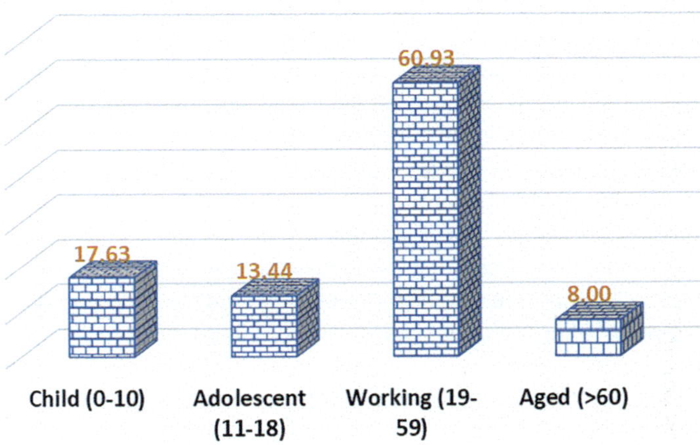

Fig. 2.9 Age structure in the slum neighbourhoods (Source: Field study, 2022)

Fig. 2.10 Housing structure of slum neighbourhoods (Source: Field study, 2022)

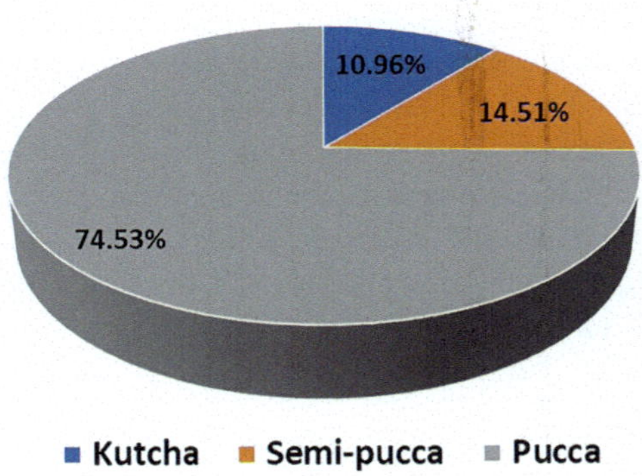

2.6 Social Deprivation

Deprivation means a lack of something in every life span. However, social deprivation indicates the extent to which an individual or group in a community does not meet the necessities for their daily standard of living. In other words, when a person or community is deprived of fundamental rights and privileges everyone holds in society, it is called social deprivation. In developing countries, deprivation is a more common phenomenon, where poor households are deprived of basic amenities, such as work, money, housing, and accessibility to healthcare facilities and services. They are also excluded from society as they are treated as an unprivileged section. In urban areas, due to high population density and congestion, the demands of the community's basic needs are more significant in the sense of living in a limited space. Therefore, the study needs to explore the evaluation of deprivation conditions in the community for proper planning and promoting a decent standard of living in the interest of building a developed nation.

For the present study, 19 relevant variables have been chosen based on careful evaluation of the previous studies (Cutter et al., 2003; Kirby et al., 2019; Aksha et al., 2019; Roy, 2016) and author perception from the field survey. The principal component analysis (PCA) through varimax with Kaiser normalization method converged in six iterations technique was employed to determine the significant variable for constructing social deprivation index (SoDI) through Eq. 2.1.

$$\mathrm{SoDI}_{\mathrm{slum}} = PC_{i-1} \times V_i \tag{2.1}$$

where PC_i is the principal component and V_i is the variable for each element. Six components have been extracted and used for index construction (Table 2.6). The SoDI value of each slum neighborhood was plotted on the map through ArcGIS v10.4 based on the SoDI value. The spatial distribution of slum neighborhoods is shown in Fig. 2.11, which reveals that most of the slum neighborhoods are moderately to highly deprived in terms of basic needs. In contrast, a few slum neighborhoods exhibit the most and least deprived areas.

Based on Getis-ord-Gi* statistics, hot spot geospatial analysis was also employed to determine the spatial grouping or pattern of slum neighborhoods in the context of SoDI. The map reveals no significant clustering as a spatial pattern among neighborhoods in the study area, meaning that slum neighborhoods are randomly distributed across the study area in the context of social deprivation (Fig. 2.12).

2.7 Poverty

Poverty is the inability to earn or the inefficiency of doing work (Sen, 1981). Poverty is a severe issue in the world, especially in the developing countries. Generally, poverty is measured in two ways—absolute and relative. However, the method of

Table 2.6 Variables and PCA for SoDI

Variables	Description	Components					
		1	2	3	4	5	6
Electricity	Percentage of households having electricity	−0.151	0.024	−0.124	−0.036	−0.018	−0.698
Separate kitchen	Percentage of households having a separate kitchen	−0.165	−0.727	−0.012	0.200	−0.049	−0.063
Separate bathroom	Percentage of households having a separate bathroom	−0.121	−0.838	−0.133	0.000	−0.083	−0.155
Toilet	Percentage of households facing poor toilet facility	0.007	0.700	0.174	0.173	−0.114	0.062
Ownership of house	Percentage of households having own house	0.104	0.300	0.162	−0.611	−0.045	−0.019
Room dampness	Percentage of household facing room dampness during the rainy season	0.007	0.157	−0.102	0.014	−0.157	0.780
Roof	Percentage of households having a pucca roof	0.780	0.077	0.071	−0.071	−0.175	−0.008
Wall	Percentage of households having a pucca wall	0.901	0.082	0.046	0.038	−0.049	0.089
Floor	Percentage of households having pucca floor	0.879	0.095	0.077	0.099	0.014	0.089
Assets	Percentage of households having a lack of basic household assets	−0.064	−0.009	−0.020	−0.038	0.721	0.001
Literacy rate	Percentage of the population who can read and write and above	0.165	0.159	0.193	0.505	−0.222	0.012
Food expenditure	Monthly per capita food expenditure (Rs.)	0.022	0.567	−0.118	0.185	−0.233	−0.281
Fuel used for cooking	Percentage of households using LPG for cooking	0.357	0.337	0.035	0.261	−0.148	0.196
Overcrowded	Percentage of households facing overcrowding	−0.201	−0.226	−0.050	−0.529	0.381	0.143
Water purifier	Percentage of households having a water purifier	−0.079	0.041	−0.101	0.020	0.725	−0.158
Unemployment	Percentage of unemployed persons per household	−0.098	−0.115	−0.968	0.045	0.015	−0.026
Employment	Percentage of employed persons per household	0.098	0.115	0.968	−0.045	−0.015	0.026
Family size	Average number of people per household	−0.017	0.079	−0.125	0.805	0.051	0.076
Income	Average per capita income	−0.059	−0.270	0.350	−0.282	0.602	0.021

Extraction method: principal component analysis. Rotation method: varimax with Kaiser normalization. (a) Rotation converged in six iterations

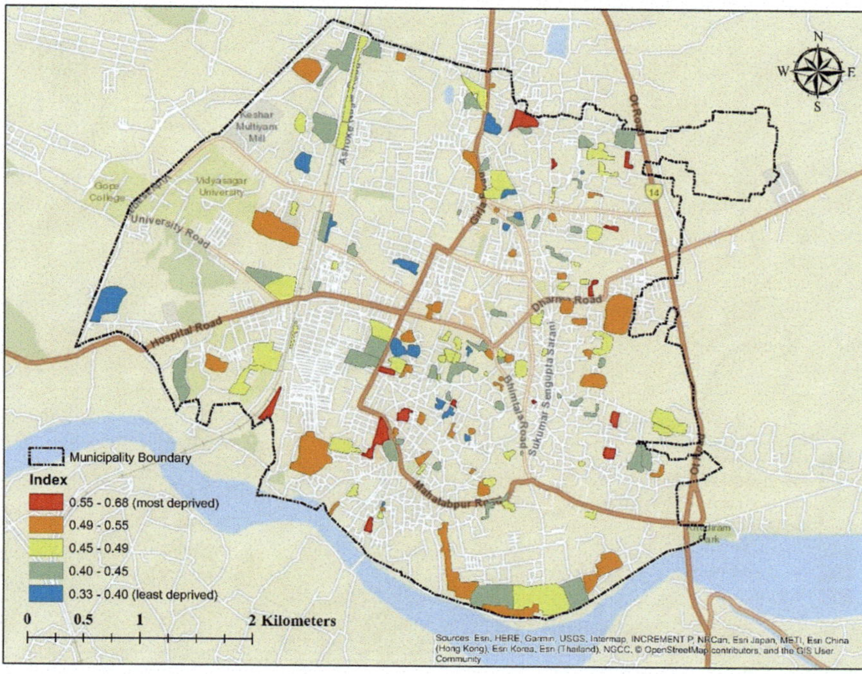

Fig. 2.11 SoDI of slum neighbourhoods by index score (Source: Author)

measuring poverty differs from country to country. Poverty indicates income and household assets as a measuring variable, as well as the state of well-being and standard of living. Therefore, many variables are significant for measuring poverty across space, even at the neighborhood level. In these perspectives, poverty is multidimensional rather than absolute or relative. A multidimensional poverty index is designed to use for identifying the poverty pattern and vulnerable people using accessible microdata from the households as well as measuring the multiple deprivations at the household level in health, education, standard of living, etc. (Alkire et al., 2015; Ravallion, 2011; Cavapozzi et al., 2015; Das et al., 2020). The weighted deprivation score (Ci) measures multiple deprivations using Eq. 2.2.

$$Ci = \sum_{j=1}^{d} gij \times wj \tag{2.2}$$

The weighted deprivation score (C_i) is estimated by multiplying the deprivation matrix (*gij*) and the weighted value (equal) allocated for each indicator or dimension (*wj*). If the household deprivation score exceeds a given threshold value (e.g., 1/3), a household is considered "multiply deprived" or simply "poor" (Alkire et al., 2015; Das et al., 2020). Weighted deprivation score ranges between 0 (non-deprived) and 1 (deprived), where $Xij > Zj$. The deprivation matrix (*gij*) is recognized by the order of achievement matrix (n x d) denoted as Xij and deprivation cutoff (*Zj*)

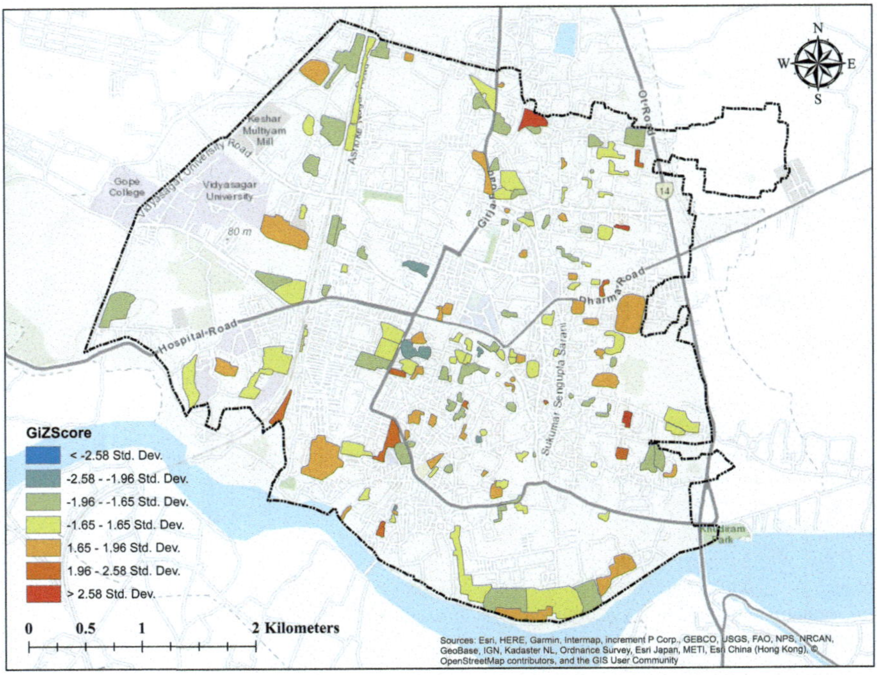

Fig. 2.12 SoDI of slum neighbourhoods by Hotspot Analysis (Getis-Ord Gi*) (Source: Author)

standard score, indicating whether the households are deprived. If the weighted deprivation score (C_i) is more than the poverty cutoff ($Ci > Pk$), then the households are considered to be as multidimensionally poor (q). The headcount ratio (HCR) is estimated by Eq. 2.3.

$$\text{HCR} = \frac{\text{Number of multidimensionally poor households}(q)}{\text{Total number of household}(n)} \tag{2.3}$$

The average number of multidimensionally poor households (A) is the intensity of the multidimensionally poor condition. Finally, the multidimensional poverty index (MPI) was derived (Alkire et al., 2015; Das et al., 2020) using Eq. 2.4.

$$\text{MPI} = \text{HCR} \times A \tag{2.4}$$

The study used micro-level household data of slum communities with their standard of living, expenditure, income per head, fuel used for cooking, ownership of land and house, electricity and education level of the households and household assets, etc., in every municipal ward, and results are evaluated using MPIs shown in Fig. 2.13. In most cases, the estimated MPI values are limited to 0.5, which indicates that most of the slum neighborhoods are multidimensionally poor and deprived of basic amenities and other services. The least number of slum neighborhoods is

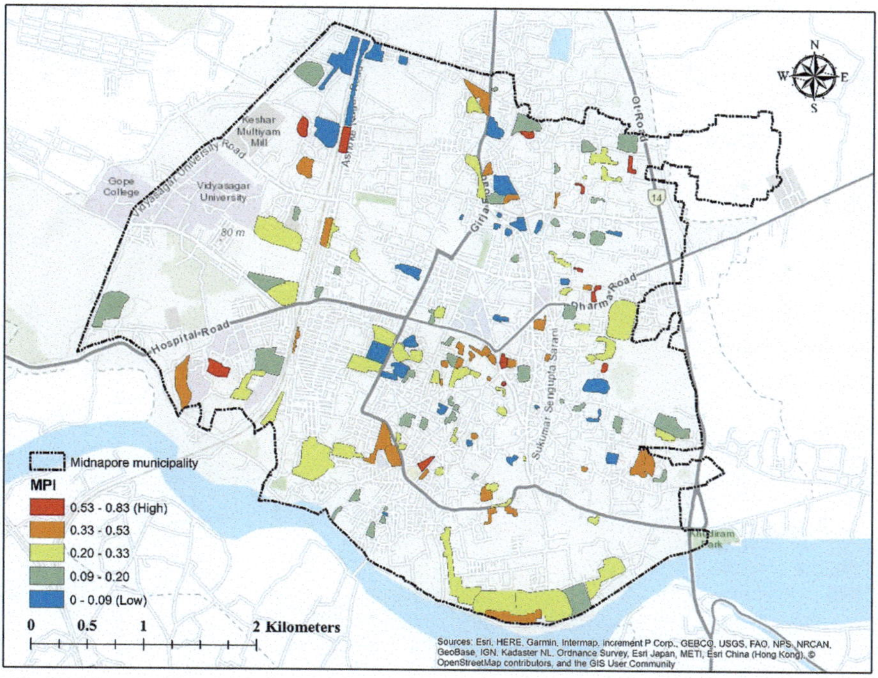

Fig. 2.13 MPI of slum neighbourhoods by Index score(Source: Author)

higher than the 0.5 MPI value. The hot spot analysis was employed based on the G_i* statistic, i.e., Z-score, to identify spatial clustering of the slum neighborhoods in the study area (Fig. 2.14), which have high or low values depending on statistically significant Z-score. The positive and negative Z-scores are more intense to high (hot spot) or low (cold spot) values, respectively. The Z-score in the vicinity of zero indicates no significant spatial clustering. The Z-score of the MPI value is 0.21 at a P-value of 0.83, indicating the random distribution pattern (Fig. 2.15) of slum neighborhoods in the context of multidimensionally poor conditions.

2.8 Quality of Life

Quality of life is the perception of human beings about their living in a healthy and comfortable environment within the physical, socioeconomic, and psychological built-up environment. The concept of quality of life is multidimensional and dynamic, and it varies across space, even at the trim neighborhood level. Moreover, quality of life varies from person to person due to physical and mental health, emotion, recreation, leisure time, freedom, security, employment and income, accessibility to basic amenities and healthcare services, and the individual perception of

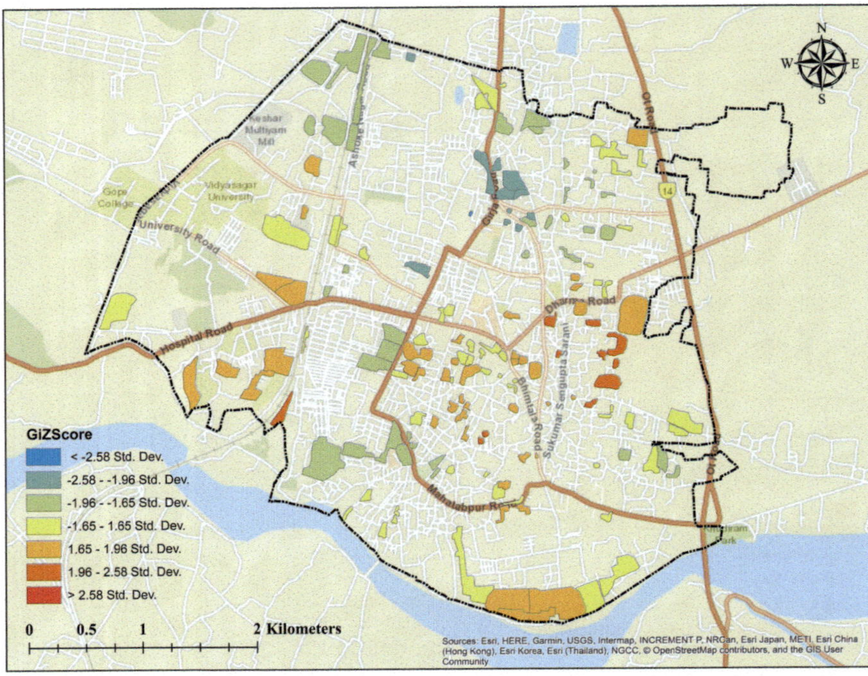

Fig. 2.14 MPI slum neighbourhoods by Hotspot Analysis (Getis-Ord Gi*) (Source: Author)

the standard of living. Quality of life is closely related to vulnerability and health security as a whole. For example, low standard quality of life intrudes people into health vulnerability as well as their mental and psychological domain. Quality of life generally depends on what households have, such as housing condition, access to drinking water, economic condition, occupation, availability of food and nutrition, ownership of the house, households' assets, education and social care, and sanitation, in addition to accessibility to basic amenities and healthcare. In India, pseudo-urbanization leads to disparities among households due to population explosion within a limited space, which results in the growth of slum areas characterized by poverty, unemployment, and disparities in access to basic needs due to inadequate provision of infrastructure and services. This condition leads to an overall degradation of the quality of life, perpetuating generation to generation and pushing toward the vulnerable group in the urban landscape. It is found that 40% of the population lives in class I cities in India without fulfilling their basic needs (Mudey et al., 2017; Komalawati & Lim, 2020). Midnapore town is a class I city with 156 notified slum neighborhoods that are assumed to be deprived groups in distinct physical and socioeconomic environments. Therefore, for assessing the quality of life among the slum neighborhoods, both subjective and objective sets of indicators have been included in the study based on the distinct characteristics of the study area, which is beyond the WHO's recommendation. The variables are selected

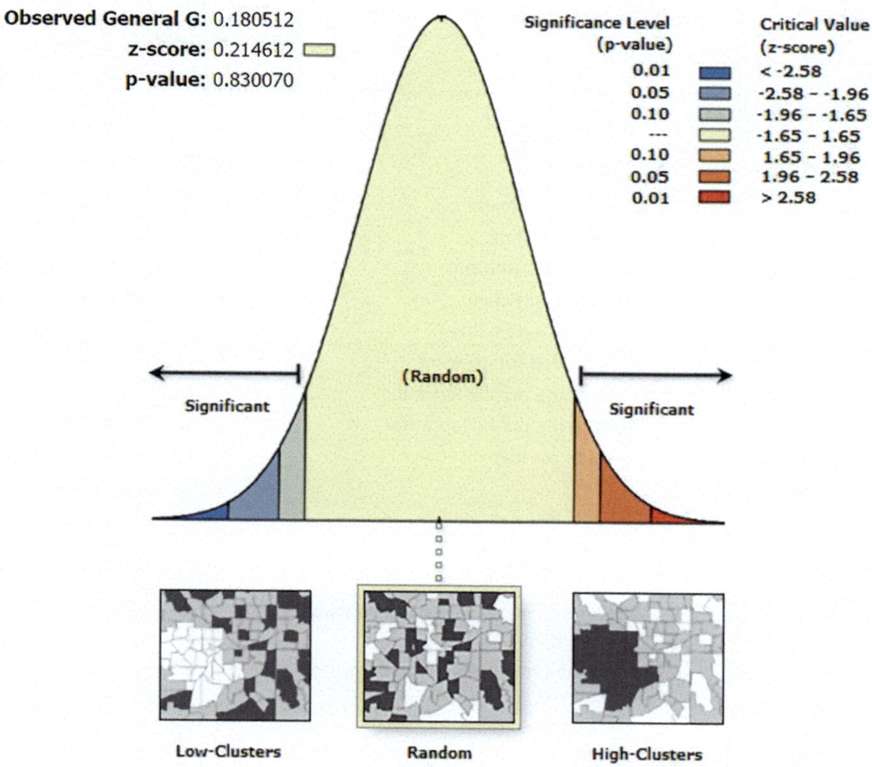

Fig. 2.15 Distribution pattern of MPI of slum neighbourhoods by Gi* statistic (Source: Author)

based on empirical observation and focus group discussion. Table 2.7 represents the set of indicators that were included in the analysis for quality of life.

Furthermore, inductive and deductive techniques were employed to breed an index of quality of life. We included all those households with heterogeneous characteristics regarding caste, religion, socioeconomic status, etc. PCA of inductive approach (in SPSS v26) and standardized method (Z-scores, the ratio of value, min-max, rescaling transformation) of deductive approach have been used to develop the index of quality of life so far as to the better understanding of the quality of life for the elimination of methodological error. The results of all methods are again compared using the Pearson correlation coefficient and mapping with the GIS algorithm. The Pearson correlation matrix (Table 2.8) shows a strong positive correlation among the index scores of employed methods. The PCA was accomplished in SPSS v26, which extracts the components with their cardinality based on the influences of the components (Table 2.9). The quality of life index formula is computed using Eq. 2.5.

$$I_{\text{slum}} = \sum PC_{i\_i} \times V_i \tag{2.5}$$

Table 2.7 Variables used for analyzing the quality of life

Variables		Mean	Standard deviation
Housing condition	Pucca floor	70.83	22.689
	Pucca wall	79.04	35.338
	Pucca roof	56.15	26.772
	Room dampness	45.58	23.314
	Overcrowding	49.57	12.594
	Separate bathroom	73.59	23.027
	Separate kitchen	63.59	25.118
	Ventilation facility	77.63	21.285
	Fuel used for cooking	74.42	22.353
Access to drinking water	Drinking private tap water	11.54	16.071
	Drinking public tap water	85.13	17.579
	Water purifier	7.50	12.579
Economic condition	BPL	30.06	22.041
	Per capita income	31.17	19.115
	Any loans taken	44.23	25.094
	Unemployment rate	47.30	11.098
Occupation	Earning members associated with the service sector	14.77	16.661
	Earning members in small business and informal sector	36.61	22.721
	Earning members in daily wage labor	36.43	27.773
Availability of food and nutrition	Per capita food expenditure	70.71	20.794
House ownership	Migrated	13.21	17.960
	Own house	81.15	18.067
Asset	Owning TV	79.68	21.325
	Owning cycle	87.50	16.796
	Owning vehicle	36.22	24.133
	Owning phone	93.27	12.295
Education	Distance to the nearest school/institution	49.74	24.573
	Literacy rate	72.23	15.701
Health and social care	Distance to the nearest healthcare center	35.71	22.321
	Use sanitary napkin	39.94	18.474
	Leisure time with family	5.71	11.535
	Leisure time watching TV	25.77	18.701
Sanitation	Sewerage system	11.41	18.012
	Availability of public toilet	87.56	16.279
	Condition of drain	78.46	20.888
	Solid waste management	25.38	24.531
	Facilitated solid waste collection	10.45	13.835

Source: Field survey (2022)

Table 2.8 Comparison of all indices of the quality of life by Pearson correlation matrix

	Principal component analysis	Ratio of value	Z-score	Min-max rescaling
Principal component analysis	1			
Ratio of value	0.945**	1		
Z-score	0.945**	1.000**	1	
Min-max rescaling	0.945**	1.000**	1.000**	1

**$p < 0.01$

where I_{slum} indicates the index score of the slums, PC_i is the principal component, and V_i is the variance of each component. The explained variance of principal components of quality of life is 67.37%. Components are extracted based on cardinality and eigenvalue, influencing quality of life. Each component is multiplied by its explanatory variance and then added to get its index score (Cutter et al., 2003; Maria et al., 2016; Aksha et al., 2019). Furthermore, cluster-outlier analysis based on Anselin Local Moran's I statistic tool in ArcGIS v10.4 was used to identify spatial clusters or grouping whether slum neighborhoods in the context of the quality of life are in the spatial pattern (clustered or random) or other forms based on Z-score of each neighborhood. The modified quality of life index has been developed by combining the score of four models and mapping in ArcGIS v10.4.

The modified QoLI for Midnapore City was estimated for each spatial unit (156 neighborhoods) by accumulating the scores derived from the principal component values, Z-score transformation, maximum value transformation (ratio of value), and min-max rescaling transformation. The total score for the modified QoLI has been developed by combining four model scores and mapping in ArcGIS v10.4. The Z-score values derived from Local Moran's I statistics were grouped into five quintiles from high (>1.5 SD) to low (<−1.5 SD). Total scores were mapped through ArcMap v10.4 (Fig. 2.17) as well, and the value of each component was also visualized for the quality of life in Midnapore (Fig. 2.16). For the quality of life, the Z-score value varies between ±1.5 S.D. The map of the quality of life (Fig. 2.17) reveals that the slum neighborhoods with a decent quality of life, such as Ram Mohan Nagar, Ashoknagar Nepali Para, Udaypally bastee, Porabanglo bastee, Chatarpara bastee, Beharapara bastee, Singpara bastee, Beltala bastee, backside of commerce college bastee, backside of Muslim burial ground and fire brigade, Subhasnagar gargi bastee, backside of Mohanananda Sevashram, Hatisalagoli bastee, are situated in the eastern, the northwest, and the central part of the city. Slum neighborhoods categorized as having a degraded quality of life are Berballavpur bastee and Bauripara bastee, primarily positioned in the north and southeast parts of the city.

The results from Figs. 2.16 and 2.17 reveal that low quality of life characterizes fewer slum neighborhoods. The reason behind the phenomenon is that most of the slum neighborhoods are under vulnerable socioeconomic status, which involuntarily falls into high health risks such as lack of housing accommodation, lack of access to drinking water, lack of proper food and nutrition, as well as lack of

Table 2.9 Results of PCA for quality of life

Principal component	Cardinality	Variables	Loadings	Explained variance (%)	Eigenvalue
PC1	(+)	Separate bathroom	0.720	12.717	4.705
		Separate kitchen	0.655		
		Sewerage system	−0.434		
		Ventilation facility	0.517		
		Fuel used for cooking	0.616		
		Per capita food expenditure	0.342		
PC2	(+)	Pucca floor	0.750	6.658	2.464
		Pucca wall	0.785		
		Pucca roof	0.783		
PC3	(+)	Drinking private tap water	0.934	6.374	2.358
		Drinking public tap water	−0.909		
PC4	(+)	Overcrowding	0.758	5.427	2.008
		Per capita income	0.726		
PC5	(+)	Water purifier	0.347	4.833	1.788
		Any loans taken	−0.688		
PC6	(+)	Migrated	0.596	4.404	1.630
		Own house	−0.787		
		Owning phone	0.506		
		Literacy rate	0.364		
PC7	(+)	Availability of public toilet	0.661	3.987	1.475
		Solid waste management	−0.666		
PC8	(+)	Condition of drain	0.526	3.825	1.415
		Earning members in small business and informal sector	0.748		
		Earning members in daily wage labor	−0.507		
PC9	(+)	Distance to the nearest school	0.799	3.716	1.375
		Facilitated solid waste collection	0.689		
PC10	(+)	Unemployment	0.452	3.449	1.276
		Leisure time with family	0.798		
PC11	(+)	Earning members associated with the service sector	0.474	3.275	1.212
		Leisure time watching TV	0.696		
PC12	(+)	Use sanitary napkin	0.804	3.028	1.120

(continued)

Table 2.9 (continued)

Principal component	Cardinality	Variables	Loadings	Explained variance (%)	Eigenvalue
PC13	(+)	Owning TV	0.560	2.943	1.089
		Owning cycle	0.875		
PC14	(+)	Room dampness	−0.437	2.374	1.012
		BPL	0.686		
		Owning vehicle	0.375		
		Distance to the nearest healthcare center	0.623		

Extraction method: principal component analysis. Rotation method: varimax with Kaiser normalization. (a) Rotation converged in 29 iterations

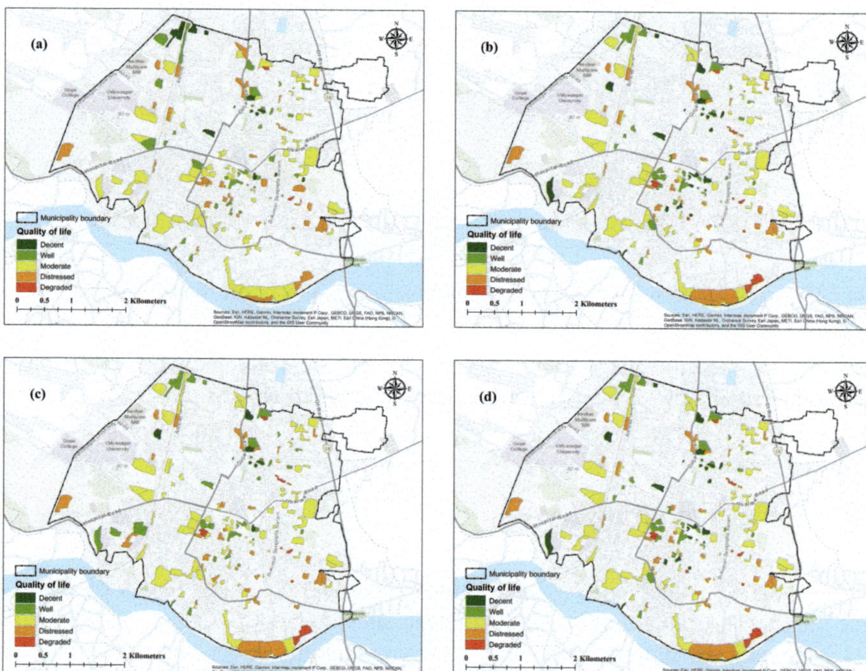

Fig. 2.16 Spatial distribution slums based on different index scores for Quality of life (**a**) Principal component analysis (**b**) Z-score transformation (**c**) Maximum value transformation (ratio of value) (**d**) Min-max rescaling transformation

healthcare facilities. Moreover, high family size and fewer literate and employed persons may contribute to their health security, living standard, and overall quality of life. Clearly, "slum neighborhoods are more vulnerable than non-slum neighborhoods" in the milieu of the quality of life. However, the methods applied in this study do not necessarily yield the same outcomes for the variables and do not spread

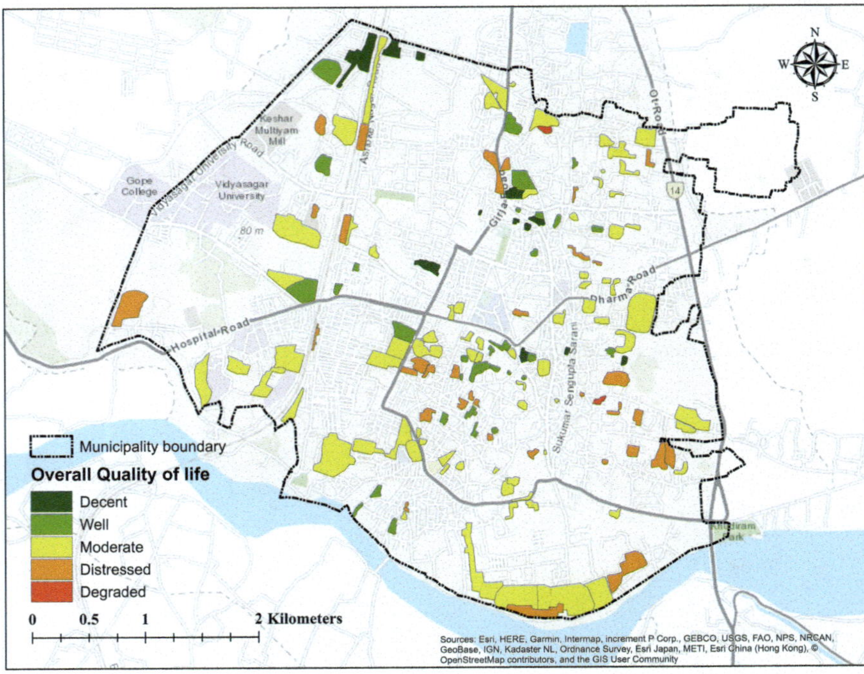

Fig. 2.17 Spatial distribution of slums as per the degree of quality of life (Source: Author)

over in the same time-space framework. Still, it can be evaluated whether the most vulnerable people of Midnapore municipality live in the worst slums.

2.9 Social Vulnerability

The concept of vulnerability is multidimensional and changing across the space-time matrix. The vulnerability may be in the context of health, society, environment, job, education, or daily livelihood. Social vulnerability is an important issue in human geography, particularly in health geography. Social vulnerability refers to the possible adverse impact on the health and well-being of an individual due to natural or human-induced hazards or disasters in the socio-physical living framework of the society (Cutter et al., 2003; HVRI 2019; Aksha et al., 2019; Birkmann, 2007; Fekete, 2009; Dunning & Durden, 2013; Beccari, 2016). Social vulnerability persists in society due to several factors that expose people to natural and artificial hazards, abuse, social exclusion, or socioeconomic deprivation, which hinder preparation, coping, and recovery from human suffering and loss, such as economic, health, society, and accessibility to resources. Social vulnerability varies from place to place and is influenced by population growth, rate of urbanization, and economic vitality associated with economic, demographic, social, and human-made environment (Birkmann & Wisner, 2006; Cutter & Finch, 2008; Montz & Tobin, 2011; Aksha et al., 2019). It is difficult to quantify social vulnerability at the microscale due to the availability of vulnerability-related data. Different methods have been used to measure social vulnerability in the socioeconomic, political, and geographical context. Social vulnerability measures the multiple burdens and shocks and accessibility to essential resources and services such as food, water, healthcare services, and transportation, all of which contribute to health and well-being. These measures can reduce the suffering and losses and are directed toward coping, recovery, and resilience from the coming natural and social hazards.

Social vulnerability is closely related to household health shocks. The increasing burden of health shock, i.e., diseases, death, etc., increases the social vulnerability of poor people, especially in urban areas. Continuous livelihood burden in the limited urban space leads to poor people becoming more vulnerable, which is gradually challenging due to the influx of rural migrants living in unrecognized or unauthorized illegal urban areas. Urban poor are facing deprivation in basic facilities and healthcare services, which fall into a poverty trap and health shocks. Thus, the urban poor are confronted with health shocks and do not maintain a standard of living (Jankowska et al., 2011; Wratten, 1995; Moser, 1998; Beall & Fox, 2009). Such phenomena are common in most slum areas in developing countries, especially Indian cities. However, the magnitude of social vulnerability may differ in the context of physical, social, demographic, and living environment and health behavior where they live. Therefore, a simple modified social vulnerability index (MSoVI) has been adopted, developed by Cutter et al. (2003), and incorporated with the study area's physical, socioeconomic, and demographic setup and health aspects. All the slum neighborhoods were taken as a unit of study, and information related to vulnerability variables was obtained, as represented in Table 2.10. PCA has been used to extract the components (Table 2.11) for constructing MSoVI (Eq. 2.6) with 70.832% of explained variance. All the variables were normalized through mean, percentage, and weightage via SPSS v26. For each component, cardinality was assigned to extract latent variables.

$$\text{SoVI}_{\text{Slum}} = \sum PC_{i-i} \times V_i \tag{2.6}$$

where PC_i is the principal component and V_i is the variance of each component. SoVI scores for each slum neighborhood were mapped through ArcGIS v10.4 to explore whether the neighborhoods are high or low vulnerable based on the Z-score. The Local Moran's I statistic has been used to find the statistical cluster or random pattern of social vulnerability at the neighborhood level (Fig. 2.18).

Vulnerability scores have been grouped into five quintiles based on estimated SD values from high (>1.5 SD) to low (<−1.5 SD) vulnerability. Afterward, the total SoVI scores were mapped through ArcMap (Fig. 2.20), and the value of each component was used to envision social vulnerability across Midnapore (Fig. 2.19). The map of total SoVI scores (Fig. 2.20) exhibits the highly socially vulnerable slum neighborhoods such as Udaypally bastee, Ashok Nagar Nepalipara, Rammohan

Table 2.10 Variables used for SoVI

Variables		Description	Mean	SD	References
Age	Child population	Percentage of child population (under 6 years)	16.74	6.9995	1, 2, 3
	Aged population	Percentage of the elderly population (60+ years)	7.01	5.825	1, 2, 3
Education and marital status	Literacy	Percentage of the population who can read and write	81.31	10.019	1, 2, 3, 4
	Marriage before 18 years	Percentage of woman's marriage age (<18 years) to the total married	11.61	7.732	4
Occupation	Earning members in the service sector	Percentage of earning members associated with the service sector	14.78	16.661	n/a
	Earning members in small business and informal sector	Percentage of earning members associated with small business and informal sector	36.61	22.721	n/a
	Earning members in daily wage labor	Percentage of earning members associated with daily wage labor	36.43	27.773	n/a
Health behavior	Timely food intake	Percentage of households taking food timely	45.83	18.455	n/a
	Skipping meal	Percentage of households skipping their meal	24.36	13.353	n/a
	Physical workout	Percentage of households with regular exercise	2.18	7.029	n/a
	Health card	Percentage of households having a health card	50.96	23.162	n/a
	Health insurance	Percentage of households having health insurance	22.76	19.824	n/a
	Interval of sickness	Often suffering from the prevalence of diseases	15.90	17.260	n/a
	Women violence	Percentage of households facing women violence	16.92	14.881	n/a

(continued)

Table 2.10 (continued)

Variables		Description	Mean	SD	References
Family structure	Total female	Percentage of females in a household	51.11	7.2794	1, 2
	Female-headed households	Percentage of female-headed households	25.00	25.00	4
	Family size	Average no. of people per household	5.076	1.0875	4
	Acquaintances	Percentage of acquaintances	48.59	22.532	n/a
	Closeness with acquaintances	Percentage of households with closeness with acquaintances	51.67	25.749	n/a
Asset	No phone	Percentage of household having no phone	6.73	12.295	1, 2, 3, 4
	Vehicle	Percentage of household having no vehicle	63.78	24.133	1, 2, 3, 4
Built environment	Electricity	Percentage of household having electricity	98.46	5.347	1, 2, 3, 4
	Sewerage system	Percentage of household having a poor sewerage system	11.41	18.012	1, 2, 3, 4
	Separate bathroom	Percentage of household having a separate bathroom	73.59	23.027	4
	Separate kitchen	Percentage of household having a separate kitchen	63.59	25.118	4
	Source of drinking water	Percentage of household using public tap water	85.13	17.579	n/a
	Water purifier	Percentage of household having a water purifier	7.50	12.579	1, 2, 3, 4
	Facilitated solid waste collection	Percentage of household facilitating solid waste collection daily	25.38	24.531	4
	Poor toilet	Percentage of households facing poor toilet facility	87.56	16.279	4
	Fuel used for cooking	Percentage of households using LPG for cooking	74.42	22.353	4

(continued)

Table 2.10 (continued)

Variables		Description	Mean	SD	References
Housing condition	Pucca floor	Percentage of households having pucca floor	70.83	22.689	2, 4
	Pucca wall	Percentage of households having pucca wall	79.04	35.338	2, 4
	Pucca roof	Percentage of households having a pucca roof	56.15	26.772	2, 4
	Overcrowding	Percentage of households facing overcrowding	49.57	12.594	n/a
	Room dampness	Percentage of household facing room dampness	45.58	23.314	n/a
Financial	Employment	Percentage of employment	52.693	11.0975	1, 2, 3, 4
	BPL	Percentage of households having BPL card	30.06	22.041	4
	Taken loans	Percentage of households carrying the burden of the loan	44.23	25.094	4
	Income	Average per capita income	2526.72	899.32	1, 2, 3, 4
Ethnicity	Minority	Percentage of households belonging to the minority	26.28	38.340	1, 2, 3, 4

Referenced studies: 1, Cutter et al. (2003); 2, Kirby et al. (2019); 3, Aksha et al. (2019); 4, Roy (2016)
Source: Field survey (2022)

Nagar, Iswarpur bastee, Hamalpara, Jharnadanga bastee, Natunpally, Kamarpara, Patnabazar bauripara, Patnabazar alokbisoi club bastee, Astana Pukur bastee, Rajibnagar bastee, Dudhar Pukur bastee, Abasanpara bastee, Pyendange tetultola bastee, etc., which are located in the eastern, the southwest, and the northwestern part of the city. The low socially vulnerable slum neighborhoods like Mirza Mahalla Jhangir bastee, Taltola bastee, Harijon Para bastee, Muchi para-Sangbad Bazaar, Kustho Colony, etc. are typically located in the central, eastern, northeast, and northeast part of the city. The slum neighborhoods, for example, Saildanga Gandhi ghat bastee, Palbari Vivekananda Pally, Mahatabpur srijonibaba ashram bastee, Brajakishor pally bastee, Mandal Mahalla Adibasi Para, Hosnabad bastee-1, Dograpukurbastee, Chatarparabastee, Sudam pukurpar bastee, Nabinabag bastee, Sarada Nagar, Hanspukur, Muslimpara bastee, Napitpara bastee, Goalapara bastee, Mitraconpound kamarpara, Sekhpura bastee, Jelapara bastee, Jhangir colony, etc., around the city core (mainly CBD) are recognized as moderate to high and low to

Table 2.11 PCA for SoVI

Principal component	Cardinality	Variables	Loadings	Explained variance (%)	Eigenvalue
PC1	(+)	Literacy	0.924	11.116	4.669
PC2	(+)	Employment	−0.956	6.105	2.564
PC3	(−)	Pucca floor	0.704	5.410	2.272
		Pucca wall	0.714		
		Pucca roof	0.853		
PC4	(+)	Separate bathroom	0.746	4.899	2.057
		Separate kitchen	0.838		
		Fuel used for cooking	0.448		
PC5	(+)	Overcrowding	0.752	4.693	1.971
		Family size	−0.701		
		Income	0.467		
PC6	(+)	Water purifier	0.474	4.172	1.752
		Earning members in the service sector	0.787		
		Minority	−0.491		
PC7	(+)	Skipping meal	−0.421	4.116	1.729
		Earning members in daily wage labor	0.732		
		Earning members in small business and informal sector	−0.677		
PC8	(+)	Sewerage system	0.402	3.713	1.559
		Poor toilet	0.710		
		Room dampness	0.400		
		Interval of sickness	−0.588		
PC9	(+)	Electricity	0.699	3.698	1.553
		No phone	−0.704		
PC10	(+)	Health card	0.733	3.391	1.424
		Health insurance	0.759		
PC11	(+)	Facilitated solid waste daily	0.510	3.199	1.344
		Marriage before 18 years	0.778		
PC12	(+)	Women violence	0.498	3.090	1.298
		Closeness with acquaintances	−0.726		
		Sources of drinking water	−0.391		
PC13	(+)	Female-headed households	−0.314	2.863	1.202
		Child population	−0.326		
		Aged population	0.802		
PC14	(+)	Acquaintance	−0.572	2.816	1.183
		Taken loans	0.814		

(continued)

Table 2.11 (continued)

Principal component	Cardinality	Variables	Loadings	Explained variance (%)	Eigenvalue
PC15	(+)	BPL	0.781	2.592	1.089
PC16	(−)	Physical workout	0.789	2.552	1.072
		No vehicle	0.395		
		Total female	0.313		
PC17	(−)	Timely food intake	−0.804	2.407	1.011

Extraction method: principal component analysis. Rotation method: varimax with Kaiser normalization. (a) Rotation converged in 20 iterations

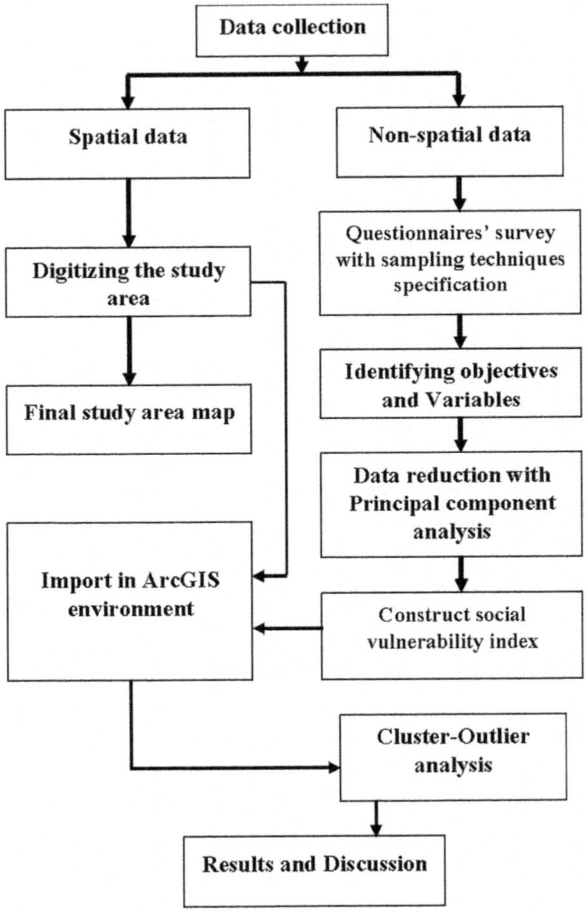

Fig. 2.18 Methodological flowchart for SoVI

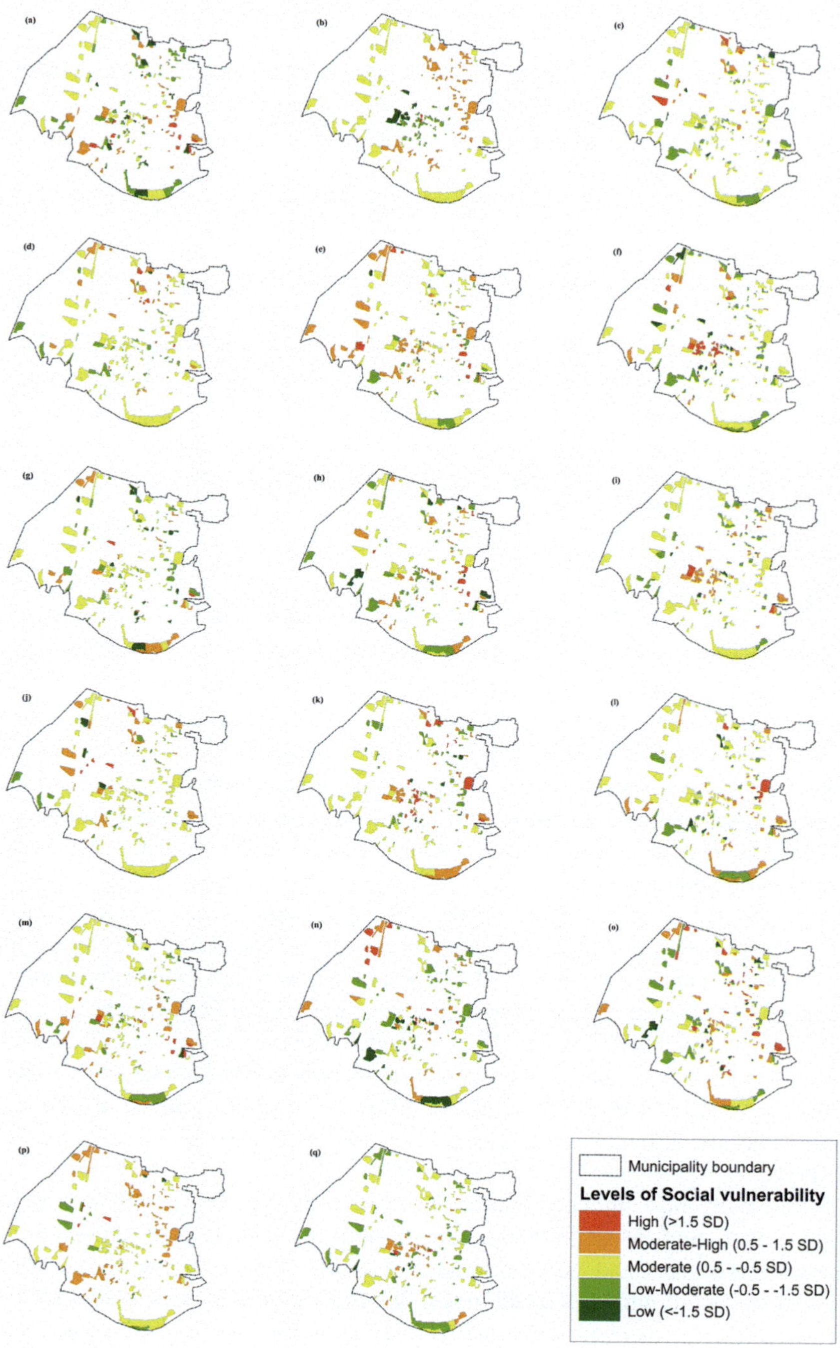

Fig. 2.19 Geographic distribution of slums based on Principal Component values (PC1 to PC 17 denoted as a to q letter) of the modified Social Vulnerability Index.(Source: Author)

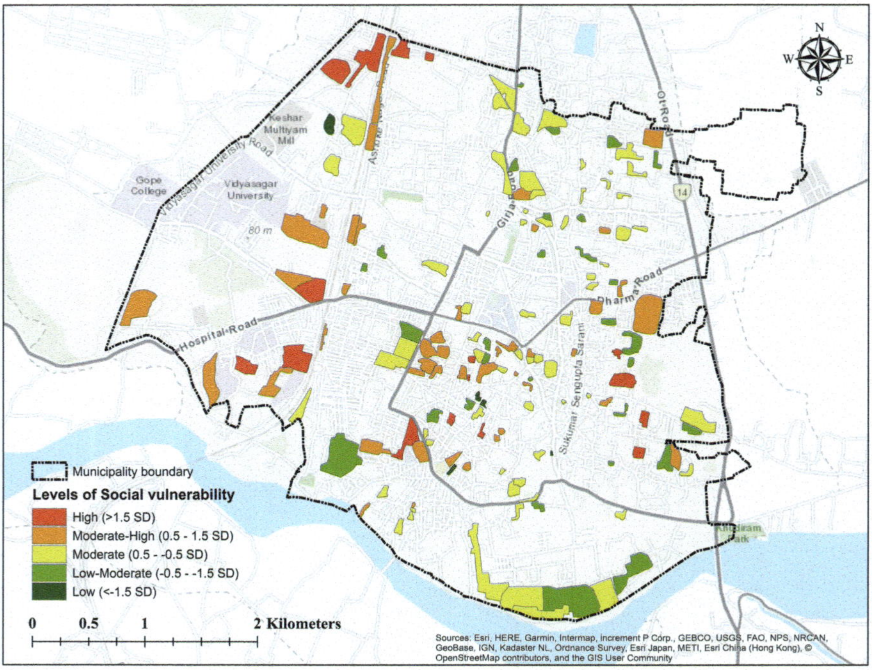

Fig. 2.20 Spatial distribution of slums based on the level of social vulnerability(Source: Author)

moderate socially vulnerable areas. This study employed Moran's *I* test to estimate the positive spatial autocorrelation with a value of 0.018 ($p < 0.18$) and a Z-score of 1.33. The value of spatial autocorrelation represents the random spatial distribution of slum neighborhoods.

The geographic distribution of the principal components is visualized in Fig. 2.19. Highly vulnerable slum neighborhoods are located in the eastern and southwest parts of the city for PC1, PC13, and PC16. Moderate to high vulnerability exhibits for PC2 toward the northwest compared to PC3 and PC4, which show relatively low vulnerability values in the north and northwest parts of the city. Besides, randomly distributed slum neighborhoods show moderate to high levels of vulnerability for PC5, PC7, PC8, PC14, and PC15, respectively. In contrast, a high level of vulnerability is found in the central and northwest parts of the city for PC6 and PC10.

In summary, most slum neighborhoods are socially vulnerable, with an intensity of moderate to high. Most probably, slum dwellers are socially deprived in terms of accessible resources and going through vulnerable socioeconomic status, which forces them to fall into high health risks. The monthly low income (<Rs. 5000) is intrinsically associated with vulnerability and health risk. Due to low-income levels, slum dwellers face difficulties in their livelihood beyond meeting their health expenditures. The observation through field survey reveals that the occupation status of

the dwellers is volatile and vulnerable as most males are rickshaw or toto pullers. At the same time, the female is engaged in insecure servant work with a low salary. The loss of jobs and lack of self-employed workers push households toward taking an informal loan or mortgaging household assets, which may influence their daily livelihood and social vulnerability and lead them into a poverty trap and a vicious cycle of health shocks. Consequently, to meet the expenditure of health shock, they again become susceptible to vulnerability.

As a whole, literacy has a significant impact on health vulnerability and daily livelihoods. Generally, a low literacy level has high social vulnerability and vice versa in the context of knowledge, perception, and management skills. Education may help change the family's superstitious nature, strengthening the family's economic planning and culture, thereby reducing health shock and economic and social vulnerability. Moreover, an educated or literate person can decide what ought to be or what is wrong in the family, thereby taking strategies to reduce the family tension and burden. Women's literacy can help to mitigate gynecological problems during the menstruation period, and they can express their problems to the doctor as most educated families do, thus reducing the mortality rate. The study investigated the fact that in the context of personal hygiene, more women use clothes during their menstruation than sanitary napkins (23%) in the present era. Low education may also affect healthcare practices for children and the aged, reducing morbidity and mortality.

The socioeconomic status of the households also influences maternal health and adolescent health. Housing infrastructure, accommodation, spacing between houses, number of rooms, separate kitchens, and bathrooms are considered a health indicator, especially for the child and women. For example, no separate kitchen results in indoor pollution, which may cause respiratory diseases among the child and women. Low-standard rooms and accommodations severely impact the breed of different infectious diseases, especially during the monsoon period. A flawed sewerage system and no separate bathroom also breed dengue mosquitoes. Most of the slum neighborhoods mostly use tap water, and without purifiers, stagnant water and dampness may cause different microbes' generation and related diseases. Therefore, overall accommodation and inaccessibility to basic amenities increase the health shock and vulnerability.

Moreover, large family size, female-headed family, and closeness to acquaintances significantly impact the family's overall socioeconomic status. For example, meeting expenditures is difficult if family members have more earners and female-headed families than small family size and male-headed families, respectively. Closeness to acquaintances is essential from a social perspective, as the family may get support mentally, economically, and socially. Ethnicity, religion, and caste also contribute to social exclusion and deprivation of the resources of society. The overall result of the study is evident that disparities among the slum neighborhoods in the context of socioeconomic, demographic, and health behaviors may contribute to the overall quality of life, social vulnerability, and health shocks. Therefore, in the context of different socioeconomic, demographic, and historical backgrounds, the place-based integrated structural mitigation program should be implemented for the overall development of slum neighborhoods and the prosperity of the city's entity.

2.10 Conclusion

In developing countries, social deprivation and poverty are common phenomena where poor households are deprived of basic amenities, work, money, housing, and accessibility to healthcare facilities and services. Due to high population density and congestion in urban areas, the community's basic needs are more significant when living in a limited space. Also, quality of life varies from person to person due to physical and mental health, emotion, recreation, leisure time, freedom, security, employment and income, accessibility to basic amenities and healthcare services, and the individual perception of the standard of living. Quality of life is closely related to vulnerability and health security as a whole. After employing principal component analysis (PCA), data was reduced into variables, and social indexes were developed using the adopted methodology. The deliberated index score joins smaller spatial units as slum neighborhoods with the spatial data in the GIS algorithm. Different GIS-based analyses like hot spot analysis and Anselin Local Moran's analysis find the spatial variations of different social aspects. This analysis also finds the spatial pattern of slum neighborhoods in the study area. The GIS-based analysis can easily find the deprived and decent social elements of the discussion in this chapter. Most of the slum neighborhoods are vulnerable, with moderate to high intensity. Most probably, slum dwellers are socially deprived in terms of accessible resources and go through vulnerable socioeconomic status, which forces them to fall into high health risks.

References

Aksha, S. K., Juran, L., Resler, L. M., & Zhang, Y. (2019). An analysis of social vulnerability to natural hazards in Nepal using a modified social vulnerability index. *International Journal of Disaster Risk Science, 10*(1), 103–116. https://doi.org/10.1007/s13753-018-0192-7

Alkire, S., et al. (2015). *Multidimensional poverty measurement*. Oxford University Press.

Beall, J., & Fox, S. (2009). *Cities and development*. (Routledge Perspectives on Development). Routledge. http://www.amazon.co.uk/Cities-Development-Routledge-Perspectives/dp/0415390990

Beccari, B. (2016). A comparative analysis of disaster risk, vulnerability and resilience composite indicators. *PLoS Curr, 8*.: ecurrents.dis.453df025e34b682e9737f95070f9b970. https://doi.org/10.1371/currents.dis.453df025e34b682e9737f95070f9b970. Erratum in: PLoS Curr. 2017 Jun 29;9: PMID: 27066298; PMCID: PMC4807925.

Birkmann, J. (2007). Risk and vulnerability indicators at different scales: Applicability, usefulness and policy implications. *Environmental Hazards, 7*(1), 20–31. https://doi.org/10.1016/j.envhaz.2007.04.002

Birkmann, J., & Wisner, B. (2006). *Measuring the un-measurable the challenge of vulnerability Jörn Birkmann & Ben Wisner* (Issue April).

Cavapozzi, D., Han, W., & Miniach, R. (2015). Alternative weighting structures for multidimensional poverty assessment. *Journal of Economic Inequality, 13*(3), 425–447.

Census of India 1961. (1961, 2011).
Cutter, S. L., & Finch, C. (2008). Temporal and spatial changes in social vulnerability to natural hazards. *Proceedings of the National Academy of Sciences of the United States of America, 105*(7), 2301–2306. https://doi.org/10.1073/pnas.0710375105
Cutter, S. L., Carolina, S., Boruff, B. J., Carolina, S., Shirley, W. L., Carolina, S. (2003). Cutter et al 2003. In Soc. Sci (Vol. 84, pp. 242–261).
Das, P., Paria, B., & Firdaush, S. (2020). Consumption poverty and multidimensional poverty: A study in Indian context. *Social Indicators Research*. https://doi.org/10.1007/s11205-020-02519-0
Dunning, C. M., & Durden, G. C. (2013). Social Vulnerability Analysis: A Comparison of Tools (IWR White Paper). Institute for Water Resources, U.S. Army Corps of Engineers.
Fekete, A. (2009). Validation of a social vulnerability index in context to river-floods in Germany. *Natural Hazards and Earth System Science, 9*(2), 393–403. https://doi.org/10.5194/nhess-9-393-2009
Hazards & Vulnerability Research Institute. (2019). Social Vulnerability Index (SoVI) 2019. University of South Carolina.
Jankowska, M. M., Weeks, J. R., & Engstrom, R. (2011). Do the most vulnerable people live in the worst slums? A spatial analysis of Accra, Ghana. *Annals of GIS, 17*(4), 221–235. https://doi.org/10.1080/19475683.2011.625976
Kirby, R. H., Reams, M. A., Lam, N. S. N., Zou, L., Fundter, G. G. J. D. D. Q. P., & Kirby, R. H. (2019). Assessing social vulnerability to flood hazards in the Dutch Province of Zeeland. *International Journal of Disaster Risk Science, 10*(2), 233–243. https://doi.org/10.1007/s13753-019-0222-0
Komalawati, R. A., & Lim, J. (2020). The reality of compact development in a developing country : Focusing on the perceived quality of life in Jakarta, Indonesia. *International Journal of Urban Sciences*, 1–32. https://doi.org/10.1080/12265934.2020.1803106
Mandal, K., Chatterjee, S., & Chatterjee, N. D. (2016). Morphological analysis of a historical urban landscape: The case of Midnapore, an Early Urban Center of Eastern India. Vol. 4(12), 28–37, December 2016. The International Research Journal of Earth Science. E-ISSN 2321–2527.
Maria, B., Hummell, D. L., Cutter, S. L., & Emrich, C. T. (2016). Social vulnerability to natural hazards in Brazil. *International Journal of Disaster Risk Science*. https://doi.org/10.1007/s13753-016-0090-9
Midnapore Municipality (2022). https://www.midnaporemunicipality.com/
Midnapore Municipality. (2012). Audit report 2011-2012. https://midnaporemunicipality.com/mun_auditreport.php
Montz, B. E., & Tobin, G. A. (2011). Social vulnerability to disasters. CRC Press.
Moser, O. N. C. (1998). Reassessing urban poverty reduction strategies: The asset vulnerability framework. *World Development, 26*(1), 1–19.
Mudey, A., Ambekar, S., Goyal, R. C., & Agarekar, S. (2017). Studies on ethno-medicine assessment of quality of life among rural and urban elderly population of Wardha District, assessment of quality of life among rural and urban elderly. 5070. https://doi.org/10.1080/09735070.2011.11886394
Mukherji, A. (2001). Towards a sustainable city: The problem of environment and urban poverty in Medinipur. http://www.anthrobase.com/txt/M/Mukherji_A_01.htm.%0A%0A https://doi.org/10.1016/j.envint.2005.10.003.
Ravallion, M. (2011). On multidimensional indices of poverty. *The Journal of Economic Inequality, 9*(2), 235–248. https://doi.org/10.1007/s10888-011-9173-4.hdl:10986/3346
Roy, U. (2016). Causative factors of social inequality and its impact on community health: A neighbourhood level study in Midnapore Municipal Area, West Bengal, India. The International Archives of the Photogrammetry, Remote Sensing and Spatial Information Sciences, XLI-B8 201. https://doi.org/10.5194/isprs-archives-XLI-B8-1417-2016.
Sen, A. (1981). Poverty and famines: An essay on entitlement and deprivation. Clarendon Press.
Wratten, E. (1995). Conceptualizing urban poverty. *Environment & Urbanization, 7*(1), 11–38. https://doi.org/10.1177/09562478950070

Chapter 3
Lifestyle Behavior and Health Awareness

3.1 Introduction

Lifestyle comprises cultural and behavioral habits of an individual lifelong socialization process, which may be harmful or promoted to health and well-being. Lifestyle changes may cause many health problems that need special care and a healthy lifestyle. Population growth with resulting rapid urbanization would face challenges of health vulnerability associated with the local environment, water, air, violence, accidents, and non-communicable diseases that are influenced by risk factors such as personal hygiene, unhealthy food habits, chewing and tobacco use, smoking, alcoholism, lack of physical exercises, crowding, malnutrition, and disease prevalence (Saikia & Das, 2014). These risk factors, directly or indirectly, affect health-seeking behavior and well-being for long periods. Since childhood, we have continuously learned lifestyle behavior from family and society. In the same way, awareness is a learning process through understanding the self and situation in a particular space and time. Health awareness is the increasing possibility of people being aware of their self-health, families, environment, and close acquaintances where they live or are engaged in services. However, family norms, values, superstitions, perceptions, attitudes toward illness, etc. strongly influence the community's health awareness. Therefore, socioeconomic and cultural factors, knowledge, and consciousness about health and disease are the significant resilience to awareness in public health.

Rapid urbanization has failed to provide essential city infrastructure and space, especially in developing countries, leading to health hazards (Montgomery, 2009) such as the COVID-19 pandemic. Seven in ten people will live in cities by 2050 (Saikia & Das, 2014), which may increase urban poverty. The continuous influx of rural migrants to urban areas brings about the growth of slum populations in the cities, which reduces urban space and increases health hazards. The reason is that most of the slum dwellers often lack basic personal hygiene and healthcare such as

© The Author(s), under exclusive license to Springer Nature Switzerland AG 2025

B. Dutta et al., *Using GIS to Identify Opportunities for Improved Health in Urban Slums*, Global Perspectives on Health Geography,
https://doi.org/10.1007/978-3-031-89617-0_3

boiling and purifying drinking water, handwashing before and after a meal and toilet, clean dress daily, etc. Family rules, superstitions, norms, and social traditions sometimes act as barriers to using modern healthcare services. This often leads to unconsciousness, carelessness, and abstinence from availing healthcare facilities in due time, which then increases morbidity, mortality, sickness, etc. This chapter, therefore, is an attempt to explore the health-seeking behavior and health awareness among the slum dwellers that would help to draw recommendations for further improvement of their health, healthcare facilities, and services, thereby increasing their coping skills in the slum pockets in the urban area.

3.2 Loan and Savings for Health Expenses

Nowadays, household expenses are increasing for the standard of living or price inflation and healthcare. The loans and savings are used during medical emergencies and offer financial coverage for expenses paid during hospitalization bills, medicine, surgeries, chemotherapy, and other similar healthcare treatments. The tremendous population growth and resulting environmental contamination-related vulnerability and poverty conditions lead people to health shocks. The increasing burden of health shocks is common in those who are living in slums and below the poverty line, the so-called urban poor. Generally, middle- and high-income groups are in health insurance and have savings for health expenses. However, the urban poor are involved in meeting their daily livelihood rather than saving for health expenses. Therefore, the study tries to understand what the urban poor are doing for the provision of expenditure. The survey from the households revealed that they are taking loans and savings to meet their expenditure, i.e., 44.23% of households took loans from informal, e.g., money lenders or mortgaged household assets. In comparison, 55.77% of households took no loans or savings for health expenditures (Table 3.1). However, the West Bengal government launched "Swasthya Sathi" in December 2016, which covers the essential health for secondary and tertiary healthcare up to 5 lakh per annum per family, covering all pre-existing diseases without the contribution of the households. During the survey (December 2022), only 50% of households had "Swasthya Sathi" health cards. During the post-survey, households reported that the "Duare Sarkar" (services at home) launched by the West Bengal government has been provisioned a "Swasthya Sathi" card at home. Hence, in the context of health expenditure, it is expected that households will have to reduce informal loans and increase savings to develop quality of life, thereby reducing health shock incidents caused by the low standard of living.

Table 3.1 Loan and savings status of the households

Loan and savings	Percentage of household
Yes	44.23
No	55.77

Source: Field survey (2022)

3.3 Process of Solid Waste Management

Solid waste management is the process of collecting, dumping, burning, and treating solid wastes generated from households and commercial, institutional, and industrial activities. There are different ways of collecting solid waste, such as daily, weekly, monthly, and never. Similarly, solid wastes are treated differently, especially in developing countries, such as remaining scattered, dumping on nearby pits or away from habitation daily on mobile collecting vans, and burning. In the study area, solid wastes collected daily, weekly, and monthly are 8%, 28%, and 12%, respectively, while never collected accounts for 52%, which led to dumping solid waste nearby pits (50.64%), away from habitation (9.1%), as well as remaining scattered (10.45%) and burning (4.17%) the solid waste (Fig. 3.1). The overall picture of solid waste management in the study area is mostly scattered and never collected from most of the slum neighborhoods, which has a severe health impact due to the breeding of microorganisms and pathogens (zoonotic diseases), mainly affecting children, women, and elderly members of the households. Moreover, solid waste from the households to and from the neighborhood produces a bad smell, which may create mental irritation, breathing problems, and susceptibility to asthma patients. Therefore, special care and management procedures should be implemented to protect and prevent communicable diseases.

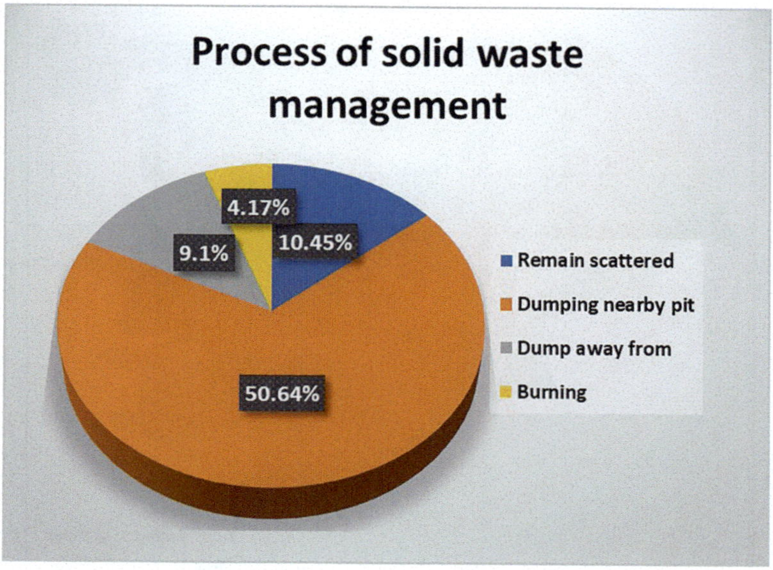

Fig. 3.1 Process of Solid waste management (Field survey, 2022)

3.4 Waterlogging Condition

Waterlogging is the condition of stagnant water for a short or long period in a particular place caused by heavy or average rainfall, poor drainage, and defective waste disposal systems. Waterlogging is a normal phenomenon on the surface of depression storage without human habitation. However, waterlogging in residential, institutional, or industrial areas badly affects houses, living conditions, and activities. The southern side of the study area is surrounded by the Kasai River, with stagnant water presence throughout the year due to the presence of the Aniket Dam. The analysis of flood buffering was estimated in ArcGIS v10.4 based on an average level of the stagnant water of the Kasai River (23 m) with a 5-m interval of water level (Fig. 3.2). If the water level increases around 20 m (during massive flood), then 50%

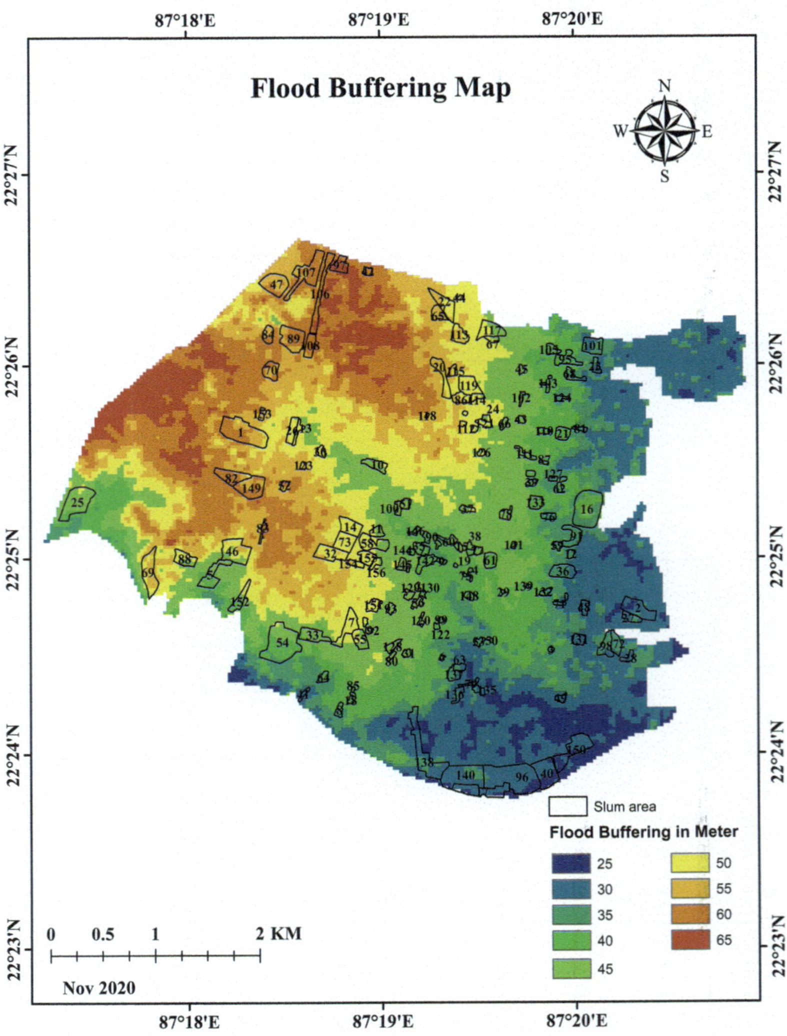

Fig. 3.2 Flood buffering for the study area (Source: USGS)

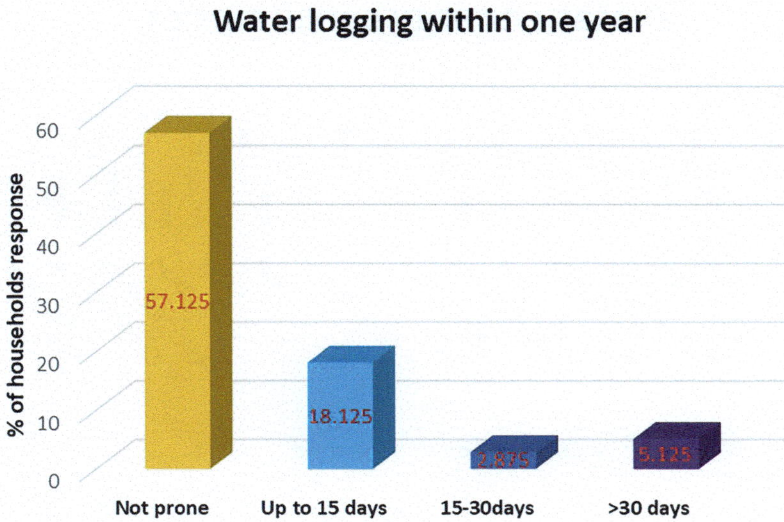

Fig. 3.3 Water logging condition (Field survey, 2022)

of the area will be waterlogged, especially in the northeast, east, and southeast parts of the city; if the water level increases by 2 m (up to 25 m), then five to ten slum neighborhoods will be inundated during an average rain. So, one can imagine the waterlogging conditions during the flood. From the interview survey, the slum dwellers residing near the Kasai riverbank reported that during the rainy season, they are mostly living in their relatives' houses (around 60 days) due to long-day waterlogging conditions. Due to the encroachment of urban spaces, major drains are occupied by building houses or informal shops filled with garbage.

During heavy rains, water overflows and enters the houses, and sometimes, this dirty water stays for more than 15 days. About 18.13% of slum dwellers reported that water stayed up to 15 days, 8% reported more than 15 days, and 57.13% of respondents did not face waterlogging during the rainy season (Fig. 3.3).

3.5 Dietary Energy Consumption

Dietary energy consumption, such as protein, fat, vitamins, etc., is the basic need of every human body, which keeps the body healthy and prevents the prevalence of malnutrition, diseases, etc. Nevertheless, malnutrition and protein and vitamin A deficiency are common in India, especially in cities' slums. The nutritional status and diets of slum dwellers among all age groups are worse than those of their rural counterparts and even lower than those of urban ones. Urban slums are mainly growing through distressing socioeconomic conditions, poverty, and malnutrition due to migration in the unprivileged part of the metropolitan area. The deficiency in dietary energy consumption may result from a lack of awareness and knowledge about proper nutrition, limited access to adequate nutritious food due to low

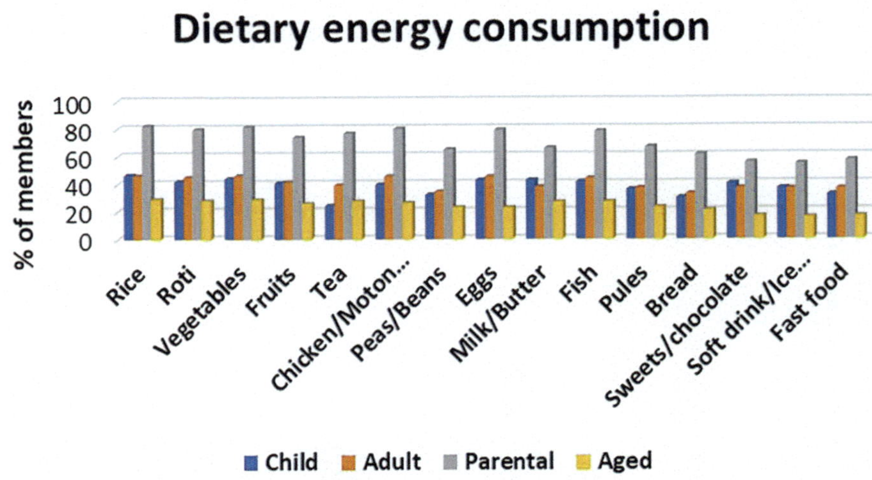

Fig. 3.4 Dietary energy consumption (Field survey, 2022)

socioeconomic status, and insufficient care or responsibility during childhood, adulthood, pregnancy, and old age. Most of the survey data revealed that the daily habitual diets of the urban poor primarily consist of cereals (mainly carbohydrates), with minimal intake of protein and fat. Protein-rich foods such as pulses, eggs, and meat are largely absent, resulting in dietary energy consumption that falls below the recommended intake (RDI). The National Institute of Nutrition revealed the dietary consumption of urban slum dwellers is worse than the rural average.

The consumption of junk or "ready-to-eat" food poses greater health risks to poor individuals, as the low quality of these foods contributes to serious health issues. This trend is especially common among the younger male and female members of low-income households. The dirty junk foods consumed by children are less nutritious, which makes malnutrition with expenditure and health problems like diarrhea, ARI, measles, etc. Underweight, stunted, and overweight also were prevalent among the slum dwellers due to their food habits. Women are suffering from malnutrition during their pregnancy period, which sometimes may cause morbidity, miscarriage, fetal death, neonatal mortality, as well as maternal mortality. The study observed that rice is more common than vegetables, fish, eggs, meat, pulses, and others (Fig. 3.4). It is also remarkable that slum dwellers now prefer fast foods, soft drinks, etc. than protein foods, which needs a more focused and integrated strategy from the nutritional and hygienic context.

3.6 Smoking, Alcohol Consumption, and Drug Addiction

The common phrase is "smoking or alcohol or drug use is injurious to health." Substance (smoking, alcoholism, or drug) use is a rising fact all over the world and particularly in India, perhaps due to the release of stresses with the changing

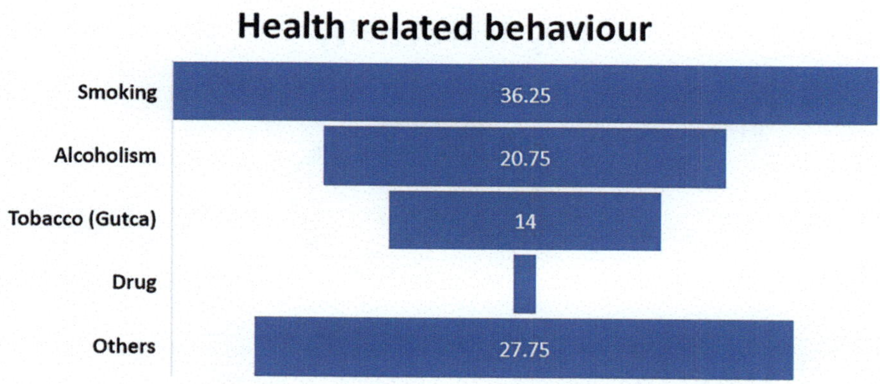

Fig. 3.5 Practices of health behavior (Field survey, 2022)

lifestyles. Substance use is cost heavily and results in significant health problems for the community.

This may be the cause of death and morbidity globally. In India, alcohol, smoking, gutka, pan, etc. have become common among people in the recent decade, albeit with differences in rural or urban places. Binge drinking, intoxication, and drug use among the urban poor may contribute to public health problems and cause injuries, violence in the family, bad relations with neighbors, etc. However, substance use varies across caste, religion, education, social groups, gender, place of residence, and standard of living to a great extent. In the context of slum areas, substance use is predominantly detrimental as they are mostly living below poverty, and the use of substances cuts their family expenditure, which leads them into the vicious cycle of poverty. Despite that, family and societal well-being are also hampered by violence, fighting, and slag language use due to binge drinking. The study observed across the all-slum neighborhoods that 36.25% of dwellers are smoking tobacco such as Biri, cigarettes, etc., 20.75% drink alcohol (frequently), 14% consume tobacco like gutka, and 27.8% consume other types of substances (Fig. 3.5). During the survey, most of the married women reported that consumption of alcohol is the leading cause of violence and poverty condition in the family. The study found a small quantity of alcohol consumption among the women. Instead, some women consumed tobacco such as gutka, pan, raw tobacco leaf, etc.

3.7 Physical and Recreational Activities for Health

Physical exercise and recreational activities are essential in every span of human life. Physical activities like exercise, yoga, playing, sleeping, dancing, and religious activities are expected to keep the body fit and healthy.

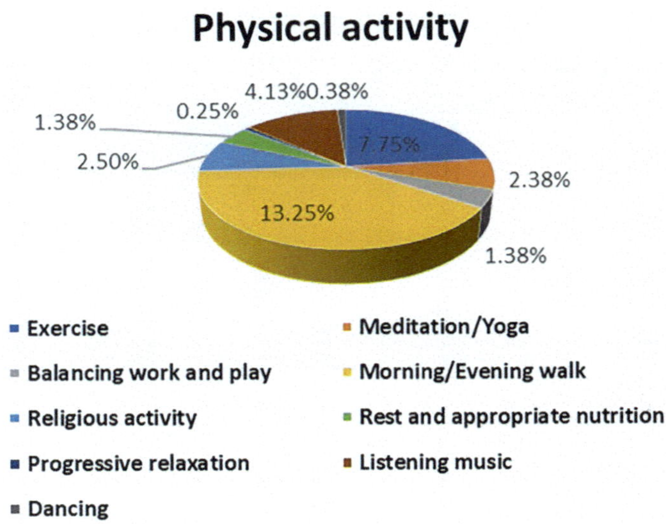

Fig. 3.6 Physical Activities for health wellbeing (Source: Field Survey, 2022)

Moreover, physical activities reduce stress, anxiety, and irritation and prevent diabetes, obesity, gastric problems, etc., across all ages. Recreational activities like watching TV or cinema, playing a game, dancing, listening to music, mooch or festivals, etc. are mental pleasure and psychological development, thereby promoting the health and well-being of the community. People living in poverty typically engage in labor-intensive work and have little extra time for health-related physical activities, as their primary focus is on meeting basic needs. Nevertheless, some do participate in various recreational activities.

The study captured responses from slum dwellers regarding their involvement in health-promoting and recreational activities. For instance, 13.25% reported participating in morning or evening walks, 7.75% engaged in physical exercise, 2.38% practiced yoga or meditation regularly, 2.50% took part in religious activities, and 4.13% regularly listened to music or participated in other similar activities (Fig. 3.6). Respondents have also reported that they are watching TV/movies at home or cinema hall (57%), listening to music (13%), playing games (5%), dancing (2%), and other activities (23%) (Fig. 3.7). Overall, health-related physical activities are not so well among the slum dwellers. Therefore, a mass awareness campaign should be needed to promote health and well-being among people experiencing poverty and reduce disease prevalence and health expenditure.

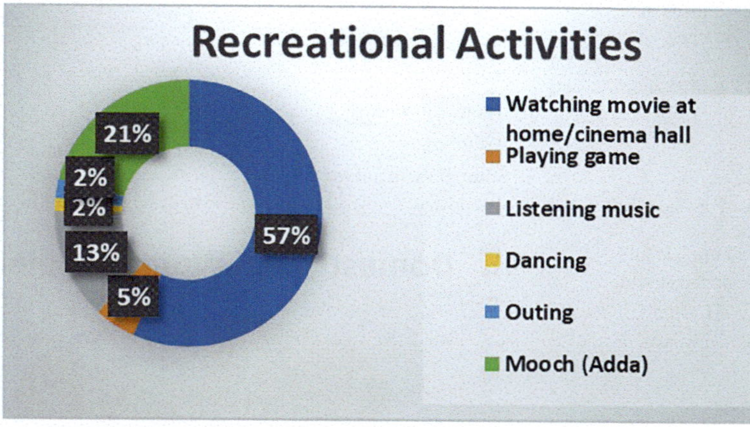

Fig. 3.7 Recreational activities for health promotion (Source: Field Survey, 2022)

3.8 Domestic Violence

Domestic violence is an aberrant behavior with low self-esteem in the adverse socioeconomic condition within the particular community or society. Domestic violence includes gender violence, especially women violence, child violence, neighbor violence, sexual violence and brutal rape, mental harassment, etc. Domestic violence in the slums of India is a common phenomenon due to the low standard of living and low socioeconomic status. Many researchers demonstrated different hidden causes of violence in the slum from various points of view, such as socioeconomic, psychological, family norms, lack of basic amenities (toilet, bathroom), patriarchal or muscularity (Chaudhuri & Morash, 2019), son preference (Shukla et al., 2017; Mahapatro et al., 2012), migration, education, caste, religion, family size, etc. Due to the lack of sanitation and resulting open defecation, young girl and married women are abducted, raped, and murdered. The menial job of men, consumption of alcohol (Sabri et al., 2014; Wagman et al., 2018), extramarital relationships, doubt in wives, eviction and job loss, frustration and stress, high expectation and low standard education (Garg et al., 2019), dowry, early marriage (Kidman, 2017; Raj et al., 2010), etc. are the leading causes of women violence. The form of women's violence is physical or sexual abuse (33.78%), beating (42.31%), mental torture (23.9%), and others (5.64%) in the study area as a whole (Table 3.2 and Fig. 3.8).The rate of women's violence varies from 1% to 45% in India (Jungari &

Table 3.2 Causes of women's violence

Causes	%
Physical/sexual abuse	33.78
Beaten	42.31
Mental torture	23.97
Others	5.64

Source: Field survey (2022)

Fig. 3.8 Domestic and women violence (Field survey, 2022)

Chinchore, 2022), while NFHS (2019–2020) reported a 23.2% prevalence of women's violence at their reproductive age in urban areas, which is gradually decreased. Women violence is strongly associated with physical, mental, and psychological health problems (Ellsberg et al., 2015; Dasgupta, 2019; Donta et al., 2016; Jeyaseelan et al., 2007). Moreover, this violence leads to low self-esteem, self-status in society, injuries, even suicides (Sabri et al., 2014), divorce, separation, etc. The prevalence of women violence not only affected women, but it also affected children in a different way, such as incomplete education, social behavior, nutritional deficiency, and adjustment in the family and neighborhoods, which perpetuate up to their adulthood (Yount et al., 2018). Therefore, women's violence is a severe issue in the present day, and there is a need to strengthen the socioeconomic status and basic amenities that can help women's empowerment and thereby reduce women's violence.

3.9 Analysis of Personal Hygiene

The World Health Organization (WHO) defines health as "a state of complete physical, mental, and social wellbeing, and not merely the absence of disease or infirmity" (WHO, 1948). The organization also defined "Hygiene refers to the conditions and practices that help to maintain health and prevent the spread of diseases." Personal hygiene, which is also referred to as "personal care, includes all of the following activities in daily life such as hair care, nail care, foot care, genital care, and dental care" (UNICEF, 2016). Personal hygiene is an art and science (Rahman et al., 2014; Adetunji et al., 2018) or a practice or strategy for acquiring healthy living and well-being for an individual in the current place of residence. It includes daily activities (Ghose et al., 2012; Rajbhandari et al., 2018) that act as disease prevention and health promotion (Odonkor et al., 2019). Standard personal hygiene helps us improve the quality of life and longevity (Pal & Pal, 2017) and prevent disease progression from environmental contamination. Standard personal hygiene practices must be adapted for future health to protect the body from respiratory infections, skin infections, and communicable diseases. Poor hygiene is a very personal subject (Dessalegn et al., 2017) and is one of the significant behavioral risk factors breeding the different infectious diseases that may affect family health burden and expenditure (Hassan et al., 2019) in varying degrees (Odonkor et al., 2019). Personal hygiene depends on socioeconomic status, family and living environment, individual awareness, healthcare delivery systems (USAID, 2004), and healthy coping behavior. Poor hygiene skills and practices are developed due to superstitious attitudes and a deficiency of proper hygiene knowledge in society or family (Odonkor et al., 2019). According to WHO (2008), to avoid the diffusion of diseases and infection, a set of conditions and practices of hygiene, such as cleanness, healthy food intake, and medicine, can help to maintain a better livelihood and promise good hygiene standards. Practices of hygiene are different from one culture to another. It is well known that "A healthy mind lives in a healthy body." Much research has reported that poor mental health is caused by unhygienic practices in developing and developed countries (Ranasinghe et al., 2016). The spreading of infectious and communicable diseases in developing countries has accounted for many illnesses and deaths worldwide (Ilesanmi, 2017). Health problems due to poor hygiene practices can be preventable by promoting proper basic and household hygiene practices or habits, improving essential services, and providing adequate health education (Hassan et al., 2019; Temitayo, 2016) from early childhood. Disease transmission cycles can be broken by the mechanisms of healthy personal hygiene practices (USAID, 2004).

The existence of slum neighborhoods is deprived chiefly of areas in cities, particularly noticeable in developing countries like India. Slum neighborhoods are associated with poor personal hygiene practices. They are more vulnerable due to high population density, compromised living situations (Pal & Pal, 2017), the spread of respiratory infection, inadequate water supply, lack of sanitary facilities, malnutrition, and vitamin deficiency, leading to rising diseases and morbidities (Temitayo, 2016). Lack of standard hygiene may also lead slum households into a poverty trap, a vicious cycle of detrimental personal hygiene practices and health shocks. As a result, disease prevalence is common among slum neighborhoods, affecting their socioeconomic condition, such as loss of savings, a mortgaged house, selling fixed assets, interrupted child education, taking another loan, etc. (Fig. 3.9). Thus, they have limited resources to cope with this condition and fall again and again into a vicious cycle.

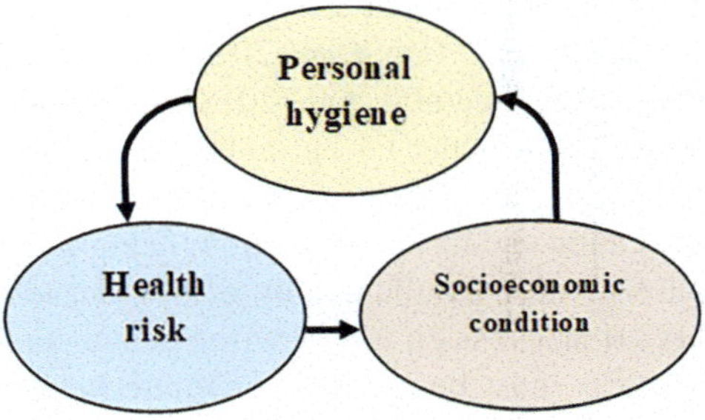

Fig. 3.9 Relation between personal hygiene and health risk

Table 3.3 Description of livelihood assets and their descriptive statistics

Asset types	Variables	Description	Mean	SD
Human	Employment capability of household	Percentage of employment capability of the household	40.02897	9.969646
	Educational level	Percentage of a literate person in the household	81.44531	10.130290
Physical	Household fixed asset	Percentage of household having minimum assets (production and appliance items)	19.23	11.943
	Housing condition	Percentage of households made with mud	20.96	19.863
	Number of rooms	Number of rooms households have	2.274	0.5211
Financial	Monthly cash income	Monthly cash income of a household	12263.65	4091.947

Source: Field survey (2022)

3.9.1 *Data and Methodology*

Personal hygiene is a critical issue worldwide, irrespective of caste, religion, ethnicity, etc. The following study relies on quantitative and qualitative data to understand the personal hygiene practices in the micro-geographical region. Therefore, in this context, four data collection methods were employed: a literature review, field visits, participatory observation, and a questionnaire survey. The study selected 15 meaningful variables (Tables 3.3 and 3.4) based on in-depth interviews with slum dwellers and some endorsement from the previous literature, which are the most influential factors for constructing the personal hygiene index (PHI). The study also collected statistical data and GIS maps from Midnapore municipality, georeferencing and digitizing the cartographic boundary of slum neighborhoods and cities in ArcGIS v10.4 (ESRI). The calculated PHI score is spatially joined with the slum shape file data. Inferential statistics were estimated based on collected data in SPSS v26 software. The methodological framework is shown in Fig. 3.10, guided by the complete study.

Table 3.4 Input data for analyzing the personal hygiene index

Types	Component	Description	Mean	Standard deviation
Basic hygiene practices	Brush at morning	Percentage of HH who brush in the morning every day	77.05	16.933
	Take a bath daily	Percentage of HHs who take baths daily	82.76	9.809
	Cleaning of dress daily	Percentage of HH cleaning of dress every day	43.91	19.363
	Handwashing before a meal and after using the toilet	Percentage of HH who clean their hand before taking a meal and after using the toilet every time	87.44	13.578
	Keep nails short	Percentage of HHs who cut their nail at the right time	51.03	18.181
	Timely food intake	Percentage of HH timely food intake	45.83	18.455
	Skipping meal	Percentage of HH leaving their meal due to various causes	24.36	13.353
	Menstruation care	Percentage of HH women who use sanitary napkins	39.94	18.474
Household hygiene practices	Separate kitchen	Percentage of HH having separate kitchen	63.72	25.355
	Separate bathroom	Percentage of HH having separate bathroom	73.59	23.027
	Prevent mosquitos and insects	Percentage of HHs who use a mosquito net	50.96	21.123
	Condition of toilet	A percentage of HHs who have a good condition of toilet	87.56	16.279
	Drinking water treatment	Percentage of HH who do treatment of drinking water	18.27	16.575
	Face overcrowding	Percentage of HH who face overcrowding	49.57	12.594
	Sewerage system	Percentage of HH not having concrete sewerage system	11.41	18.012

Source: Field survey (2022)

3.9.2 Result and Discussion

3.9.2.1 Result of Principal Component Analysis and PHI

Kaiser-Meyer-Olkin (KMO) and Bartlett's tests were conducted to determine the variables' suitability for principal component analysis (PCA). The value of the KMO is 0.662, and Bartlett's test of sphericity χ^2 (105) = 199.068, $p < 0.01$, indicating that correlations among variables were appropriate for PCA. PCA extracted components from the 15 variables, and Kaiser normalization and varimax rotation methods were employed with eigenvalues higher than 1.0 through the dimension

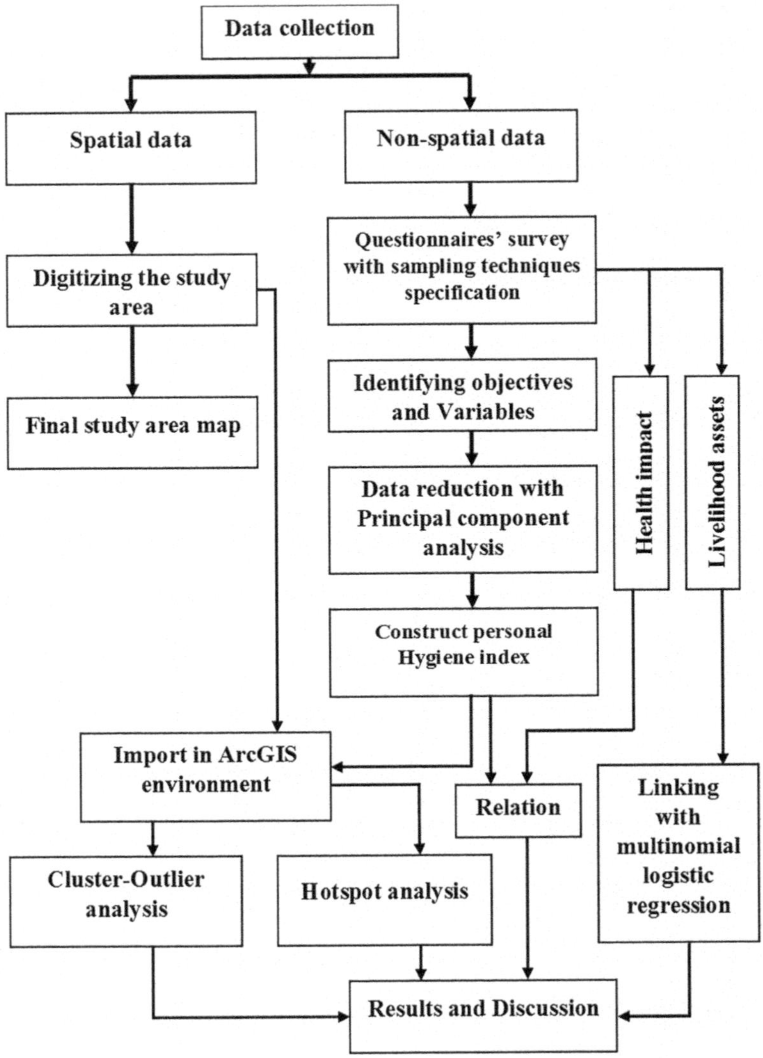

Fig. 3.10 Flowchart of the methodology for personal hygiene

reduction tool of SPSS v26. Varimax rotation with Kaiser normalization converged into ten iterations, explaining 62.895% of the variance for the components together, as summarized in Table 3.5. The first seven components with an eigenvalue greater than 1.0 were identified as the latent variables and assigned cardinality (±). The seven components were then described by signs denoting their influence on personal hygiene. For each slum neighborhood, the highest loading component was selected to calculate the final personal hygiene index (PHI). PHI

Table 3.5 Principal components (PC), cardinality, variables, loadings, and explained variance for the PHI

Component name	Cardinality	Variables	Loadings	Explained variance (%)
PC1	(+)	Separate kitchen	0.862	13.104
		Separate bathroom	0.843	
PC2	(+)	Cleaning of dress daily	0.694	10.554
		Handwashing before a meal and after using the toilet	−0.476	
		Keep nails short	0.648	
		Sewerage system	0.479	
PC3	(+)	Face overcrowding	0.819	9.593
PC4	(+)	Menstruation care	0.754	8.217
		Prevent mosquitoes and insects	0.617	
PC5	(+)	Timely food intake	−0.641	7.487
		Condition of toilet	0.758	
PC6	(+)	Brush at morning	0.811	7.202
PC7	(+)	Take a bath daily	0.681	6.738
		Skipping meal	−0.651	
		Drinking water treatment	0.470	

Extraction method: principal component analysis
Rotation method: varimax with Kaiser normalization
(a) Rotation converged in ten iterations

was calculated by adding the values of the principal component to arrive at total PHI values (Eq. 3.1).

$$PHI_{Slum} = PC_1 + PC_2 + PC_3 + PC_4 + PC_5 + PC_6 + PC_6 + PC_7 \quad (3.1)$$

Assuming that different components have equal importance in overall personal hygiene, the study did not assign any weight to the components. To identify the spatial clusters of the magnitude of personal hygiene practices, PHI values of the slum neighborhoods were mapped in ArcGIS through cluster-outlier and hot spot analysis spatial statistic tools.

3.9.2.2 Result of Cluster-Outlier Analysis

The cluster-outlier analysis is used to detect statistically significant groupings based on a fixed distance band with a Euclidean distance of 928.36 meters in the ArcGIS environment. The deployment of Anselin Local Moran's I test is to find out local cluster's coexistence such as high (HH) and low (LL) cluster and high (HL) and low (LH) outlier association (Fig. 3.11). From the result, high personal hygiene (HH) is found in Mukundasagar adibasi bastee, Panpara bastee, Beherapara, Dewannagar,

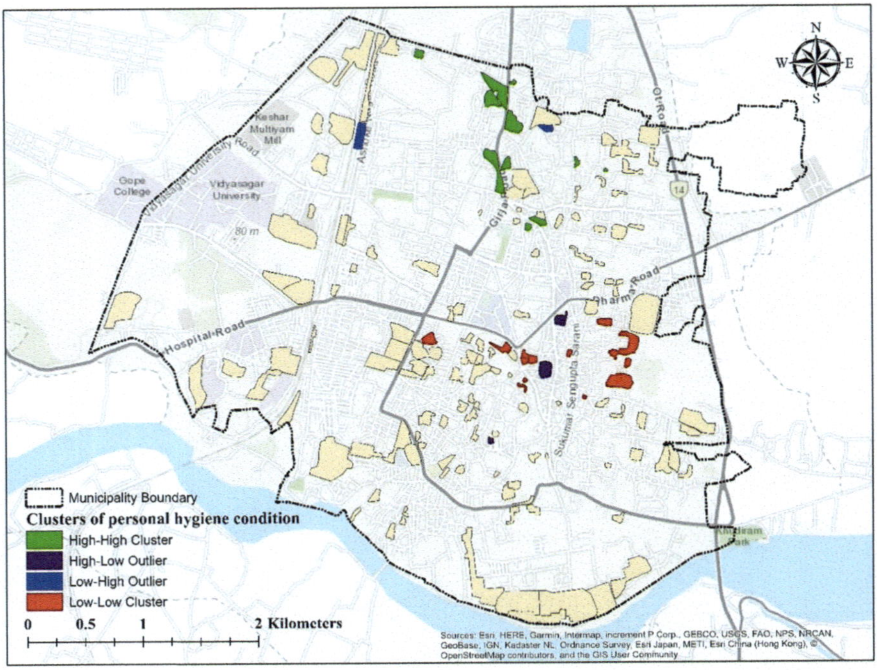

Fig. 3.11 Results of Cluster-outlier analysis for personal hygiene

Akarshanagar, Mirzamahalla Jahangir bastee, Singpara bastee, Minimarket bastee, Muchipara bastee, Rabindranagar bastee, Halugali bastee, and Chirimarsai bustee, and low personal hygiene (LL) is also found in Nepalipara bastee, Sepoybazar muslimpara bastee, Bhuniapore, Ladey para, Mallickpukur bastee, Sukumar Sengupta bastee, Promodnagar bastee, and Palbari bastee. In the case of high (HH) and low (LL) personal hygiene clusters, the adjacent polygon's value is similar. Therefore, high (HL) and low (LH) personal hygiene outliers were found in Goalapara bastee, Darjipara bastee, Chandannagar bastee, Mandal mahalla bastee, and Sitala mandir bauripara bastee, Station Road bastee, Dhunaru bastee, and Majdurnagar bastee, respectively. For the outlier, the value of the adjacent polygon is dissimilar. The Local Moran's I statistical test endorsed positive spatial autocorrelation with a value of 0.13 ($p < 0.01$) and a Z-score of 7.65, indicating a clustered pattern of the slum clusters (HH or LL) with high and low values, respectively.

3.9.2.3 Result of Hot Spot Analysis

Hot spot analysis on personal hygiene displays an apparent spatial pattern of the statistically significant hot spot (shown in red) or cold spot (shown in blue) with critical values >±1.65 SD in Z-score using the fixed distance of 928.36 meters (Fig. 3.12). The result shows that specific slum neighborhoods are under hot spots of standard (good) personal hygiene as Bara Astana, Sirisbagan bastee, Jharnadanga

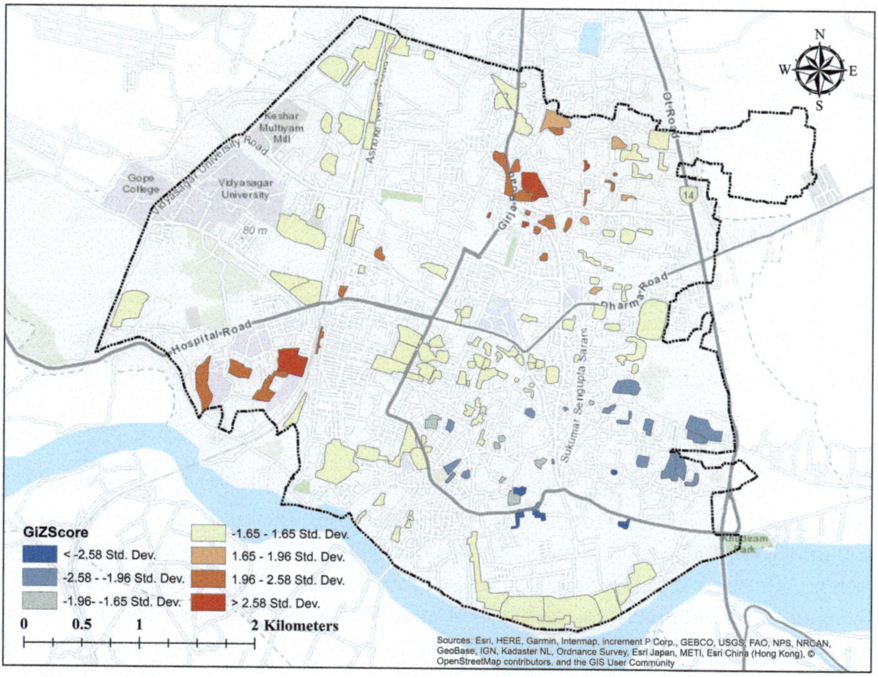

Fig. 3.12 Results of Hotspot analysis for personal hygiene

bastee, Sepoybazar bastee, backside of Muslim burial ground and fire brigade, Subhasanagar Garage gali, and Boroastana bastee. Nevertheless, there is a conspicuous cold spot in the study area characterized by low personal hygiene slum neighborhoods such as Brajakishorpally bastee, Minimarket bastee, Gandhipally bastee, Mahatabpur shainji baba asram bastee, and Mahatabpur sasanghat colony. Some slum neighborhoods are not significant or not intense, making clusters that lie between ±1.65 SD as a non-clustered class, probably scattered or randomly varying in the study area. The estimated Getis-Ord Gi* local statistic for the slum clusters with high or low PHI values in an ArcGIS environment is endorsed by observed general G-value of 0.17 ($p < 0.01$) and a Z-score of 2.277 indicating the more intense spatial clustering or location of slums with high PHI values (hot spot) and low PHI values (cold spot) in a low-cluster pattern. Except for hot spots or cold spots, many non-significant slum neighborhoods are observed to be characterized by an average percentage of personal hygiene. It is clear from the cold spots that there is a lack of public awareness and a decent environment.

3.9.2.4 Households Faced Various Problems Regarding Personal Hygiene

In the analysis, 85.90% and 93.58% of slum neighborhoods face problems with basic and household hygiene practices, respectively. In the case of basic hygiene practices, slum households have predominantly faced problems during

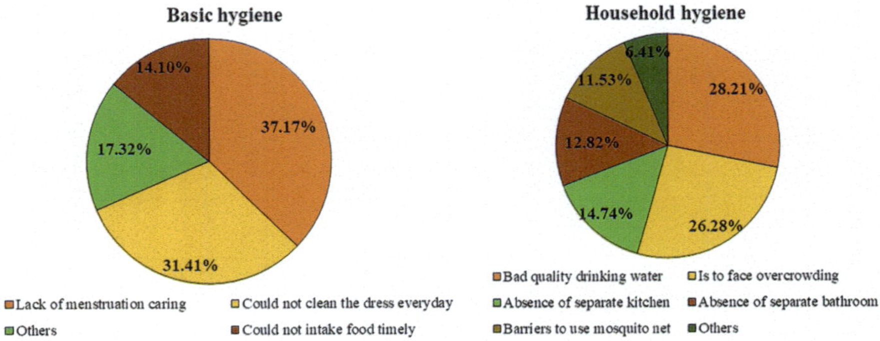

Fig. 3.13 Mostly faced problem by households regarding personal hygiene (Source: Field survey, 2022)

menstruation care (37.17%), could not clean their dress daily (31.41%), and timely food intake (17.32%) and others (14.10%). Therefore, the study selected the most problems from these three dependent variables among the eight variables of basic hygiene. Furthermore, slum households have also faced problems during their household hygiene practices, such as bad quality of drinking water (28.21%), overcrowding (26.28%), absence of separate kitchens (14.74%) and separate bathrooms (12.82%), use of mosquito net (11.53%), and others (6.41%). For this reason, these challenging five variables among the seven variables of household hygiene have been nominated. The variables were selected based on their relative importance in each slum through the independence of observations. The eight dependent variables of personal hygiene practices are lack of menstruation care, inability to clean dress daily, inability to take in food timely, bad quality use of drinking water, overcrowding, absence of a separate kitchen, and absence of a separate bathroom shown in Fig. 3.13. The dwellers could not take in food in a timely manner, and use of mosquito nets was used as a reference item for the model. Also, livelihood assets as the predictor or independent variables were used based on their linear relationship with personal hygiene practices: employment, literacy, household fixed assets, house type, number of rooms, and household monthly income (Table 3.6). Then, the multinomial logistic regression model was used to estimate the most driving predictors affecting basic and household hygiene practices of the slum neighborhoods.

3.9.2.5 Role of Livelihood Assets to Involve in Personal Hygiene

Livelihood assets have been the driving components for personal hygiene practices. Table 3.6 shows the role of assets in basic hygiene practices estimated from the multinomial logistic regression model. In the analysis, the "housing condition" is the only significant component affecting slums' basic hygiene practices of "lack of menstruation care." The regression coefficient (*B*) for this component (−0.350) is significantly negative at the 0.01 level, and the exponentiation of the coefficients

Table 3.6 Role of assets in basic hygiene practices (parameter) estimated from the multinomial logistic regression model

Dependent variable	Independent variable	*B*	S.E.	Wald	Exp (*B*)
Lack of menstruation care (*n* = 58)	Intercept	−1.186	2.544	0.023	–
	Employment	−0.003	0.022	0.013	0.997
	Literacy	−0.004	0.022	0.032	0.996
	Household fixed asset	−0.030	0.020	2.318	0.971
	Housing condition	−0.350***	0.013	7.131	0.965
	Number of rooms	0.446	0.502	0.008	1.047
	Household monthly income	0.000**	0.000	5.935	1.000
Could not clean dress daily (*n* = 49)	Intercept	−1.819	3.407	0.285	–
	Employment	−0.032	0.031	1.084	0.969
	Literacy	0.002	0.028	0.003	1.002
	Household fixed assets	−0.045*	0.024	3.570	0.956
	Housing condition	0.241*	0.014	3.042	1.025
	Number of rooms	0.392	0.626	0.022	1.097
	Household monthly income	0.000**	0.000	4.061	1.000

*Significant at the 0.10 level
**Significant at the 0.05 level
***Significant at the 0.01 level
The reference category is food that could not intake timely
Cox and Snell R-squared is 0.227, Nagelkerke R-squared is 0.260, and McFadden is 0.125
Regarding goodness-of-fit, the Pearson chi-square is 262.520, df is 244, and Sig. = 0.198

(Exp (*B*)) is <1. This means that in households with bad housing conditions, women are less likely to care during the menstruation period compared to could not take food timely. Bad housing condition and household fixed assets are shown to be two components that influence the slum's basic hygiene practices of "could not clean of dress daily" in which the regression coefficient of "bad housing condition" is significantly positive at the 0.10 level with Exp (*B*) >1, whereas "household fixed assets" are significantly negative with Exp (*B*) <1 at the same level. The results indicate that households with poorer housing conditions are more likely to neglect daily cleanliness of clothing and basic hygiene practices, even more so than skipping meals. In contrast, households with more fixed assets are more likely to maintain daily cleanliness of clothing and uphold basic hygiene, reflecting greater stability and self-care.

Table 3.7 shows the role of assets in household hygiene practices estimated from the multinomial logistic regression model. In the analysis, the "number of rooms" is the only significant component affecting slum household hygiene practices of "use of bad quality drinking water." The value of *B* for this component is significantly negative at the 0.10 level, with Exp (*B*) being <1. This means that households with more rooms are less likely to "use bad quality drinking water" than "use of mosquito nets." Additionally, "bad housing condition" is the only significant component

Table 3.7 Role of assets in household hygiene practices (parameter) estimated from the multinomial logistic regression model

Dependent variable	Independent variable	*B*	S.E.	Wald	Exp (*B*)
Use of bad quality drinking water (*n* = 44)	Intercept	1.148	3.356	0.117	–
	Employment	−0.043	0.032	1.790	0.958
	Literacy	0.017	0.029	0.345	1.017
	Household fixed asset	0.002	0.026	0.005	1.002
	Housing condition	−0.010	0.017	0.387	0.990
	Number of rooms	−1.128*	0.638	3.123	0.324
	Household monthly income	0.000**	0.000	5.712	1.000
Is to face overcrowding (*n* = 41)	Intercept	0.235	3.267	0.005	–
	Employment	−0.009	0.030	0.095	0.991
	Literacy	0.012	0.028	0.200	1.013
	Household fixed asset	−0.006	0.026	0.047	0.994
	Housing condition	−0.007	0.016	0.219	0.993
	Number of rooms	−0.629	0.610	1.063	0.533
	Household monthly income	0.000	0.000	2.220	1.000
The absence of a separate kitchen (*n* = 23)	Intercept	−4.938	3.981	1.538	–
	Employment	0.031	0.034	0.855	1.032
	Literacy	0.023	0.034	0.490	1.024
	Household fixed asset	−0.041	0.029	1.984	0.960
	Housing condition	0.035*	0.018	3.683	1.035
	Number of rooms	−0.625	0.682	0.840	0.535
	Household monthly income	0.000**	0.000	5.985	1.000
Absence of a separate bathroom (*n* = 20)	Intercept	11.238**	4.512	6.204	–
	Employment	−0.059	0.041	2.072	0.943
	Literacy	−0.015	0.034	0.202	0.985
	Household fixed asset	0.007	0.034	0.042	1.007
	Housing condition	−0.008	0.019	0.192	0.992
	Number of rooms	−1.409*	0.834	2.856	0.244
	Household monthly income	0.000***	0.000	7.156	1.000

*Significant at the 0.10 level
**Significant at the 0.05 level
***Significant at the 0.01 level
The reference category is the use of mosquito nets
Cox and Snell R-squared is 0.368, Nagelkerke R-squared is 0.386, and McFadden is 0.149
Regarding goodness-of-fit, the Pearson chi-square is 779.137, df is 556, and Sig. = 0.000

affecting slum household hygiene practices of "absence of separate kitchen." The *B*-value for this component is significantly positive at the 0.10 level with Exp (*B*) >1. This demonstrates that households have more "bad housing conditions" the more they have "absence of a separate kitchen" while maintaining their household

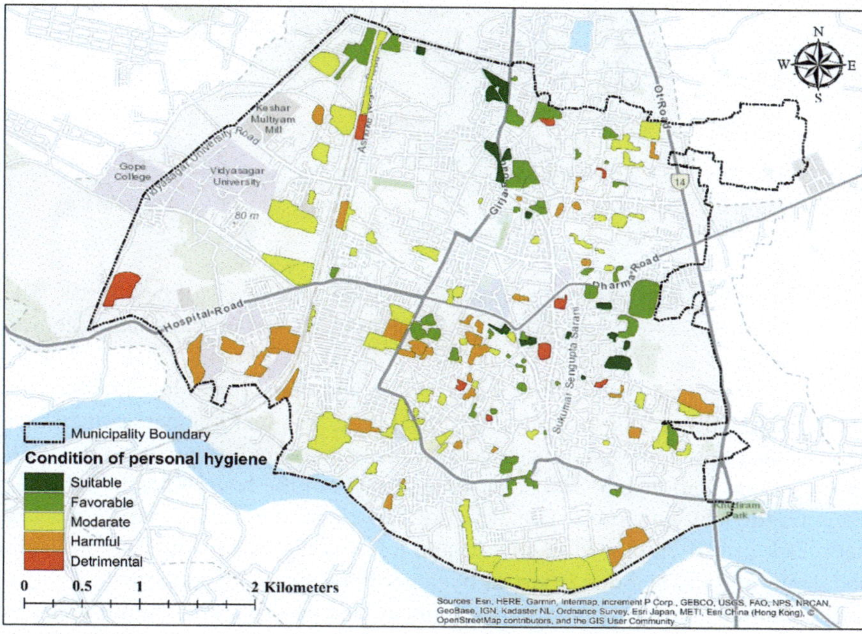

Fig. 3.14 Spatial distribution of the index score of personal hygiene

hygiene practices than the "use of mosquito nets." The study also found that the number of rooms is the only significant component affecting slum household hygiene practices, such as the "absence of separate bathrooms." The *B*-value for this component is significantly negative at the 0.10 level, with Exp (*B*) being <1. This reveals that households with more "rooms" are less likely to have an absence of a "separate bathroom" to maintain their household hygiene practices compared to the "use of a mosquito net." The study could not find any significant predictor for households faced with overcrowding. Household monthly income is a significant predictor at 0.01 and 0.05 levels, respectively, but does not affect basic and household hygiene practices as Exp (*B*) is absolute 1.0.

A major finding of this study is that marked differences were observed in the spatial distribution of personal hygiene among the slum neighborhoods. The slum neighborhoods in central and northwestern, western, and southeastern parts of the municipality exhibit relatively suitable to favorable personal hygiene showing the highest PHI scores with a low level of health risk, while slum neighborhoods with comparatively harmful to detrimental personal hygiene represent low PHI scores characterized by a high level of health risk distributed randomly across the municipality. Similarly, slum neighborhoods with moderate personal hygiene exhibit a moderate level of health risk that prevails in all parts of the municipality (Fig. 3.14 and Table 3.8). This finding is also stimulating because slum neighborhoods residing within similar geospatial environments exhibit a marked difference in personal hygiene and exposure to different levels of health risk. However, social, economic, and built environmental attributes make the slum neighborhoods different in terms

Table 3.8 Magnitude of personal hygiene and health risk level

LMiZ Score Fixed 928	Level of personal hygiene	Number of slums	Health risk level
3.738683–8.806771	Suitable	15	1 (low)
1.112794–3.738683	Favorable	41	2
−0.411173–1.112794	Moderate	57	3
−2.113245 to −0.411173	Harmful	34	4
−4.028007 to −2.113245	Detrimental	9	5 (high)

of the level of personal hygiene, which in turn affects the levels of health risk. The driving livelihood assets that affect personal hygiene are employment, literacy, household fixed assets, housing conditions, number of rooms, and household monthly income. One contributing factor is that households with low income (Rs. <5000 per month) are intrinsically associated with personal hygiene as they face various difficulties and cannot meet basic needs such as land ownership, job stability, proper sanitization, etc. The occupational status of the households in the study area is fragile and vulnerable. Female work participation as a servant and at a low salary, lack of self-employed members, and job loss may also contribute to livelihood assets and maintaining personal hygiene. In recent years, vulnerable occupational conditions have become more prominent, and people are facing difficulty in recovering the extra cost of maintaining personal hygiene to overcome COVID-19. Another contributing factor is assumed to be low personal hygiene due to low literacy and vice versa. The superstitious nature in the family can be reduced if the family has high literacy and increases the skill of maintaining personal hygiene. Moreover, a society with low literacy has superstitions during the menstruation period of the women in the family. High literacy can reduce this superstitious nature and make women aware that it is a biological process and needs extra care during this period. Literacy can also reduce local environmental pollution as it is substantiated that households with high literacy society use dust bins and throw waste material into particular places, thereby reducing the spreading of diseases. Literacy immensely impacts child and aged care and increases public awareness, as seen during this pandemic. Most of the research found that environmental conditions and accommodation of the households and family indoor environment are directly linked with personal hygiene as they would face a range of difficulties during the monsoon and summer due to urban flash floods and hot temperatures. During monsoon season with excessive rainfall, urban flash floods cause dampness in the room and stagnant water, which may cause breeding grounds for different infectious microbes (such as dengue mosquito), as most slum households developed across urban sewerage channels. Besides, during the hot, dry summer season, excessive temperature may cause sweating and dehydration because the roof is mostly covered by asbestos, tin, or tiles. Furthermore, members of the slum households bathed in open places as they didn't have separate bathrooms, as well as they used community toilets. Moreover, lower room density and lack of separate kitchens may cause congestion, overcrowding, and indoor pollution, respectively, severely affecting the health status of family members, especially women and children. A bad

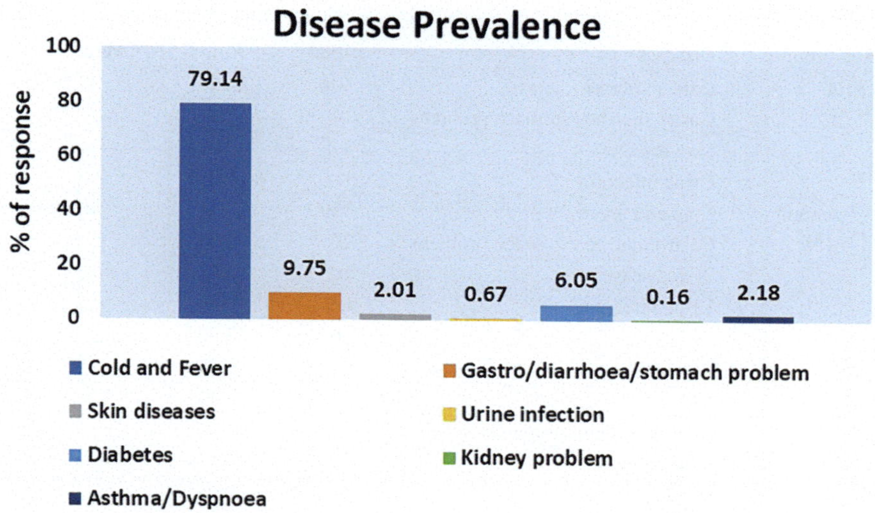

Fig. 3.15 Overall common diseases reported by the slum dwellers (Source: Field survey, 2022)

sewerage system and no separate bathroom can inevitably spread diseases, and mosquitoes (such as dengue) have a dreadful impact on personal hygiene. Household fixed assets were the most consistently associated with personal hygiene. Major household assets include ownership of a house, toilet facilities, separate bathroom, refrigerator, mobile phone, electric fan, TV, bicycle, car, sewing machine, water pump, computer, air conditioner or cooler, washing machine, motorcycle or scooter, etc. These assets may affect the personal hygiene practices of the slum dwellers; for instance, mobile phones can provide information through different health applications (LeFevre et al., 2019), such as Aarogya Setu Mobile App for Indian people, which may guide to fight against coronavirus. Furthermore, a motorcycle or scooter may facilitate travel without physical hard work and easily connect with healthcare facilities. Toilet facilities and water sources may protect against disease exposure and maintain standard hygiene practices.

3.9.2.6 Disease Prevalence With Relation to Personal Hygiene

Due to a lack of hygiene practices and skills, people in the study area reported different types of human diseases (chronic, infectious, and hiding). Due to the negligence of their hygiene, diseases spread rapidly among them. Cold and fevers are the common diseases, accounting for 79.14% (Fig. 3.15), and are affected by age group, i.e., child and adolescent to 25.56% and 28.51%, respectively (Table 3.9). Diarrheal and stomach problems are the most common initiates in the adolescent and old age groups, with a percentage of 5.88% and 1.68%, respectively. Even out of it, the overall percentage shows that adolescent and child groups (40.20%, 27.50%)

Table 3.9 Diseases that were reported by the different age groups in the study area

Age group	Diseases	Total	%	Overall %
Child (0–10)	Cold and fever	304	25.56	27.50
	Gastro/diarrhea/stomach problem	14	1.17	
	Skin diseases	8	0.67	
	Urine infection	1	0.08	
Adolescent (11–18)	Cold and fever	339	28.51	40.20
	Gastro/diarrhea/stomach problem	70	5.88	
	Skin diseases	10	0.84	
	Urine infection	6	0.50	
	Diabetes	36	3.02	
	Asthma/dyspnea	17	1.42	
Parental (19–59)	Cold and fever	232	19.51	21.19
	Gastro/diarrhea/stomach problem	12	1.00	
	Skin diseases	5	0.42	
	Diabetes	1	0.08	
	Kidney problem	1	0.08	
	Asthma/dyspnea	1	0.08	
Old age (>60)	Cold and fever	66	5.55	11.10
	Gastro/diarrhea/stomach problem	20	1.68	
	Skin diseases	1	0.08	
	Urine infection	1	0.08	
	Diabetes	35	2.94	
	Kidney problem	1	0.08	
	Asthma/dyspnea	8	0.67	

Source: Field survey (2022)

Table 3.10 Linear regression between PHI and health shock incident

Model		Unstandardized coefficients		Standardized coefficients	t	Sig.
		B	Std. error	Beta		
1	(Constant)	32.923	0.997		33.028	0.000
	PHI	−0.571	0.027	−0.862	−21.149	0.000

Coefficients[a]
(a) Dependent variable: health shock incident

(Table 3.9) are the most affected group of human diseases. A simple Pearson correlation was applied to estimate the relation between PHI and health shock incidents, which displays a strong negative correlation (−0.862) at a highly significant level. Table 3.10 shows the beta value (−0.862) at a highly significant level, and the adjusted R^2 value (0.744) with SE (3.275) signifies that there is a severe lack of hygienic practices, which contributes to the high intensity of health shock incidents in the study area.

3.10 Health Risk Analysis

The health risk is the product of the hazards and vulnerability of people with respect to capacity building. It is argued that failure resistance to environmental threats may cause a health risk. Hazards and vulnerability also cause environmental threats. Hazards and vulnerabilities arising from social, human-induced, and natural factors cause numerous chronic, communicable, and non-communicable diseases that can negatively impact people's mental, physical, and socioeconomic health. A slum is known to be the hot spot of a health hazard as they are in deprived, poor socioeconomic groups worldwide, particularly in India. Slums have been proliferating within a limited space in the cities for the past decade with a lack of basic amenities such as safe drinking water, sanitation, minimum living space, housing accommodation, and secure tenure. Moreover, the slum is deprived of access to healthcare services, formal jobs, and proper education, as well as a lack of personal hygiene. Due to meeting the basic livelihood, they lack awareness about health, hygiene, and the environment. Therefore, they have faced multiple stressors, health shocks, and risks in their daily livelihood. Poverty is strongly related to the health consequences of the poor, as they are unable to meet basic healthcare expenditures. Despite that, many factors are determinants of the health risk of the urban poor. Moreover, little is known in slums regarding the magnitude, distribution pattern, and risk factors of illness before the manifestation of havoc health consequences such as stroke, kidney failure, TB, heart disease, nerve problems, AIDS, etc. Furthermore, slum dwellers have reported both infectious and non-infectious chronic diseases such as asthma, hypertension, diabetes, mental illness, cancer, skin diseases, diarrhea, obesity, underweight, alcohol use, reproductive health problems, etc. Therefore, the present study aims to explore the major health risk and underlying factors across the slum neighborhood with distinct socioeconomic entities.

3.10.1 Data and Methods

Different factors have contributed to the health risk of the community. The study was conducted by examining factors such as air quality, average annual temperature, socioeconomic vulnerability, personal hygiene, social vulnerability, and proximity to existing health care centers (HCF) (Fig. 3.16). These factors were then categorized into four model maps: distance to HCF, environmental vulnerability, personal hygiene, and socioeconomic vulnerability. Data has been collected from the West Bengal Pollution Control Board (WBCPB, https://www.wbpcb.gov.in), Landsat OLI-8 data from USGS (https://earthexplorer.usgs.gov) for land surface temperature (LST), and data of personal hygiene and socioeconomic vulnerability collected through questionnaire survey schedule and location of the major healthcare center. PCA was used to find the PHI and socioeconomic vulnerability through

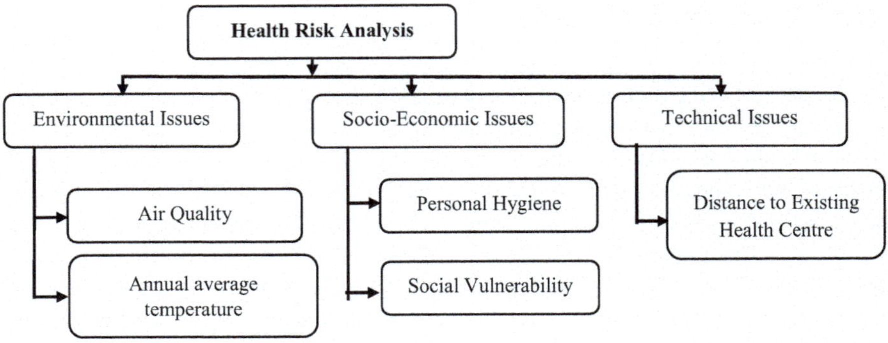

Fig. 3.16 Variables and model map used for health risk analysis

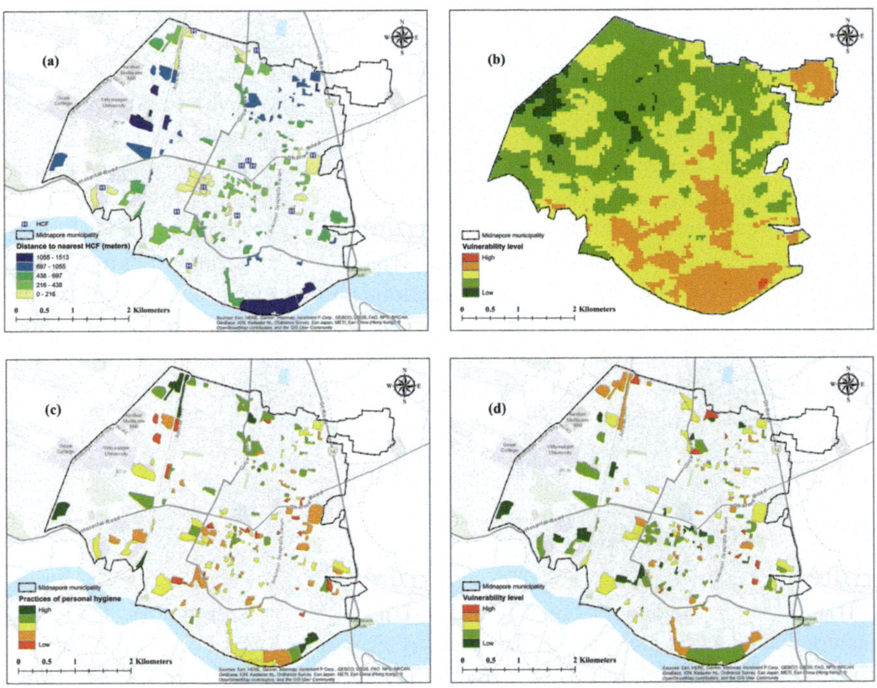

Fig. 3.17 Health Risk variables (**a**) Distance to HCF (**b**) Environmental issue or vulnerability (**c**) Personal hygiene (**d**) Socio-economic vulnerability

SPSS v26 and mapping in ArcGIS v10.4, while environmental vulnerability and distance to HCF were mapped in ArcGIS v10.4 (Fig. 3.17). All the model maps were then overlaid to find the health risk zone of the slum neighborhoods and the magnitude of health risk.

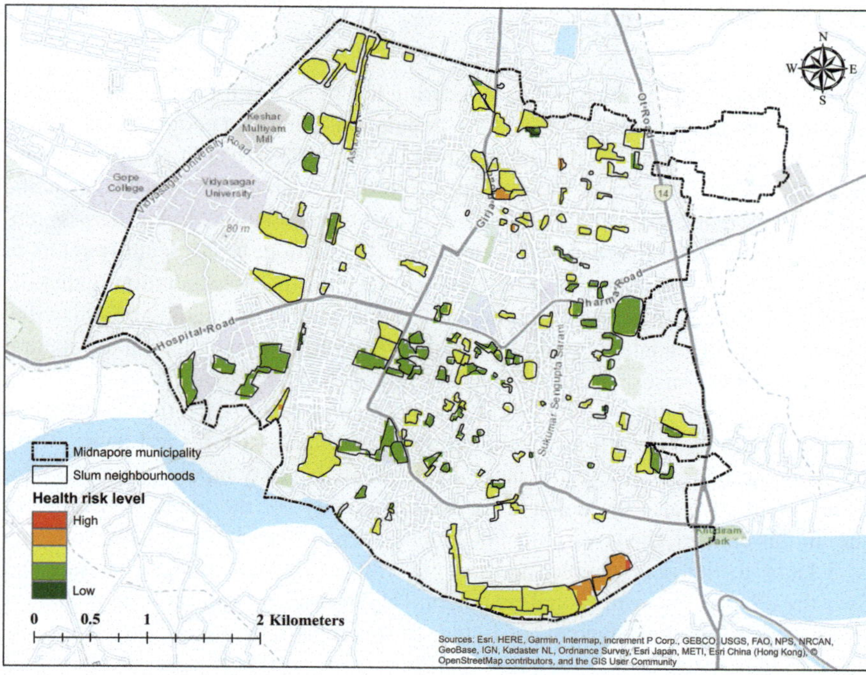

Fig. 3.18 Spatial Distribution of slums as per magnitude of health risk atial Distribution of slums as per magnitude of health risk

3.10.2 Result and Discussion

Health risks are common in slums in India. Based on the employed criteria, the magnitude of health risk was shown in Fig. 3.18, which revealed that most of the slum is undergoing moderate health risk. However, many slum neighborhoods are at high health risks, such as Palbari, Bara Astana, Sirisbagan, Beltala, backside of commerce college, Sialdanga, and Nibeditapally bastee. Therefore, special care should be given to those slums with high health risks and to reduce the burden of health expenditures. The study also suggests that various underlying socioeconomic, demographic, and basic amenity-related factors may contribute to health risks; however, these were not included due to limitations in data availability and economic feasibility. Moreover, further refinement and research need to focus on health risk factors. Based on the collected data, the author argues that the health risks in the study area are primarily due to several factors, including lack of awareness, inadequate primary healthcare services, poor housing conditions, insufficient waste management, lack of sanitation and regular drainage (sewerage) maintenance, limited access to nutritious food, excessive consumption of junk food by children and adolescents, alcoholism, smoking, and poor personal hygiene.

3.11 Health Security

Public health security is the process of securing health and well-being from the environmental threats and vulnerability of the population living in a geographical region. According to the World Health Organization (WHO, 2020), health security encompasses the "activities required to minimize the danger and impact of acute public health events that endanger the collective health of populations living across geographical regions and international boundaries." In addition, health security refers to the actions and measures aimed at preventing infectious diseases and minimizing the impact of acute health incidents, particularly in the context of ongoing urbanization, environmental degradation, rapid population growth, overexploitation of natural resources, and the misuse of antimicrobials (Stoeva, 2020; Paranjape & Franz, 2015). The objective of health security worldwide is to enhance the detection, prevention, and response to diseases through public health surveillance (Blazes & Lewis, 2016). The changing human society and natural calamities that may emerge from the pandemic potentially raise the demand for health security for the population of every state. In this context, the government of India launched "Ayushman Bharat," offering Rs. 5 lakhs to the poor people, and the government of West Bengal has initiated "Swasthya Sathi" for institutional treatment. Health insecurities prevail in the slums of India due to the low standard of living. In this regard, health security is closely related to the quality of life. Pseudo-urbanization (Ray, 2017; Kundu 1983) induced growing imbalance and inequalities in the slums of India, which emerging lack of hygiene, safety, and security (Mudey et al., 2017; Komalawati & Lim, 2020). The deprived group is slum exposed to different communicable and non-communicable diseases, necessitating a focus on prevention through geospatial analysis.

3.11.1 *Data and Methodology*

The concept of health security is multidimensional; for that reason, both subjective and objective groups of variables have been selected, which were collected through a questionnaire survey and focus group discussion. For the distinctive nature of the study area, variables were selected (Table 3.11) which, beyond the recommendation of WHO, as well as inductive and deductive approaches, have been employed to construct the health security index (HSI) (Eq. 3.2). The Pearson correlation coefficient was used to validate the result derived from inductive (PCA) and deductive (standardization) methods. PCA extracted the 20 variables among the selected 37 variables with 66.04% explained variable with the help of SPSS v26 (Table 3.12). Cluster-outlier analysis in ArcGIS v10.4 was also used to find the grouping or spatial cluster based on the Z-score among the slum neighborhood in the context of HSI. Similarly, OLS was used to find the degree of relationship between the quality of life and health security.

$$I_{\text{slum}} = \sum \text{PC}_{i\_i} \times V_i \tag{3.2}$$

Table 3.11 Variables used for analyzing health security

Variables		Description	Mean	Standard deviation
Accessibility of healthcare center	Distance to the nearest healthcare center	Whether there is a healthcare center within 2 km	35.71	22.321
	Source of transport reaching the healthcare center	Transportation to healthcare center arranged by family	81.79	23.514
Awareness about health	Satisfaction level of health infrastructure	Percentage of households satisfied with health infrastructure facility	55.71	27.241
	Awareness of health and nutrition	Percentage of households being aware of health and nutrition	33.72	20.921
	Taken regular health checkups and medicine	Percentage of households taking regular health checkups and medicine	11.41	14.567
	Maintains personal hygiene	Percentage of households maintaining personal hygiene	55.91	6.020
	Disease prevention equipment	Percentage of households using disease prevention equipment	24.62	8.652
	Perception of family health	Percentage of people realizing about family health	76.67	18.917
Food and nutrition	Prevalence of malnutrition	Percentage of people being malnourished	13.654	6.6864
	Food insecurity	Percentage of people suffering from food insecurity	70.71	20.794
Accommodation and poverty	BPL	Percentage of households below the poverty line (BPL)	30.06	22.041
	Drinking water treatment	Percentage of household caring for drinking water	18.27	16.575
	The environment of housing condition	The housing environment of the households	58.55	8.974
Prevalence of disease	Interval of sickness	Often suffering from the prevalence of diseases	15.90	17.260
	Infectious disease	Percentage of infectious disease	38.61	23.033
	Chronic disease	Percentage of chronic disease	10.52	16.445
	Hiding disease	Percentage of hiding disease	1.78	5.234

(continued)

Table 3.11 (continued)

Variables		Description	Mean	Standard deviation
Health facility	Health card	Percentage of households having a health card	50.96	23.162
	Health insurance	Percentage of households having health insurance	22.76	19.824
Gender-based violence	Women violence	Percentage of households facing women violence	16.92	14.881

Source: Field survey (2022)

Table 3.12 Results of principal component analysis for health security

Principal component	Cardinality	Variables	Loadings	Explained variance (%)	Eigenvalue
PC1	(+)	Distance to the nearest healthcare center	−0.477	11.417	2.283
		Women violence	0.724		
		Prevalence of malnutrition	0.751		
		Satisfaction level of health infrastructure	0.443		
PC2	(+)	Interval of sickness	−0.617	10.408	2.082
		Infectious disease	0.755		
PC3	(+)	Drinking water treatment	0.513	8.529	1.706
		Taken regular health checkups and medicine	0.447		
		Disease prevention equipment	0.545		
		Chronic disease	0.639		
PC4	(+)	The environment of housing condition	0.525	7.436	1.487
		Food insecurity	0.762		
PC5	(+)	Health card	0.770	6.444	1.289
		Health insurance	0.794		
PC6	(+)	BPL	−0.844	6.033	1.207
		Perception of family health	0.509		
PC7	(+)	Maintaining personal hygiene	0.848	5.593	1.119
PC8	(+)	Awareness of health and nutrition	0.867	5.165	1.033
PC9	(+)	Source of transport reaching the healthcare center	−0.469	5.024	1.005
		Hiding disease	0.809		

Extraction method: principal component analysis
Rotation method: varimax with Kaiser normalization
(a) Rotation converged in 23 iterations

Table 3.13 Comparison of all indices by Pearson correlation matrix

Correlations	Principal component analysis (PCA)	Ratio of value	Z-score	Min-max rescaling
Principal component analysis (PCA)	1			
Ratio of value	0.945**	1		
Z-score	0.945**	1.000**	1	
Min-max rescaling	0.945**	1.000**	1.000**	1

**$p < 0.01$

3.11.2 *Comparison of all Indices*

The Pearson correlation coefficients were used to find the correlation among the four indices of aggregated scores. The correlation value indicates the strong positive correlation between the three standardized methods and PCA for the health security index (Table 3.13). Thus, there is no significant difference between inductive and deductive methods indicating validation for index construction.

3.11.3 *Result and Discussion*

The health security index has been formulated by adding the score of each component by-product of variance based on the cardinality. In the same way, a composite score was prepared by the standardization methods (Z-score transformation, ratio of value, min-max rescaling transformation). The score has been grouped into five classes from high (>1.5SD) to low (<−1.5SD) and mapped with ArcGIS v10.4 (Fig. 3.19). The value between ±0.5 represents the moderately prepared, and the value between 0.5 to 1.5 and >1.5 denoted the more and most prepared for health security, respectively, while −0.5 to −1.5 and >−1.5 represents the least prepared for health security and vulnerable, respectively. The final HSI has been developed by combining the four-model score and mapping through ArcGIS v10.4 (Fig. 3.20) of all slum neighborhoods of Midnapore city. Map of health security shows the slum neighborhoods, for example, Sitalamandir bauripara bastee, Porabanglo bastee, Kamarpara basteee, Amtala khas bastee, Judgecourt horijan pally, Goalapara bastee, Patnabazar bauripara bastee, boxibazar tantipara, Pakija basteee, Ghusipara bastee, backside of commerce college, backside of Muslim burial ground and fire brigade, Mitracompound kamarpara, and Talpukur basteee, are mostly prepared for their health security distributed toward the southwest part of the city. Besides, slum neighborhoods, for example, Bhunia Para Station Road, Baburchipara bastee, Chandan bastee, Hatharmath bastee, Talpukur colony bastee, Berballavpur bastee, and Jelepara bastee, are more vulnerable due to their lack of preparedness against forthcoming health shocks.

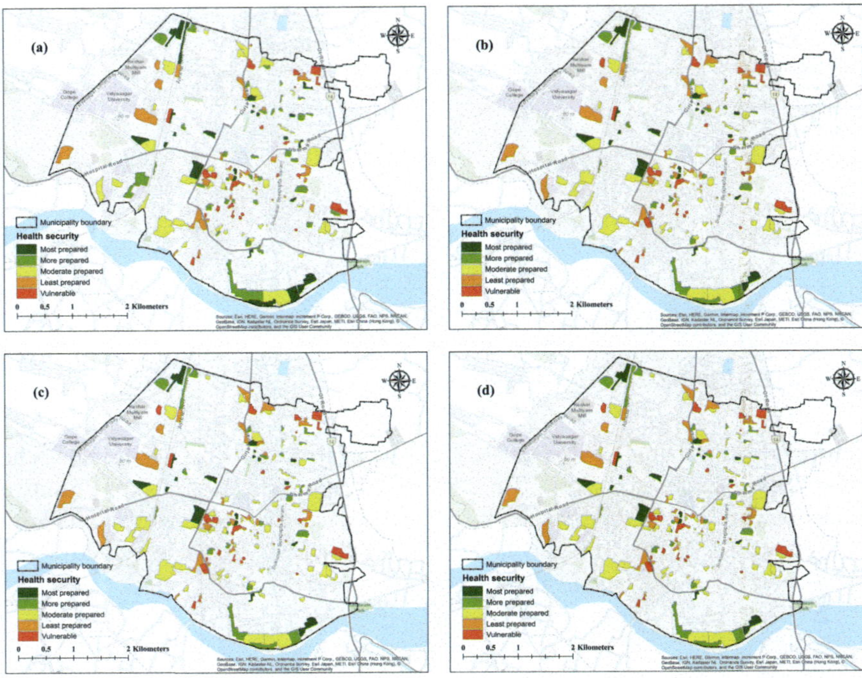

Fig. 3.19 Spatial distribution of different index scores for Health security (**a**) Principal component analysis (**b**) *Z*-score transformation (**c**) Maximum value transformation (ratio of value) (**d**) Min–max rescaling transformation

The map revealed that a maximum number of slum neighborhoods are unprepared and vulnerable to health shocks. The main factor leading to high vulnerability in slum neighborhoods may be low socioeconomic status, which pushes them into a high health risk. The model also estimated the significant variables that may contribute to the vulnerable health security condition of the slum neighborhoods at $p < 0.01$ level (Tables 3.14 and 3.15), such as lack of housing accommodation, drinking water, nutrient foods, social care, and healthcare services. In addition, lack of education, large family size, and few earning members in the family may also play an important role in health security among the slum neighborhoods, even if they are living with moderate quality of life. The survey also estimated that yearly 2.45 people per household experience health shocks. The spatial distribution of slum neighborhoods in the context of health security is more random than a cluster, which indicates that heterogeneous entities characterize slum neighborhoods and account for variability.

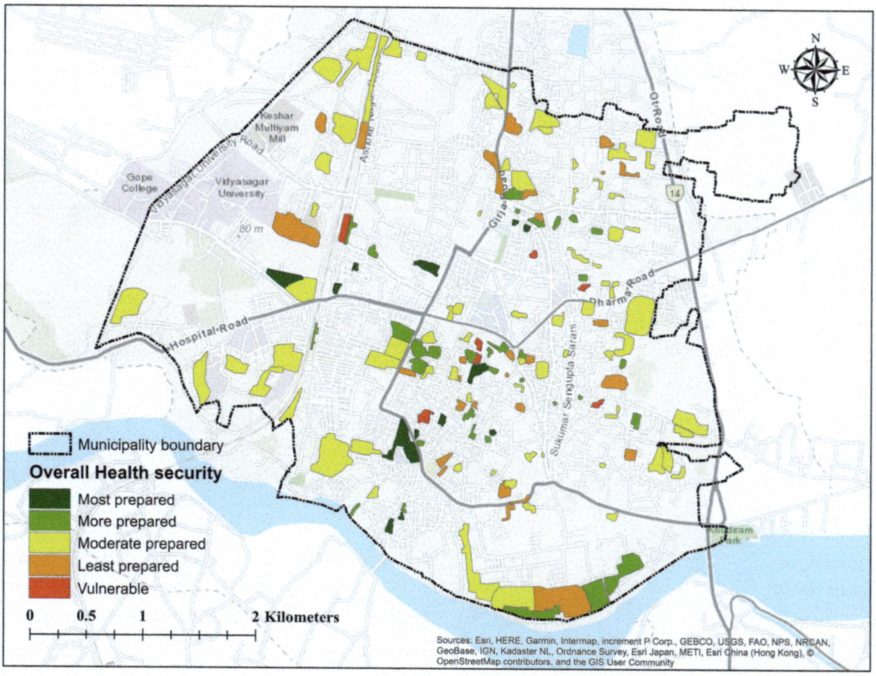

Fig. 3.20 Overall spatial distribution of slums for Health security

3.11.4 Relationship between the Quality of Life and Health Security

The general thought is to believe that health security is coherently related to the quality of life of the households. The ordinary least squares (OLS) regression was applied to estimate the relationship between the quality of life and the indicators of health security and the relationship between HSI and the indicators of quality of life. Similarly, this method was used to explore the impact of the variables on the quality of life and health security, as well as the similarity among the four models with the impact of the indicators shown in Tables 3.14 and 3.15. The result exhibits (Table 3.14) that PC3 and PC4 of health security significantly ($p < 0.01$) have an impact on the quality of life such as housing condition, taking regular health check-ups and medicine, chronic disease, disease prevention equipment, as well as drinking water treatment and food insecurity. Furthermore, three standardized models have been exploring the factors that are significantly ($p < 0.01$) related to the quality

Table 3.14 Relation between QoLIs and indicators of health security

	Model 1		Model 2		Model 3		Model 4	
	Principal component analysis		Z-score		Ratio of the value		Min-max rescaling	
Variables	β	*t*	β	*t*	β	*t*	β	*t*
PC1	−0.059	−0.859						
PC2	−0.038	−0.576						
PC3	0.223**	3.331						
PC4	0.507**	7.712						
PC5	0.088	1.365						
PC6	0.148*	2.303						
PC7	0.025	0.379						
PC8	0.040	0.618						
PC9	0.112	1.656						
Accessibility of healthcare center			2.536**	27.110	2.504**	27.141	0.153*	2.296
Awareness about health			1.762**	25.796	1.750**	25.985	0.109	1.581
Food and nutrition			3.545**	27.663	3.511**	27.780	0.214**	3.166
Accommo-dation and poverty			7.567**	28.134	7.496**	28.258	0.462**	6.871
Prevalence of disease			0.712**	18.575	0.681**	18.017	0.047	0.699
Health security			2.098**	26.433	2.066**	26.401	0.128	1.890
Gender-based violence			−1.127**	−23.121	−1.122**	−23.348	−0.068	−1.011
Constant	6314.955**		22.767**		5.173**		1148.346**	
Number	156		156		156		156	
R-squared	0.412		0.894		0.897		0.380	
Adjusted R-squared	0.376		0.888		0.891		0.351	

$^{*}p < 0.05$, $^{**}p < 0.01$

of life, such as nutritious food, accommodation and poverty, and accessibility of healthcare centers. However, there are no substantial variables of quality of life found to be put forward on the health security shown in Table 3.15. Nevertheless, some variables that may contribute to health security are consistent with the three standardized models ($p < 0.01$): food and nutrition, access to drinking water, housing conditions, health, social care, etc.

Table 3.15 Relation between HSIs and indicators of quality of life

	Model 1		Model 2		Model 3		Model 4	
	Principal component analysis		Z-score		Ratio of the value		Min-max rescaling	
Variables	B	*t*	β	*t*	β	*t*	β	*t*
PC1	−0.097	−0.903						
PC2	0.056	0.649						
PC3	−0.012	−0.133						
PC4	−0.018	−0.175						
PC5	0.073	0.864						
PC6	−0.036	−0.380						
PC7	−0.122	−1.390						
PC8	0.075	0.827						
PC9	−0.092	−1.065						
PC10	0.043	0.472						
PC11	0.023	0.242						
PC12	−0.116	−1.364						
PC13	0.046	0.518						
PC14	−0.018	−0.184						
Housing condition			0.214**	2.786	0.214**	2.786	0.214**	2.786
Access to drinking water			0.185**	2.718	0.185**	2.718	0.185**	2.718
Economic condition			0.071	0.990	0.071	0.990	0.071	0.990
Occupation			0.125	1.833	0.125	1.833	0.125	1.833
Food and nutrition			0.334**	4.863	0.334**	4.863	0.334**	4.863
House ownership			−0.055	−0.798	−0.055	−0.798	−0.055	−0.798
Asset			−0.058	−0.775	−0.058	−0.775	−0.058	−0.775
Education			0.083	1.238	0.083	1.238	0.083	1.238
Health and social care			0.232**	3.388	0.232**	3.388	0.232**	3.388
Sanitation			0.214	2.786	−0.072	−1.066	−0.072	−1.066
Constant	6437.045**		−4.269**		0.303**		−0.426**	
Number	156		156		156		156	
R-squared	0.068		0.385		0.385		0.385	
Adjusted *R*-squared	−0.024		0.343		0.343		0.343	

*$p < 0.05$, **$p < 0.01$

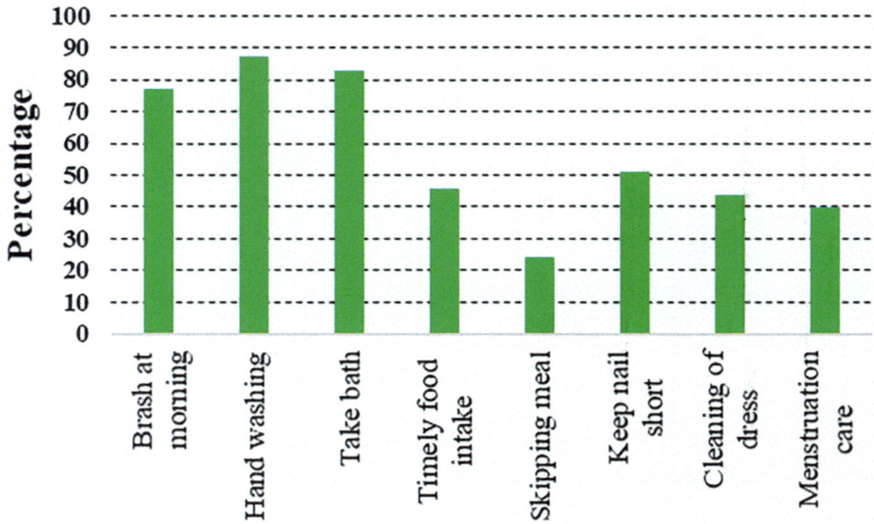

Fig. 3.21 Public awareness statistics of basic hygiene (Source: Field survey, 2022)

3.12 Public Awareness

Public awareness is the most influencing aspect to avert the extent of communicable diseases. Personal hygiene is closely linked to an individual's health status and can be improved through widespread public awareness. Due to a lack of hygienic practices and skills, the diseases spread rapidly among them. During the study period, slum dwellers have reported different types of human diseases (chronic, infectious, and hiding), their hygienic practices, and awareness of maintaining good health. We also collected age group-wise disease prevalence data in the study area during the same study period. Among the diseases reported by the people of the households, the most dominant diseases were summarized into a percentage (Table 3.9). Furthermore, public awareness statistics were aggregated in percentage, which is related to basic and household hygiene practices. The best examples of the spreading of infectious diseases are a lack of proper personal hygiene skills, knowledge, and practices during the novel coronavirus (COVID-19) in 2020. The significant statistics of public awareness of the study area were collected and analysis is shown in Fig. 3.21. The study found that over 75% of slum dwellers regularly brush their teeth in the morning, bathe daily, and wash their hands before and after meals, while also rarely skipping meals. However, at the same time, more than 50% of slum dwellers do not eat meals on time, do not regularly trim their nails, or clean their clothes daily. Additionally, many women have limited awareness regarding menstrual hygiene care. It is clear from the observation that slum dwellers have a piece of knowledge about personal hygiene to a certain extent; rather they don't know about standard personal hygiene practices. Overall, they have less frequent awareness about their health and personal hygiene. The study investigated that most households do not have health insurance and health scheme. Few people have government and private health insurance and

Table 3.16 Health insurance card of the households

Health insurance card	Percentage of households
Yes	50.87
No	49.12

Source: Field survey (2022)

Fig. 3.22 Dwelling place and Housing structure

scheme. Roughly, 50.88% of households have health cards such as "Swasthya Sathi" or "Ayushman Bharat," and 49.12% of households do not have any health insurance card still at the end of 2022 (Table 3.16).

3.13 Conclusion

Lifestyle behavior depends on cultural and behavioral habits directly allied to health awareness, which may lead to a healthy life and society. Unhealthy lifestyle behaviors contribute to various health problems and are further exacerbated by economic and environmental challenges. All cities and megacities in developing countries contain huge slum areas with very high population growth. As a result, they faced different vulnerability conditions such as local environment; wired housing conditions (Fig. 3.22); huge water, air, and soil pollution; monsoonal waterlogging; violence; accidents; and very high prevalence of diseases, which increased their health risk as soon as possible. Due to the economic crisis, individuals have adopted poor hygiene practices, unhealthy eating habits, smoking, alcoholism, malnutrition, and a lack of physical activity. Most of the slum neighborhoods of the study area face health shocks during the whole year, and they are forced to take a loan or mortgage loan or sell their household property to recover from health shocks. Their poor mental health is influenced by harmful housing and environmental conditions. As a result, physical abuse or beating and mental torture are the common types of women violence among the family in slum neighborhoods. Slum dwellers have limited awareness of menstrual hygiene due to their inability to afford sanitary napkins and the lack of a consistent water supply, which also hinders daily cleaning of clothes. Finally, slum neighbors are recognized as having moderate to low health risk levels. The employed model would have been useful to explore the significant factors in this concern.

References

Adetunji, H., Baothman, M., Alserhan, F., Almunyif, A., & Samaren, H. (2018). Knowledge, attitude, and practice (KAP) of personal hygiene among food handlers in the South Region of Makkah, Saudi Arabia. 96–102.

Blazes, D. L., & Lewis, S. H. (Eds.). (2016). Disease surveillance: Technological contributions to global health security. CRC Press. https://doi.org/10.1201/b19506

Chaudhuri, S., & Morash, M. (2019). Building empowerment, resisting patriarchy. *Sociology of Development, 5*(4), 360–380. https://doi.org/10.1525/sod.2019.5.4.360

Dasgupta, S. (2019). Attitudes about wife-beating and incidence of domestic violence in India: An instrumental variables analysis. *Journal of Family and Economic Issues, 40*, 647–657. https://doi.org/10.1007/s10834-019-09630-6

Dessalegn, T., Tamiru, D., Argaw, A., Gerbaba, M., Ayana, G., Nigussie, A., Jisha, H., & Belachew, T. (2017). Enhancing personal hygiene behavior and competency of elementary school adolescents through peer-led approach and school-friendly: A quasi-experimental study. *Ethiopian Journal of Health Sciences, 27*(3), 245. https://doi.org/10.4314/ejhs.v27i3.6

Donta, B., Nair, S., Begum, S., & Prakasam, C. P. (2016). Association of domestic violence from husband and women empowerment in slum community, Mumbai. *Journal of Interpersonal Violence, 31*(12), 2227–2239. https://doi.org/10.1177/0886260515573574

Ellsberg, M., Arango, D. J., Morton, M., Gennari, F., Kiplesund, S., Contreras, M., & Watts, C. (2015). Prevention of violence against women and girls: What does the evidence say? *The Lancet, 385*(9977), 1555–1566. https://doi.org/10.1016/S0140-6736(14)61703-7

Garg, S., Singh, M. M., Rustagi, R., Engtipi, K., & Bala, I. (2019). Magnitude of domestic violence and its socio-demographic correlates among pregnant women in Delhi. *Journal of Family Medicine and Primary Care, 8*(11), 3634–3639. https://doi.org/10.4103/jfmpc.jfmpc_597_19. PMID: 31803665; PMCID: PMC6881926.

Ghose, J. K., Rahman, M. M., Hassan, J., Khan, M. S., & Alam, A. A. (2012). Knowledge and practicing behaviour related to personal hygiene among the secondary school students of Mymensingh Sadar Upazilla, Bangladesh. *Microbes and Health, 1*(1), 34–37. https://doi.org/10.3329/mh.v1i1.13712

Hassan, O. N. A. Z., et al. (2019). Knowledge and practices of personal hygiene among mentally retarded students at El-Fikrya schools in Minia governorate, Egypt. *IOSR Journal of Nursing and Health Science (IOSR-JNHS), 8*(3), 33–43. https://doi.org/10.9790/1959-0803053343

Ilesanmi, O. T. (2017). Knowledge and practices of personal hygiene among senior secondary school knowledge and practices of personal hygiene among senior secondary school students of ambassadors college, Ile- Ife, Nigeria. *Texila International Journal of Public Health, 4*, 4. Dec 2016.

Jeyaseelan, L., Kumar, S., Neelakantan, N., Peedicayil, A., Pillai, R., & Duvvury, N. (2007). Physical spousal violence against women in India: Some risk factors. *Journal of Biosocial Science, 39*(5), 657–670. https://doi.org/10.1017/S0021932007001836

Jungari, S., & Chinchore, S. (2022). Perception, prevalence, and determinants of intimate partner violence during pregnancy in urban slums of Pune, Maharashtra, India. *Journal of Interpersonal Violence, 37*(1–2), NP239–NP263. https://doi.org/10.1177/0886260520914548

Kidman, R. (2017). Child marriage and intimate partner violence: A comparative study of 34 countries. *International Journal of Epidemiology, 46*(2), 662–675. https://doi.org/10.1093/ije/dyw225

Komalawati, R. A., & Lim, J. (2020). The reality of compact development in a developing country : Focusing on the perceived quality of life in Jakarta, Indonesia. *International Journal of Urban Sciences*, 1–32. https://doi.org/10.1080/12265934.2020.1803106

Kundu, A. (1983). Theories of city size distribution and Indian urban structure: A reappraisal. *Economic and Political Weekly, 18*(31), 1361–1368.

Lefevre, A., Agarwal, S., Chamberlain, S., Scott, K., Godfrey, A., Chandra, R., Singh, A., Shah, N., Dhar, D., Labrique, A., Bhatnagar, A., & Mohan, D. (2019). Are stage-based health information messages effective and good value for money in improving maternal newborn and child health outcomes in India? Protocol for an individually randomized controlled trial. *Trials, 20*(1), 1–12. https://doi.org/10.1186/s13063-019-3369-5

Mahapatro, M., Gupta, R. N., & Gupta, V. (2012). The risk factor of domestic violence in India. *Indian Journal of Community Medicine, 37*(3), 153–157. https://doi.org/10.4103/0970-0218.99912

Montgomery, R. M. (2009). Urban poverty and health in developing countries. Population Bulletin. Washington, DC.

Mudey, A., Ambekar, S., Goyal, R. C., & Agarekar, S. (2017). Studies on ethno-medicine assessment of quality of life among rural and urban elderly population of Wardha District, assessment of quality of life among rural and urban elderly. 5070. https://doi.org/10.1080/09735070.2011.11886394.

Odonkor, S. T., Kitcher, J., Okyere, M., & Mahami, T. (2019). Self-assessment of hygiene practices towards predictive and preventive medicine intervention: A case study of University Students in Ghana.

Pal, J., & Pal, A. K. (2017). Knowledge, attitude, and practice of personal hygiene and its predictors: A school-based study among adolescent girls in an urban slum, *6*(9), 2–7. https://doi.org/10.5455/ijmsph.2017.0617706072017

Paranjape, S. M., & Franz, D. R. (2015). Implementing the Global Health Security Agenda. *Lessons from Global Health and Security Programs, 13*, 1. https://doi.org/10.1089/hs.2014.0047

Rahman, M. M., et al. (2014). Knowledge and health problems related to health behavior among secondary school children in Rural community of Dhamrai Upazila, Dhaka. *Anwer Khan Modern Medical College Journal, 5*(2), 18–22.

Raj, A., Saggurti, N., Lawrence, D., Balaiah, D., & Silverman, J. G. (2010). Association between adolescent marriage and marital violence among young adult women in India. *International Journal of Gynecology & Obstetrics, 110*(1), 35–39. https://doi.org/10.1016/j.ijgo.2010.01.022

Rajbhandari, A. K., Dhaubanjar, R., GC, K. B., & Dahal, M. (2018). Knowledge and practice of personal hygiene among secondary school students of grade nine and ten. *Journal of Patan Academy of Health Sciences, 5*(2), 107–113. https://doi.org/10.3126/jpahs.v5i2.24030.

Ranasinghe, S., Ramesh, S., & Jacobsen, K. H. (2016). Hygiene and mental health among middle school students in India and 11 other countries. *Journal of Infection and Public Health, 9*(4), 429–435. https://doi.org/10.1016/j.jiph.2015.11.007

Ray, B. (2017). Quality of life in selected slums of Kolkata: A step forward in the era of pseudo-urbanization. *Local Environment, 22*(3), 365–387. https://doi.org/10.1080/13549839.2016.1205571

Sabri, B., Renner, L. M., Stockman, J. K., Mittal, M., & Decker, M. R. (2014). Risk factors for severe intimate partner violence and violence-related injuries among women in India. *Women and Health, 54*(4), 281–300. https://doi.org/10.1080/03630242.2014.896445

Saikia, D., & Das, K. K. (2014). Status of rural health infrastructure in the north-East India. *Management in Health, 18*(2), 34–38.

Shukla, A., Kumar, V., & Jain, K. (2017). Site suitability evaluation for urban development using remote sensing, GIS, and analytic hierarchy process (AHP). Proceedings of International Conference on Computer Vision and Image Processing.

Stoeva, P. (2020). Dimensions of health security—A conceptual analysis. 1700003. https://doi.org/10.1002/gch2.201700003.

Temitayo, I. O. (2016). Knowledge and practices of personal hygiene among senior secondary school students of ambassadors college, Ile-Ife, Nigeria. *Texila International Journal of Public Health, 4*(4), 648–660. https://doi.org/10.21522/tijph.2013.04.04.art055

U.S. Agency for International Development. (2004). *USAID HIV/AIDS technical publications*. USAID.

UNICEF. (2016). School-based healthy living and HIV/AIDS prevention education for middle and high schools (Teachers' Guide): Personal Hygiene. https://www.unicef.org/lifeskills/files/5thGrade.pdf

Wagman, J. A., Donta, B., Ritter, J., Naik, D. D., Nair, S., Saggurti, N., Raj, A., & Silverman, J. G. (2018). Husband's alcohol use, intimate partner violence, and family maltreatment of low-income postpartum women in Mumbai, India. *Journal of Interpersonal Violence, 33*(14), 2241–2267. https://doi.org/10.1177/0886260515624235

World Health Organization. (1948). Preamble to the Constitution of the World Health Organization as adopted by the International Health Conference, New York, 19–22 June 1946. http://whqlibdoc.who.int/hist/official_records/constitution.pdf

World Health Organization. (2008). The world health report 2008: Primary health care (now more than ever). https://www.who.int/publications/i/item/9789241563734.

World Health Organization. (2020). Coronavirus disease (COVID-19) pandemic. Retrieved from https://www.who.int/europe/emergencies/situations/covid-19

Yount, K. M., Roof, K. A., & Naved, R. T. (2018). Multilevel influences on men's partner violence justification, control over family decisions, and partner violence perpetration in Bangladesh. *Psychology of Violence, 8*(3), 367–378. https://doi.org/10.1037/vio0000171

Chapter 4
Infrastructure and Provision of Healthcare Services

4.1 Introduction

Healthcare infrastructure and provision of the healthcare system are the backbone of societal development. As we know, health is wealth. Also, healthcare infrastructure can help to improve the healthcare system and prevent the mortality rate in society. Healthcare infrastructure includes the entire infrastructure for the operative delivery of services such as physical and non-medical equipment, transportation, expertise technology, etc. Healthcare provision refers to allocating funds, drugs, staff, and medical equipment for effective healthcare delivery. WHO (2016) defines "healthcare provision as giving inputs such as money, staff, drugs, medical equipment, doctors, nurses, etc., for delivering healthcare intervention." Healthcare provision aims to provide healthcare services to the places by reducing inequality in health, improving health and well-being, and preventing disease prevalence. Healthcare is unequal in many developing countries where inequality is more prevalent in healthcare infrastructure and provision. Inadequate accessibility to healthcare services leads to poor people's further poorer health and a vicious cycle, which further increases health risks with suffering from chronic and communicable diseases for a prolonged period. Therefore, the overall development of the poor people remains static, perpetuating their poor condition. The accessibility to healthcare services is less provident to the poor people due to a lack of available medical staff, essential drugs, and provision of medical resources, workforce, supervision, infrastructure, equipment, etc.

The provision of a healthcare system is essential for slum dwellers as they are trapped in a poverty trap because the national policy neglects slum dwellers due to an over-focus on rural counterparts in developing countries. The rate of urbanization is higher in India, and unplanned growth is due to a lack of infrastructure barriers to cope with health risks in the urban slum arena. Urban poor people are constantly engaged in meeting their basic needs, with the need for more basic

© The Author(s), under exclusive license to Springer Nature Switzerland AG 2025
B. Dutta et al., *Using GIS to Identify Opportunities for Improved Health in Urban Slums*, Global Perspectives on Health Geography,
https://doi.org/10.1007/978-3-031-89617-0_4

amenities rather than healthcare affordability, as most urban rich people do. Moreover, the urban poor are neglected from the benefits of government or NGO health programs, as both organizations focus more on rural areas. In that context, the healthcare system of the urban poor is worse than the rural population. Moreover, the need for more information regarding available healthcare services restricts the urban poor to use the healthcare facilities of both government and private healthcare services. These conditions make the urban poor dwellers more vulnerable and prone to health risks than the rural areas. Furthermore, as the slum dwellers grow with illegal structures and land tenure, local government does not tend to recognize them as a part of urban society. Therefore, the slum poor are neglected from the benefits of subsidized social schemes and healthcare. Thus, this chapter wants to look at the impact of healthcare infrastructure and the provision of the healthcare system on the slum dwellers of the study area.

4.2 Hierarchy of Healthcare Services in India and Midnapore

India has a three-tier hierarchical referral system, i.e., primary, secondary, and tertiary public healthcare service facilities (Fig. 4.1), i.e., sub-health centers (SHC) and primary health centers (PHC) have been controlled by community health centers (CHCs) and sub-district hospitals (SDH) are also regulated by district hospitals (DH) and medical colleges (MC) (Ministry of Health and Family Welfare of India, 2011; Dutta et al., 2018). Primary healthcare center covers an area with a population of 20,000 in hilly, tribal, or challenging areas and 30,000 populations in plain areas with 6 surveillance beds. Each primary healthcare unit acts as a referral unit for six sub-centers and transfers chronic cases to community healthcare (30 bedded hospitals) and hospitals at sub-district and district levels

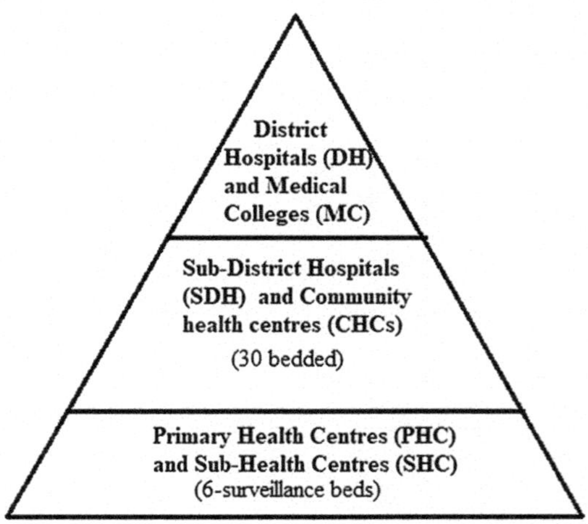

Fig. 4.1 Hierarchy of healthcare system in India

Fig. 4.2 Hierarchy of healthcare system in Midnapore Municipality

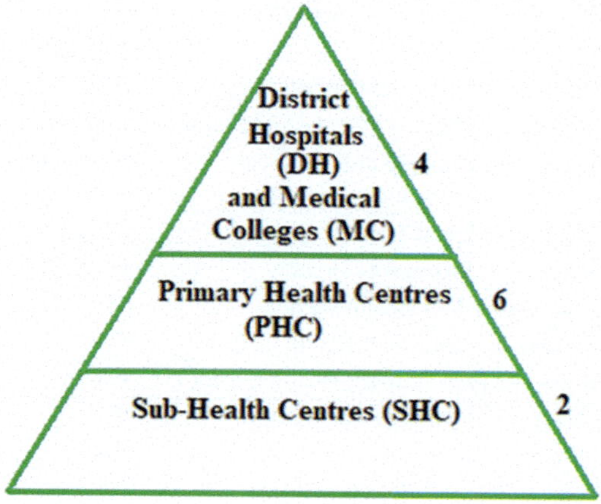

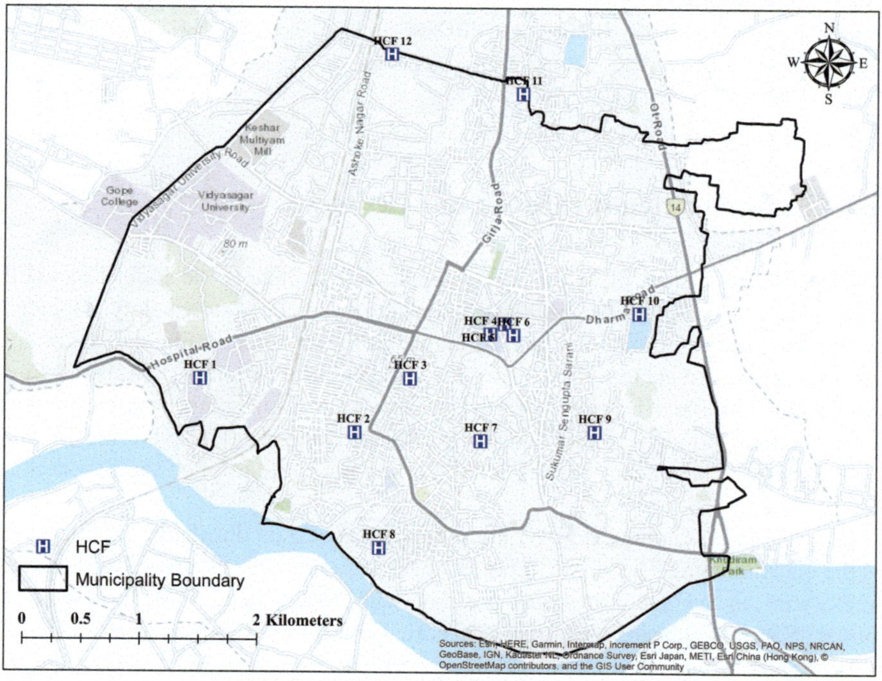

Fig. 4.3 Location of Healthcare system in Midnapore Municipality (Source: Dutta et al., 2021)

(Department of Health & Family Welfare, India, 2013; Dutta et al., 2018). The healthcare system of Midnapore municipality maintains the Indian standard (Figs. 4.2 and 4.3). However, the healthcare center needs to be allocated appropriately across the study area, which requires attention for the equitable distribution of healthcare services.

Table 4.1 Location of HCFs in the study area

Healthcare facility center (HCF)	Type of HCF	Ward no.	Latitude	Longitude
Midnapore Medical College and Hospital	Hospital	12	22°25′12.36″	87°19′ 19.92″
Midnapore Homeopathic Medical College and Hospital	Hospital	12	22°25′12.36″	87°19′30″
Indian Red Cross Society	Hospital	12	22°25′15.6″	87°19′ 22.8″
Vidyasagar Institute of Health	Hospital	24	22°25′1.2″	87°17′52.8″
Sahebpukur Sub Health Center	SHC	14	22°24′45.04″	87°19′16.85″
Ramkrishnanagar Sub Health Center	SHC	9	22°25′23.32″	87°20′13.81″
Kuikota Urban Primary Health Center	PHC	2	22°26′21.87″	87°19′32.78″
Udaypally Social Service Center	PHC	3	22°26′33.18″	87°18′50.51″
Saratpally Urban Primary Health Center	PHC	23	22°24′47.31″	87°18′39.3″
Pathanmohalla Urban Primary Health Center	PHC	13	22°25′1.5″	87°18′53.09″
Najarganj Health Post	PHC	20	22°24′15.52″	87°18′46.59″
BerBallabpur Urban Primary Health Center	PHC	16	22°24′49.47″	87°20′0.43″

Source: Author(s)

4.3 Nature of Healthcare Service Centers

The healthcare system of Midnapore municipality maintains the Indian standard (Figs. 4.2 and 4.3), but the healthcare center has not been adequately allocated across the study area; out of a total of eight healthcare centers (Fig. 4.29), six are primary healthcare centers (PHC), two are sub-healthcare centers (SHC), and four are hospitals (Table 4.1 and Fig. 4.4). The study area also has a private healthcare center (nursing home). Primary and sub-healthcare centers and hospitals are significant sources of healthcare centers for the urban poor due to poor socioeconomic conditions.

The survey was conducted over eight healthcare centers during the same period as the questionnaire. The survey reported that PHC and SHC are open 3–4 days per week, with an average of 6 hours per day, with 50% official staff, doctors, and nurses. Some healthcare centers also reported that they do not have any official staff, doctors, or nurses. These healthcare centers have been open during health awareness programs at the community level (81.81%), polio and vaccination programs (100%), and special health checkups (18.19%) at the school level as an outreach program only for the urban poor. Healthcare centers do not have ambulance facilities for referral to the Midnapore Medical College and Hospital and other hospitals with an average distance of 2–4 km (Fig. 4.4). Fifty percent of healthcare centers have laboratory facilities, and most are provided with essential healthcare services (Table 4.2) and encounter common diseases (Table 4.3) during a health checkup.

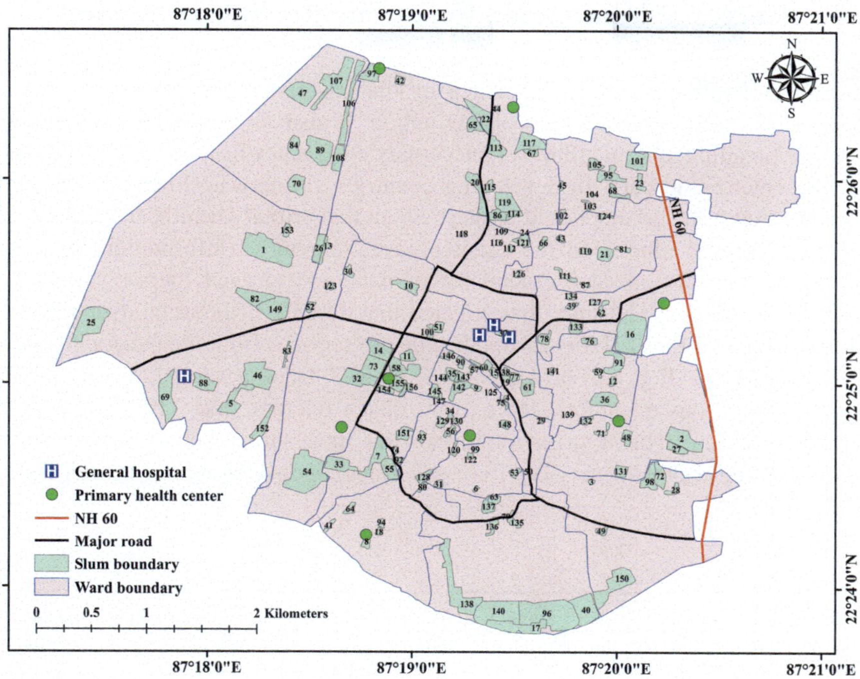

Fig. 4.4 Major road network and HCF in Midnapore Municipality (Source: Dutta et al., 2021)

Table 4.2 Type of services provided by PHC and SHC

Type of service provided	Percentage
Only outdoor	15.38
Checkup and medicine supply	5.38
Health checkup and diagnostic	7.69
Health campaign and awareness	30.76
Vaccination and polio program	23.07
ANC	7.69

Source: Field survey (2022)

Table 4.3 Common diseases encountered during health checkups

Common problems	Percentage
Fever	20
Cold and cough	20
Blood pressure	20
Blood sugar	10
Skin problems	10
During pregnancy	13.33
Others	6.67

Source: Field survey (2022)

4.4 Sources of Information Regarding Healthcare Services

A primary healthcare service provided by the government is less expensive and evident for promoted healthcare of any nation. It also saves money in preventing disease spreading, chronic illness, unnecessary emergency hospital visits, and other health center-based care. In developing countries, primary healthcare is the most critical care for the urban poor. It is also evident that primary healthcare can prevent risk and chronic illness. In the exact phrase, sources of information regarding healthcare services are also crucial for poor people to avail themselves of free healthcare services. The study investigated that 37.25% of households are getting information about availability of free healthcare services from the general hospital, 22% from ICDS (Fig. 4.5) or Anganwadi, 13.63% from private clinics or nursing homes, 10% from neighbors, 9.5% from quack doctors, and 7.62% from other sources (Fig. 4.6). This information reveals that most slum dwellers depend on the availability of healthcare services from government-subsidized healthcare facilities in the study area.

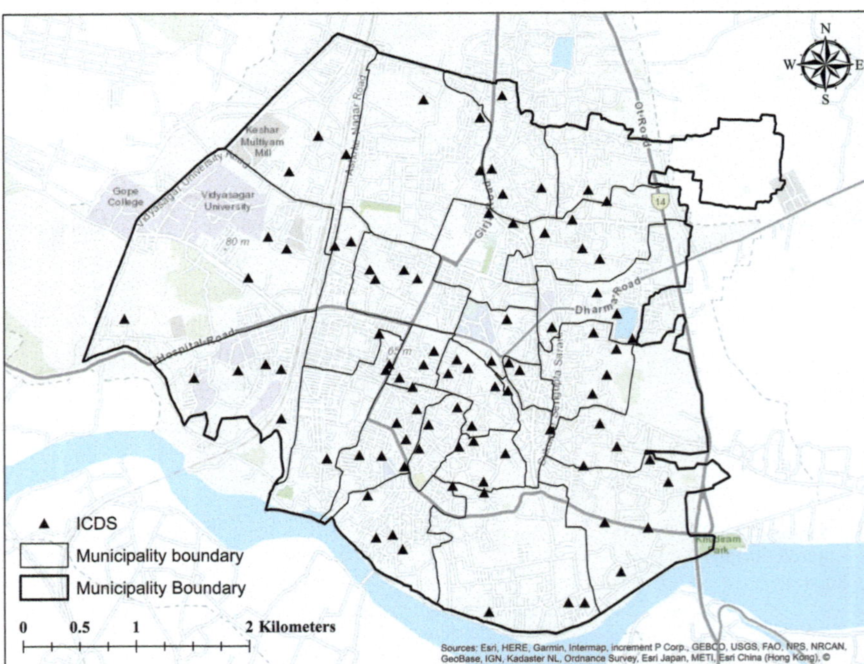

Fig. 4.5 Major ICDS center in Midnapore municipality (Source: Author)

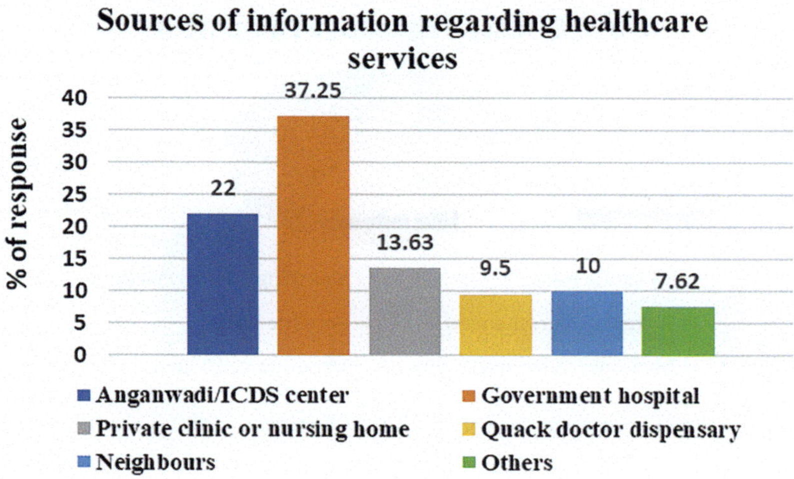

Fig. 4.6 Sources of information regarding healthcare services (Source: Field survey, 2022)

4.5 Use of Disease Prevention Equipment

Preventive healthcare or prophylaxis measures the action taken and precautions for disease prevention. Disease and morbidity are caused by environmental degradation, personal hygiene practices, socioeconomic status, disease agents, lifestyle choices, and use of disease prevention equipment. Health shocks are hidden at the primary stage and are subsequently realized by the individuals during the later stage. Therefore, disease prevention is much more important before it can spread in the form of an epidemic or at a community level, and slum neighborhoods are more vulnerable to health risks to some extent; they need to pay attention to disease prevention equipment before they are affected. In this context, disease prevention depends on prophylactic actions intended to circumvent the development of risk aspects in the first stage, early in life. It is found that mosquito nets and oil or coils, potable water, and medicine are the most useable disease prevention equipment (Fig. 4.7) among the slum dwellers in the Midnapore municipality.

4.6 Provision of Healthcare Supplements and Equipment

Healthcare supplements and equipment such as drugs, surgical supplies, and infrastructural equipment are most significant in impacting the quality of healthcare and healthcare costs. Healthcare supplements and equipment are intended for disease prevention at the primary level, such as a village, block, and municipality level. Provision of supplements and healthcare equipment is most important in India at the primary level for preventing disease progression and reducing infant and maternal

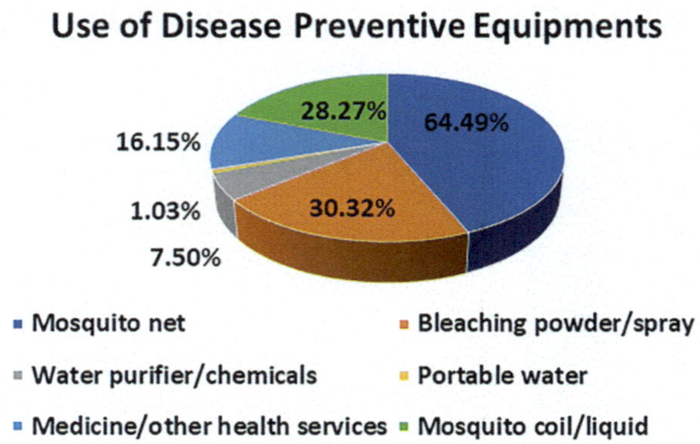

Fig. 4.7 Use of disease prevention equipment (Source: Field survey, 2022)

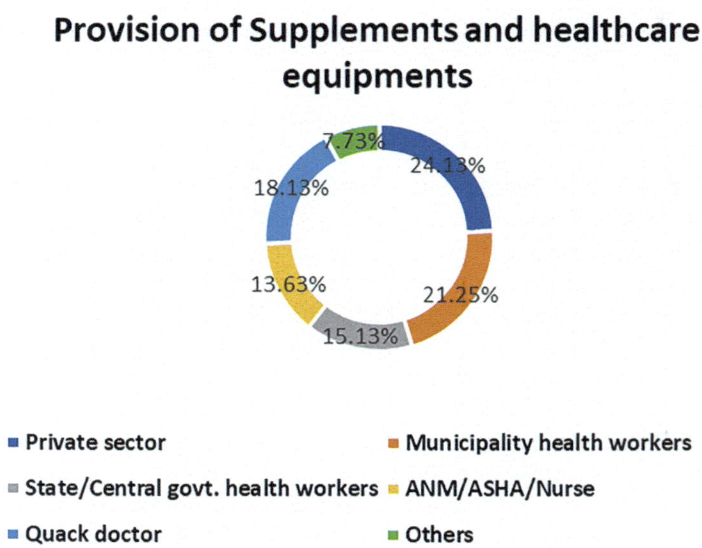

Fig. 4.8 Provision of healthcare supplement and equipment (Source: Field survey, 2022)

mortality, healthcare cost, and malnutrition in the context of health planning and management with the health and well-being of the country, thereby reducing health inequality and medical poverty trap. However, knowing how and who provided the equipment is essential. The household survey revealed that primary healthcare services and disease prevention equipment (Fig. 4.8) are mainly provided by municipal health workers (21.25%), low-cost quack doctors (18.13%), the private sector (24.13%), and NGOs. Additionally, some households place their trust in healthcare services provided by state or central government organizations (15.13%).

4.7 Sources and Provision of Healthcare Vehicles

Healthcare vehicles or transportation is a vital issue for the accessibility to healthcare centers and services in a particular geographical region. Distance to healthcare centers and the provision of healthcare vehicles are both significant factors in accessibility to healthcare services. Arcury et al. (2005) comprehended that transportation is an aiding factor for measuring geographic accessibility and spatial healthcare behavior with the distance, availability of vehicles, and activity space. Nemet and Bailey (2000) showed that utilization of healthcare services could be reduced to a provider by increased distance and outside the activity space. Therefore, the study estimates the impacts of sources and provision of transportation on the utilization of healthcare services for poor households in a small urban area. Using data collected from a survey, they considered transportation and distance as vital factors for healthcare accessibility that could impede healthcare utilization. The study estimated that the average distance to the nearest healthcare center is 2.4 km, which is mainly traveled by loto/toto or rickshaw (66%), tempo or auto-rickshaw (20.75%), and motorbike or scooter (22.75%) and arranged mainly by family members (89.19) and sometimes arranged by ASHA health workers (8.43%) typically during delivery of care services (Table 4.4 and Fig. 4.9). As the distance to the nearest healthcare center is the minimum, dwellers sometimes walk (26.63%) to the healthcare center, i.e., at the doctor's dispensary.

Table 4.4 Sources and provision of healthcare vehicles

Sources and provision of healthcare vehicles	Percentage of response
Arranged by family	89.19
Arranged by ASHA	8.43
Others	2.38

Source: Field survey (2022)

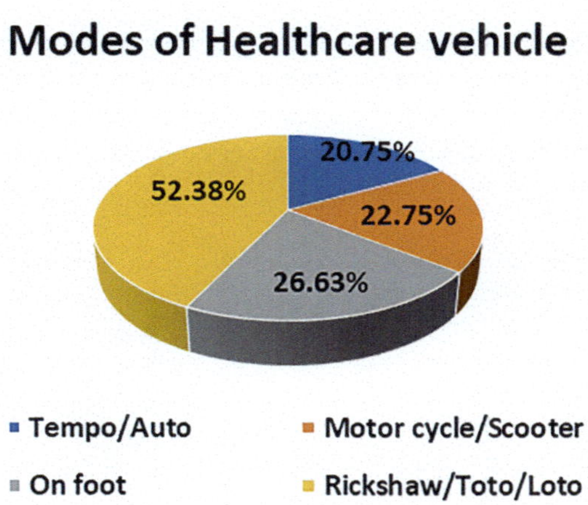

Fig. 4.9 Modes of healthcare vehicles (Source: Field survey, 2022)

4.8 Modes of Communication and Distance to Healthcare Centers

Modes of communication and distance to the nearest healthcare center are of utmost importance for patients with morbidity or associated with chronic illness. Considerable distance to the healthcare center and lack of communication may increase morbidity and mortality as well as ignorance of health consciousness, which means to say occulted the disease, which subsequently increases the health risks and medical costs. Therefore, easy and modern communication devices with a minimum distance to healthcare centers may decrease the healthcare burden and reduce morbidity and mortality trends. Nowadays, different communication modes have been developed to communicate to the healthcare center, either private, government, or trust for the aim of effective and quick health checkups and cures such as telephone, video calling, voice messages, mail, social media chat, offline booking, reference booking, campaign booking, etc. Similarly, dwellers are more conscious near the healthcare center and regularly communicate with doctors. The study estimated the average distance of the nearest healthcare center in Midnapore municipality is 2.4 km from the poor slum neighborhoods, which is mainly traveled by auto/toto or rickshaw (66%) and sometimes, walking (26.63%) to the healthcare center communicated offline, i.e., at the doctor's dispensary (50%), telephonic booking (19.13%) and by local authority reference (16.88%) (Table 4.5). However, rare modern healthcare communication devices are adequately utilized by dwellers, perhaps due to a lack of knowledge and interest to a great extent, although 40% of healthcare centers are located within a 1–2 km distance (Fig. 4.10) and choose offline booking for a health checkup.

4.9 Satisfaction Level of Provision of the Healthcare Infrastructure and Services

The satisfaction level of the patients reflects the perceptions and needs of healthcare services and their utilization. The satisfaction level of the patients is significant for assessing the allocation of healthcare facility centers and determining the indicators of the quality of healthcare services and delivery, responsiveness of the healthcare

Table 4.5 Modes of communication to the healthcare center

Modes of communication to healthcare center	Percentage of response
Telephone	19.13
Offline booking	50.38
By local authority reference	16.88
Others	9.63

Source: Field survey (2022)

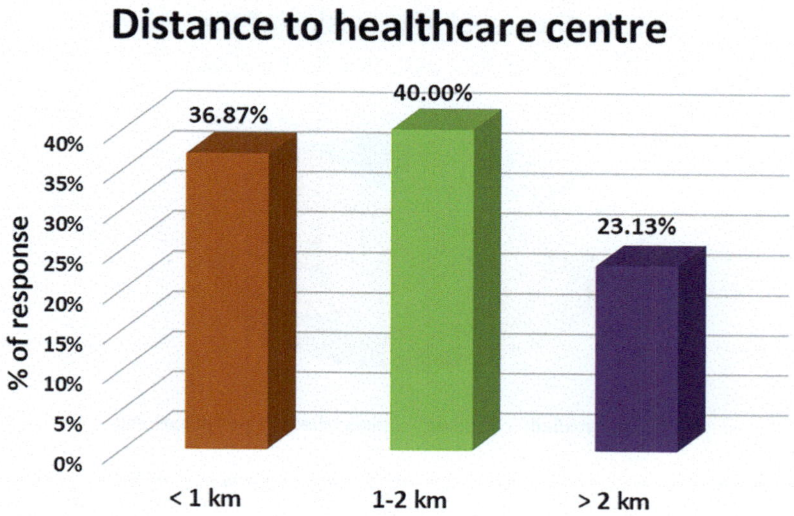

Fig. 4.10 Distance to healthcare centre (Source: Field survey, 2022)

system, and needs of healthcare services. A higher level of satisfaction includes the patient's satisfaction with received healthcare services, quality of doctors and medical staff, overall infrastructure and facilities, level of cure, medical cost, provision of medical facilities and management, etc. Moreover, measuring patient satisfaction can help to improve the service delivery system, allocate medical resources, and prioritize capacity-building needs. Thus, higher patient satisfaction indicates improved health outcomes and well-being. Many researchers have noted that in developing countries, patient satisfaction with healthcare services tends to range from moderate to low, with the exception of a few nations such as China and Japan (Kumar et al., 2012; Goel et al., 2014). In India, the data shows moderate levels of patient satisfaction based on overall facility types and levels (Shukla et al., 2017; Chahal & Mehta, 2013). Overall, around 60% (Fig. 4.11) of slum dwellers have been satisfied with their received care. Private facilities had the highest level of patient satisfaction. A relatively less proportion of dwellers were satisfied for the primary care and public facilities. However, receiving care from the quack doctor had a relatively higher level of satisfaction, i.e., 63.4%.

4.10 Site Suitability Analysis (SSA)

"Health is wealth," which was mentioned in the third Sustainable Development Goals (Lawal & Anyiam, 2019; Wang et al., 2018). Nowadays, healthcare facilities are the most critical factor in social well-being (Mansour, 2016; WHO, 2001, 2004; Alabi, 2011; Danjuma, 2015), which has been experienced by recent pandemics

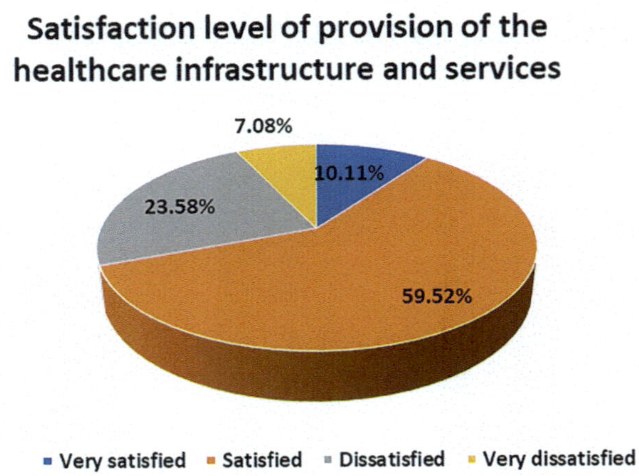

Fig. 4.11 Satisfaction level of healthcare services (Source: Field survey, 2022)

worldwide. In developing countries, small urban areas continuously absorb the population from surrounding rural areas, which demand affordable healthcare services (Lawal & Anyiam, 2019; Alabi, 2011; Danjuma, 2015). However, most developing countries have been unable to maintain the carrying capacity due to unplanned urban growth and uneven distribution of healthcare facility centers. Equitable access to healthcare facilities protects against unavoidable morbidity, illness, diseases, and deaths. Above and beyond, urban areas in most developing countries function as districts (Dutta et al., 2018) or block headquarters that provide services to their surrounding rural areas (Mansour, 2016; Christaller, 1966; Sedenu et al., 2016). Therefore, to maintain the population flux from rural areas, urban dwellers are given less importance for their healthcare (Mansour, 2016; Dinda et al., 2018; Danjuma, 2015). The rate of urbanization is high in India, especially in small cities. Thus, to achieve millennium development goals, we need to focus more on providing and properly allocating healthcare facilities. This would benefit poor people living in urban areas from the perspective of the social equity of healthcare accessibility (Kim et al., 2018). Therefore, the study seeks to assess the distribution and deficient areas of public healthcare facilities and to find the new primary healthcare sites (PHS) in the deficient healthcare facility (HCF) areas.

4.10.1 Data and Method

Spatial data such as significant healthcare facility centers (HCF), slums (156), and wards (25) boundary has been collected from the municipality. In contrast, non-spatial data, like household information, were collected through a questionnaire. The GPS device was employed to collect location points of the healthcare centers in the municipality, along with geospatial data collected from the Earth

Explorer-USGS open website (https://earthexplorer.usgs.gov). The municipality's total area is approximately 20 km^2, within which 12 governmental public healthcare centers (Midnapore Municipality, 2022), 3 hospitals, and 9 primary healthcare centers (Fig. 4.19) are dispersed across the study area. We incorporated both the statistical model-building algorithm using MS Excel v2016 (Microsoft Corporation) and SPSS v25.0 (IBM Analytics) along with raster and vector datasets in a GIS platform (ArcGIS v10.4). The total analysis is divided into two parts—spatial analysis of healthcare facilities (HCF) and site suitability for new primary healthcare facilities (PHCF)—and methodological design (Fig. 4.12) is highlighted below:

4.10.2 Result and Discussion

4.10.2.1 Spatial Analysis of HCF

Euclidean Distance

Euclidean distance is measured from any point to the center of the source point in a geographical region by a straight-line distance (meter or feet). In a GIS platform, the nearest point (Mansour, 2016) is calculated by Eq. 4.1.

$$d = \sqrt{\sum_{i=1}^{n} (X_i - Y_i)^2} \tag{4.1}$$

The Euclidean distance has been used to find the nearest or proximate accessible primary healthcare centers (PHCs) and suitable locations for PHCs. In the study area, low accessible facilities were found toward the far west, southeast, and northeast marginal parts of the municipal boundary due to the central (central and southwest) location of high-density healthcare facility centers. Therefore, the distance increases from the city center, which has low accessibility or public healthcare facilities (Fig. 4.13).

Mean Center

The mean center is a point of mean x and y coordinate among the points of a particular geographical area, measuring distance by the following Eq. 4.2:

$$\overline{X} = \frac{\sum_{i=1}^{n} x_i}{n}, \overline{Y} = \frac{\sum_{i=1}^{n} y_i}{n} \tag{4.2}$$

where x_i and y_i are the coordinates for points i and n is equal to the number of coordinate points, while $\overline{X}$ and $\overline{Y}$ define the mean center. The location of the mean

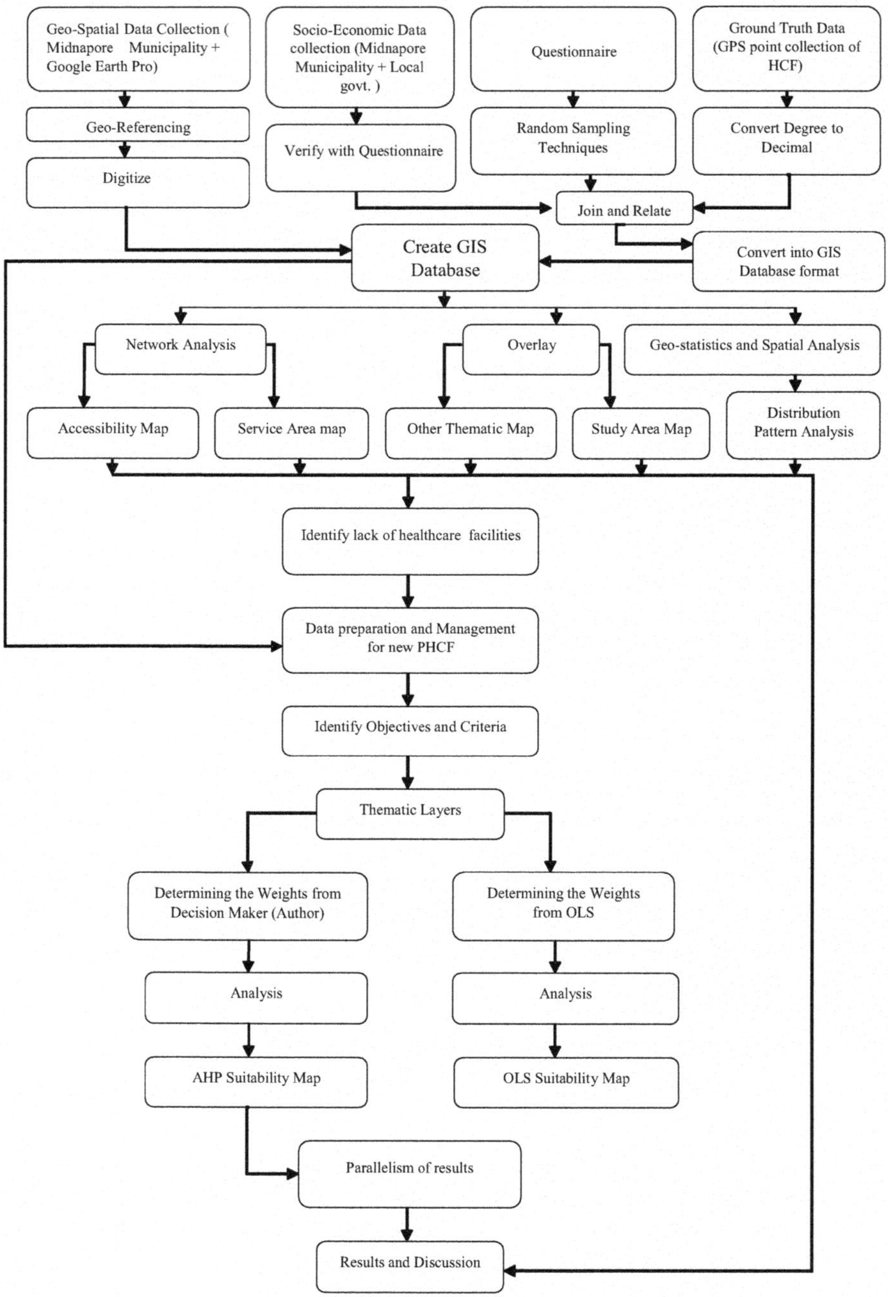

Fig. 4.12 Methodological progress of the study (Source: Author)

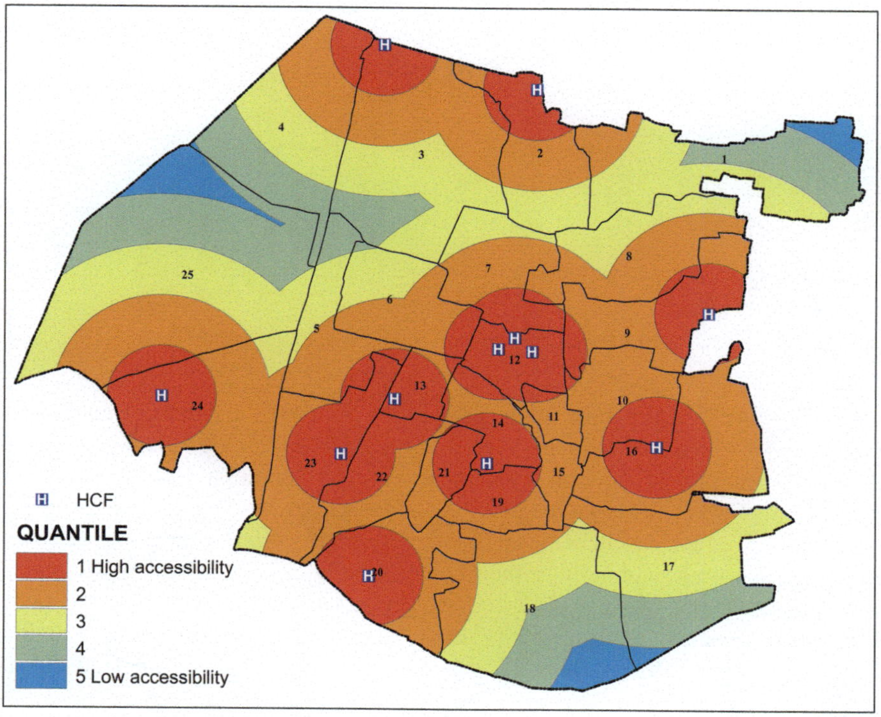

Fig. 4.13 Map showing spatial accessibility to HCF of the study area (Source: Author)

center indicates the distribution of healthcare centers or possible mean healthcare centers (Sedenu et al., 2016) in the GIS platform.

The analysis depicted (Fig. 4.14) that the mean center is located slightly in a moderate-density area acting as a geographical point, which indicates that PHCs are not located correctly in the study area and need to be more attentive.

Central Features

In the ArcGIS tool, central features identify the centrally located point among the existing points by measuring the weighted mean center (if weight is specified) from the distance (Euclidean) of all input points (Sedenu et al., 2016). This feature recognizes the most central location of the input dataset. In the study area, Midnapore Medical College and Hospital is recognized as a central location (Fig. 4.14) among the healthcare facility centers (HCFs), with highly accessible, well-connected roads and equipped medical infrastructures, and acting as district hospitals.

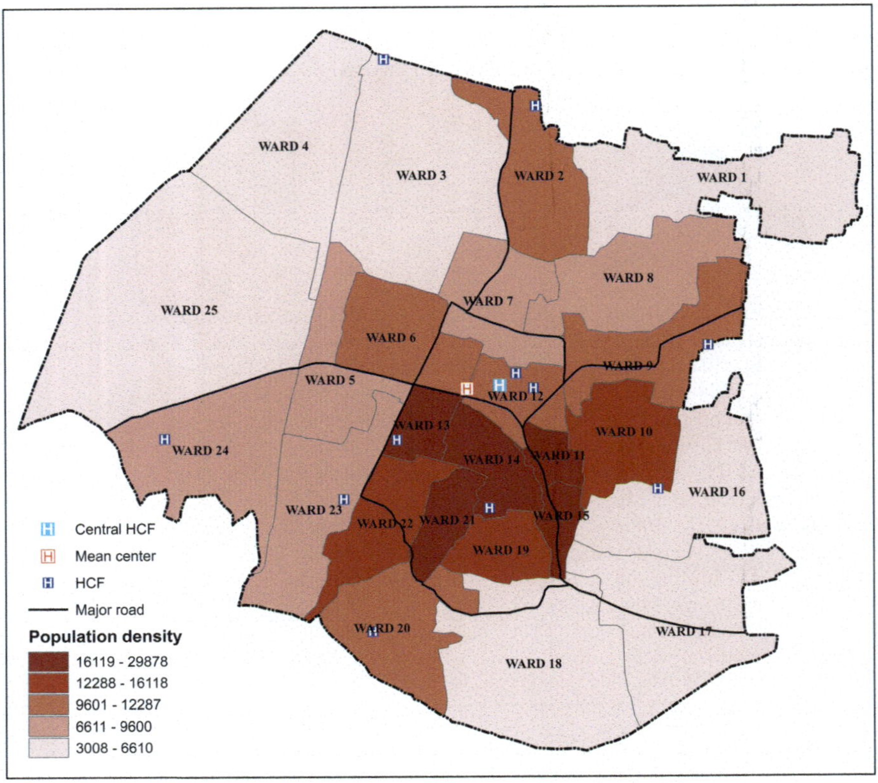

Fig. 4.14 Shows the mean centre and Central HCF of the study area (Source: Author)

Buffer Analysis

A buffer is developed around a point, line, or polygon by covering a particular distance (Euclidean) in a two-dimensional Cartesian plane in the GIS environment. A 1 km distance of buffer zone has been applied around public healthcare centers to find the spatial distribution of healthcare units (Mansour, 2016; Parry et al., 2018; Zhou & Wu, 2012; Murseli, 2014; Zhou & Daniel, 1996; Jamal, 2016; Danjuma, 2015; Wang et al., 2018). The analysis reveals (Fig. 4.15) that the maximum portion of the municipality remains outside the 1 km buffer zone. More specifically, nine municipal wards (3, 4, 5, 6, 17, 18, and 25) show 20–50% unserved areas with a lack of accessibility to the nearest HCF within a 1 km distance of the buffer radius, which needs to be additional HCFs for the unserved areas.

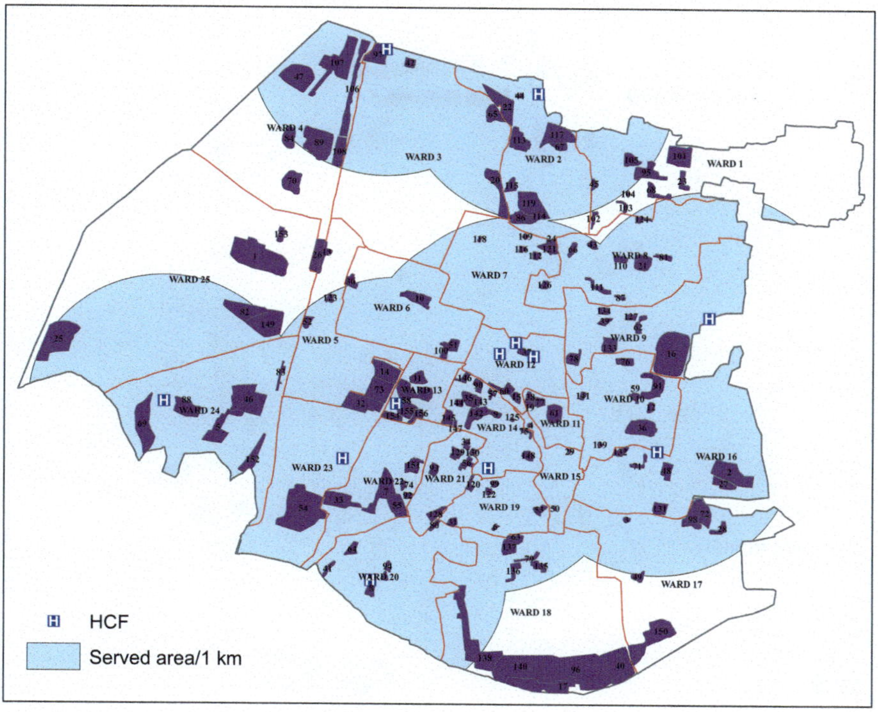

Fig. 4.15 Map showing 1 km served the area from HCF in the study area (Source: Author)

Near Analysis

Near analysis in GIS algorithm measuring the distance of nearest points (HCFs) to the input points (slum neighborhoods) through using Euclidean distance (meter or km) with specified distance or radius within a particular geographical region (Zhou & Wu, 2012; Kim et al., 2018; Wang et al., 2018; Danjuma, 2015; Wang & Luo, 2005). The analysis reveals (Fig. 4.16) that 16.5% of slum neighborhoods avail the healthcare facility within 200 m from their residence. In comparison, about 77% of slum neighborhoods avail healthcare facilities at a distance from 200 m to 1 km. In contrast, only 6.50% of slum neighborhoods avail healthcare services further from proximity to healthcare centers (HCFs) outside 1 km.

Standard Distance

The standard distance is the estimation of the degree to which a circle polygon with a particular radius equivalent to the standard distance (Sedenu et al., 2016) around the geographic mean center is distillate or diffused can be expressed as (Eq. 4.3)

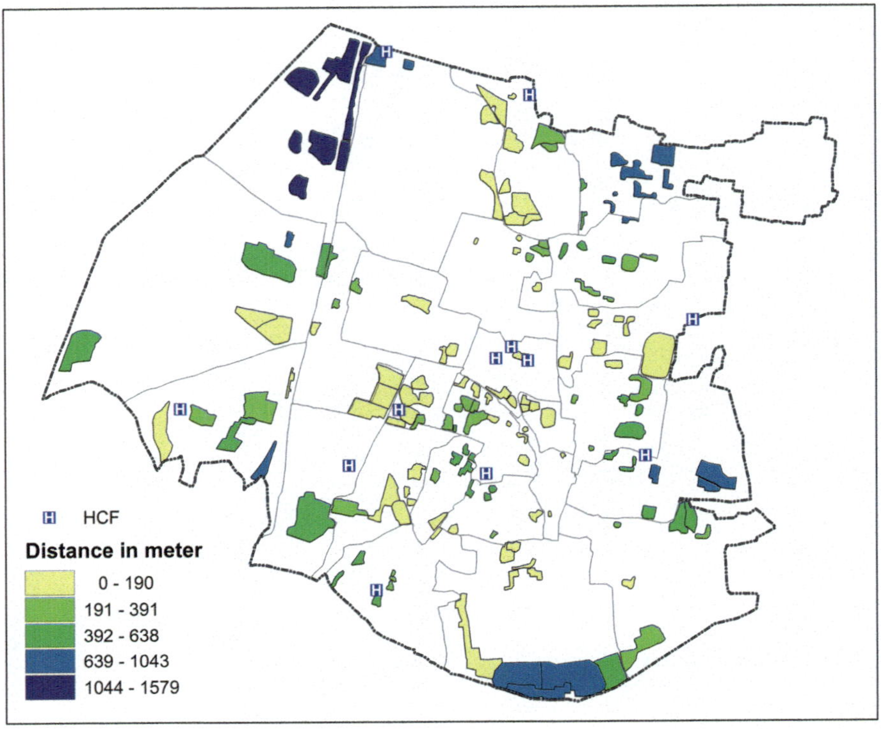

Fig. 4.16 Shows the nearest slums area from HCF in the study area (Source: Author)

$$\mathrm{SD} = \sqrt{\frac{\sum_{i=1}^{n}\left(x_i - \bar{X}\right)^2}{n} + \frac{\sum_{i=1}^{n}\left(y_i - \bar{Y}\right)^2}{n}} \tag{4.3}$$

where x_i and y_i are coordinates and $\bar{X}$ and $\bar{Y}$ define the mean center. The standard distance estimates the solidity of distributed topographies around the mean center on a map. Figure 4.17 exhibits a circle of standard distance around a mean health center, indicating that only 50% of healthcare centers deviate from a standard normal distribution (roughly 68%) and showing the dispersed pattern of healthcare services. Moreover, more than standard distance is needed to ensure service provision to the encircled area, which needs devotion from the urban authorities.

Directional Distribution

Directional distribution in the GIS platform measures the direction (x and y) of points or areas by estimating the standard distance. It is also known as the axes of an ellipse, i.e., a standard elliptical polygon that shows the distribution of features in a definite orientation, statistically expressed as (Eqs. 4.4 and 4.5)

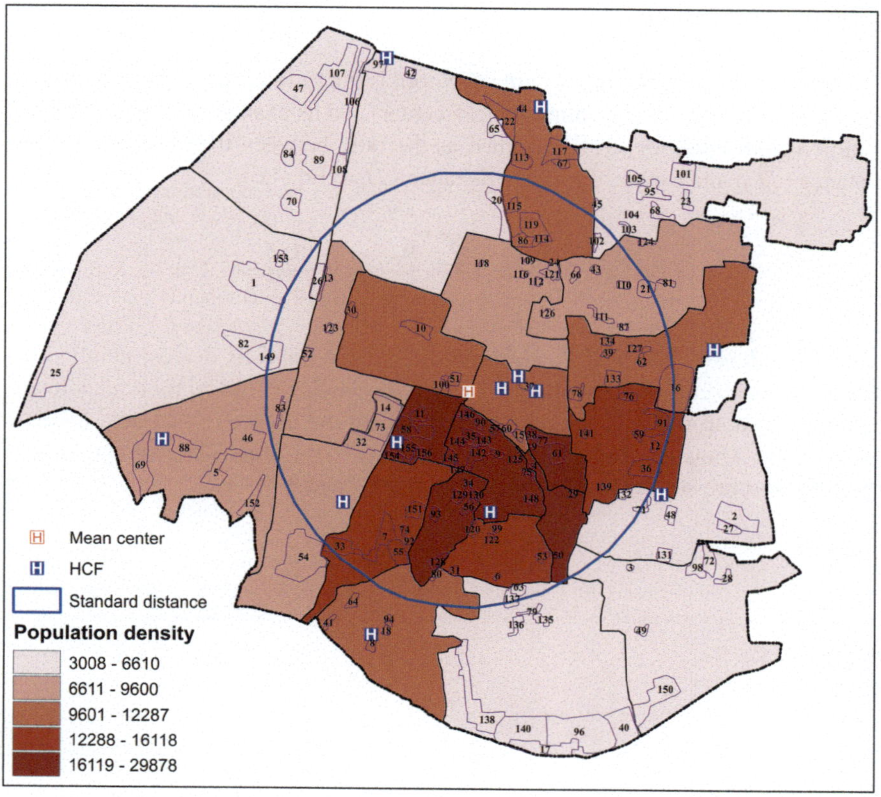

Fig. 4.17 Shows the standard distance of HCF in Midnapore Municipality (Source: Author)

$$\mathrm{SDE}x = \sqrt{\frac{\sum_{i=1}^{n}\left(x_i - \bar{X}\right)^2}{n}} \tag{4.4}$$

$$\mathrm{SDE}y = \sqrt{\frac{\sum_{i=1}^{n}\left(y_i - \bar{Y}\right)^2}{n}} \tag{4.5}$$

where x_i and y_i are coordinates, $\bar{X}$ and $\bar{Y}$ represent the mean center, and n is equal to the total quantity of features. Directional distribution within normal distribution represents the dispersion or spread of features. The result of the analysis exhibits (Fig. 4.18) the direction of the development of healthcare centers is from SW to NW, including geographic mean centers with higher density than the other parts of the town. Therefore, inequalities in locational healthcare services can be recapitulated for the public interest and further policy-making.

Average Nearest Neighbor

The Average Nearest Neighbor in the GIS tool is the ratio of the average Euclidean distance between centroid features (mean center) and its nearest neighbor's centroid (public health facilities). It is designed as the ratio between the observed average distance ($\bar{D}_0$) and expected average distance ($\bar{D}_E$) (Eq. 4.6).

$$\text{ANN} = \frac{\bar{D}_0}{\bar{D}_E} \bar{D}_0 = \frac{\sum_{i=1}^{n} \text{d}_i}{n} \bar{D}_E = \frac{0.5}{\sqrt{n / A}} \tag{4.6}$$

where d_i indicates the distance between the feature i and its nearest neighboring feature, n represents the total number of features, and A is the minimum enclosing rectangle area around all features, or the value of the user-specified area (Sedenu et al., 2016; Mansour, 2016; Danjuma, 2015) is 0.0175264. The average nearest neighbor Z-score statistics are derived as (Eqs. 4.7 and 4.8)

$$z = \frac{\bar{D}_0 - \bar{D}_E}{SE} \tag{4.7}$$

where

$$\text{SE} = \frac{0.26136}{\sqrt{n^2 / A}} \tag{4.8}$$

The Average Nearest Neighbor tool finds the value of the nearest neighbor ratio (NNR), which is 1.52 ($p < 0.0001$), and the Z-score is 3.43 ($p < 0.0001$). The value intrinsically indicates that public health facilities in the study area are spatially dispersed (Fig. 4.19) in nature, while the null hypothesis is accepted, which raises that there is an existence of spatial pattern among the public healthcare centers with riotous and bungling dispersal of public healthcare centers in the city. Naturally, public healthcare centers have been developed in the central part of the city and disseminated toward the periphery of the town. For example, the south, southeast, north, and northwest parts of the towns have the minimum number of public healthcare centers (Fig. 4.18). Correspondingly, the absence of healthcare centers has been acknowledged in the southeast and eastern parts of the town, perhaps due to pseudo-urbanization and new housing development.

Population Density

Density analysis in the GIS platform was derived by the ratio of the population (as per Census of India, 2011) and the area (km^2) of each ward (Sedenu et al., 2016) (Fig. 4.20). The equation used for estimated population density is derived by Eq. 4.9.

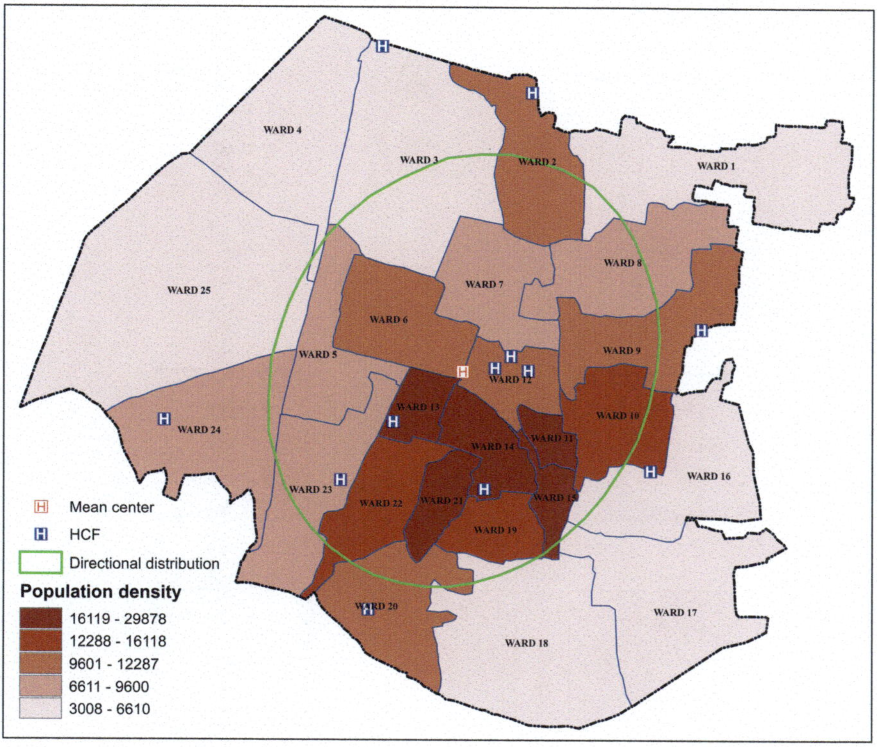

Fig. 4.18 Map showing the directional distribution of HCF in the study area (Source: Author)

$$\text{Population Density}\left(\text{Estimated}\right) = \frac{\text{Population at each ward}}{\text{Area}\left(\text{km2}\right)\text{of ward}} \tag{4.9}$$

Health Facility Ratio

In the twenty-first century, the government of India (Department of Health & Family Welfare, India, 2013) has given initiative to public healthcare facilities at each administrative unit in all states of India for the purpose of meeting the basic needs of the healthcare services based on population density and distribution (Mansour, 2016). According to WHO (2010), "the health facility ratio is the facility distribution per 10,000 populations" as a convenient, standardized indicator that estimates the magnitude of accessibility of healthcare services by population and recognizes the gaps in healthcare service coverage within a particular administrative unit (Eq. 4.10).

$$\text{Health Facility Ratio} = \frac{\text{Number of public health facility centers}}{\text{Total population of each ward}} \times 10{,}000 \tag{4.10}$$

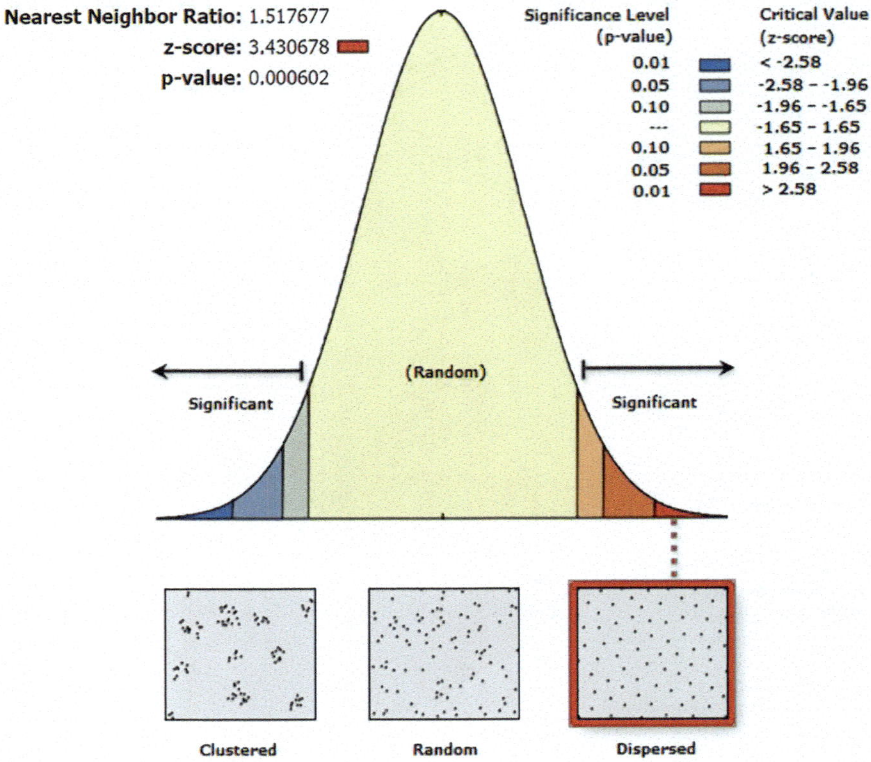

Fig. 4.19 Result of average nearest neighbour for HCF in the study area (Source: Author)

The analysis reveals (Fig. 4.21) that a maximum number of municipal wards displayed 0 ratios and less than 1 facility per 10,000 population. The central part of the city has a high healthcare facility ratio due to the presence of government hospitals and private healthcare centers. Similarly, some municipal wards in the northern, eastern, and southwest parts exhibited moderate healthcare facilities due to the provision of 24-h primary healthcare facilities. In contrast, most of the municipal wards from the city core toward the periphery exhibited limited (mostly zero) provision of public healthcare facilities. Therefore, unequal distributions of healthcare facilities persist throughout the city regarding equity in public healthcare coverage.

Service Area

A service area in a GIS Network Analyst extension is an area that provides services with distance and time from the slum neighborhoods to the nearest public healthcare center, including all access roads, expressed as (Eq. 4.11)

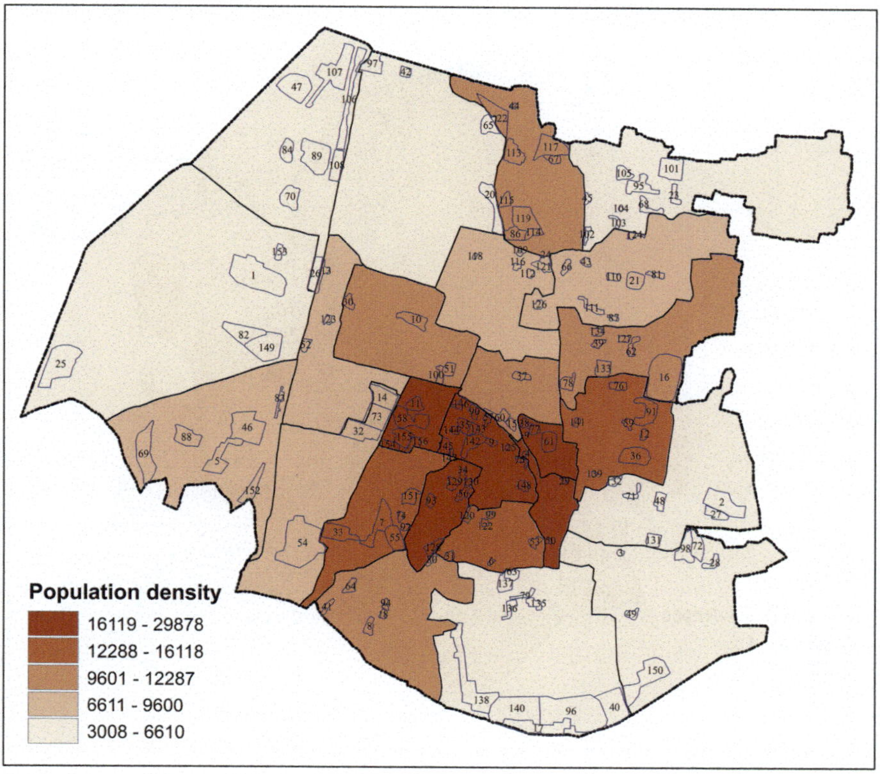

Fig. 4.20 Map showing the population density of the study area (Source: Author)

$$\text{Time} = \frac{\text{Distance}}{\text{Speed}} \tag{4.11}$$

The service area map (Fig. 4.22) shows the most accessible or the shortest or least-cost routes, roads, or streets to the nearest healthcare center at a constant speed of 40 km/hour (Sedenu et al., 2016) within a given time of 3, 5, 7, and 9 minutes. The network also estimated the total served and unserved areas of healthcare services are 12.28 and 7.72 km^2, respectively (Fig. 4.23). The map (Fig. 4.23) indicates that nine municipal wards have 90% unserved areas, four wards have 30–50% of their areas unserved, and three wards have 20% unserved areas. In contrast, eight municipal wards are fully covered by public healthcare facilities and enjoy high accessibility. Therefore, it is suggested that healthcare services be appropriately allocated to unserved areas.

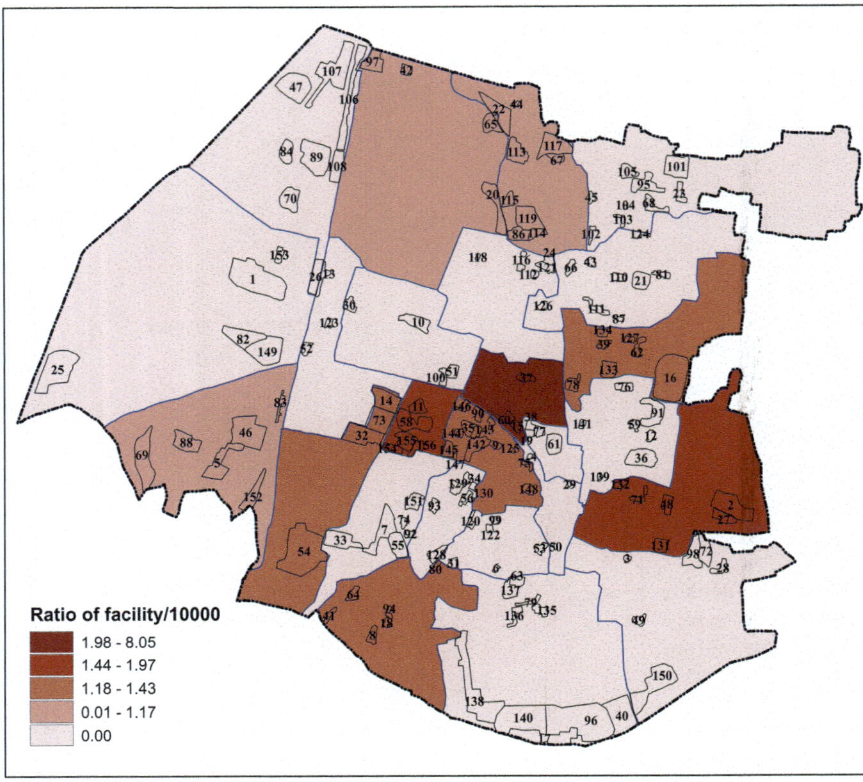

Fig. 4.21 Percentage of health facility ratio of HCF in each ward (Source: Author)

Multidimensional Poverty Index (MPI)

Multidimensional poverty indices (MPI) indicate the multiple deprivations based on micro-level data from a household survey designed to find the inclusive state of targeted vulnerable sociocultural groups living in poverty (Alkire et al., 2015; Ravallion, 2011; Cavapozzi et al., 2015; Das et al., 2020). The deprivation condition of the household is determined by a given "poverty cutoff" through a weighted deprivation score (C_i) using Eq. 4.12.

$$Ci = \sum_{j=1}^{d} gij \times wj \tag{4.12}$$

The weighted deprivation score (C_i) is the product of the deprivation matrix (*gij*) and weighted value (equal) allotted for each indicator (*wj*). A household is thought to be "multiple deprived" or simply "poor" if the household deprivation score exceeds a given threshold (e.g., 1/3) (Alkire et al., 2015; Das et al., 2020). The following Eq. 4.13 estimates the headcount ratio (HCR):

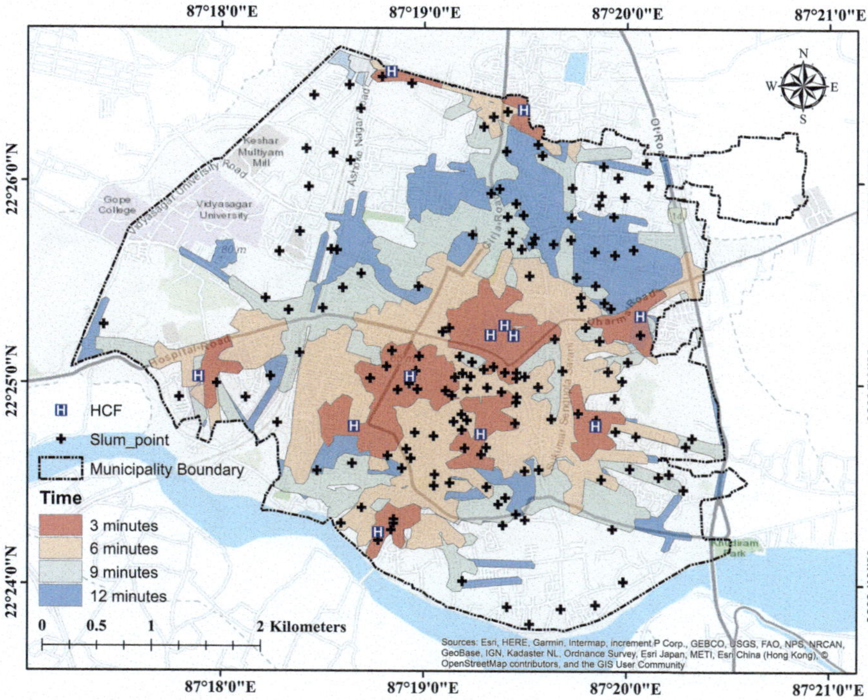

Fig. 4.22 Map showing the service area of HCF in the study area (Source: Author)

$$\text{HCR} = \frac{\text{Number of multidimensionally poor households}(q)}{\text{Total number of household}(n)} \quad (4.13)$$

Finally, the MPI is calculated by the product of the average number of multidimensionally poor households (A) and headcount ratio (HCR) designed (Alkire et al., 2015; Das et al., 2020) as (Eq. 4.14)

$$\text{MPI} = \text{HCR} \times A \quad (4.14)$$

The household data of slum neighborhoods in the Midnapore municipality has been used to find the MPI and mapped with ArcGIS v10.4, as shown in Fig. 4.24.

4.10.2.2 Site Suitability Analysis for New PHCF

Proper allocation of medical resources indicates developed societies and minimizes the health risks, cost of travel time, etc. Moreover, suitable healthcare site selection helps to augment the provision, distribution, and accessibility of healthcare

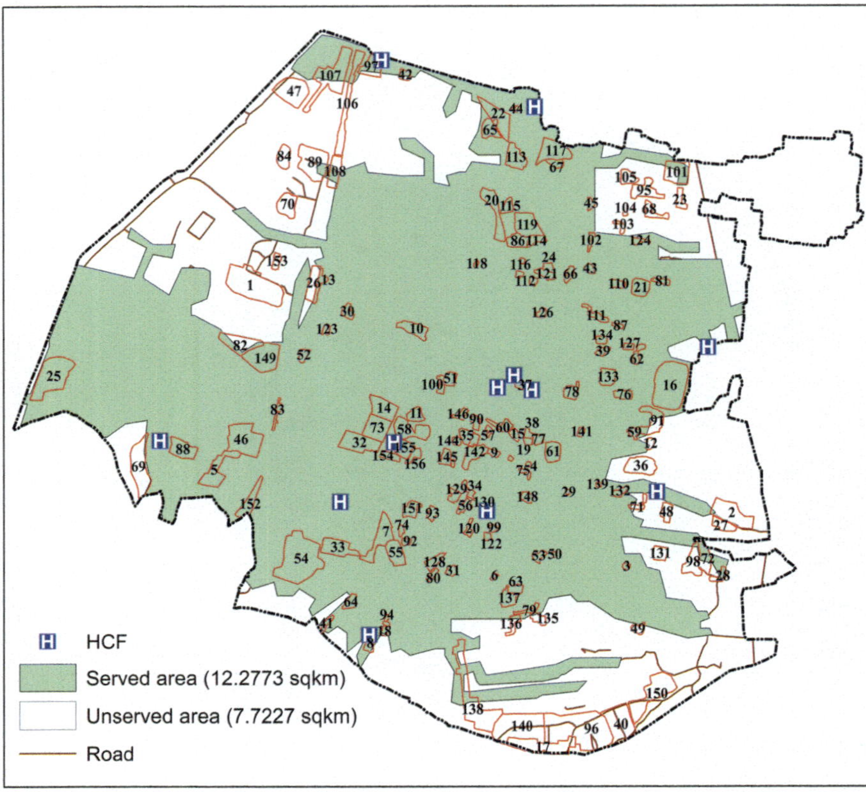

Fig. 4.23 Map showing the total served and unserved area by HCF in the study area (Source: Author)

facilities, which in turn optimize the healthcare services and planning processes (Zhou & Wu, 2012; Wang et al., 2018). Geospatial site selection (GSS) is a framework that finds suitable sites considering space availability and suitability (Malczewski, 2006; Abujayyab et al., 2016) and significant issues, for example, technical, environmental, physical, social, and many others (Lipton et al., 2000; Parry et al., 2018; Al-Shalabi et al., 2006). In this context, remote sensing, GIS, GPS, AHP, and statistical models are acting as important tools (Shukla et al., 2017; Sandipan et al., 2013; Parry et al., 2018). The study employed a multi-criteria decision support system (MCDSS) (Ahmed et al., 2016) to identify optimal locations for new primary healthcare centers (PHCs). This approach combined geographic information system (GIS) analysis and the ordinary least squares (OLS) statistical method with the analytical hierarchy process (AHP) model, addressing the shortage and improper distribution of healthcare service sites. According to the spatial analysis, a set of criteria (Fig. 4.25) concerning geophysical and socioeconomic variables such as (a) distance to slum areas, (b) distance to poverty-affected areas, (c) distance to population density areas, (d) distance to existing HCF, (e) distance to major roads have been considered for finding suitable sites for the new healthcare center.

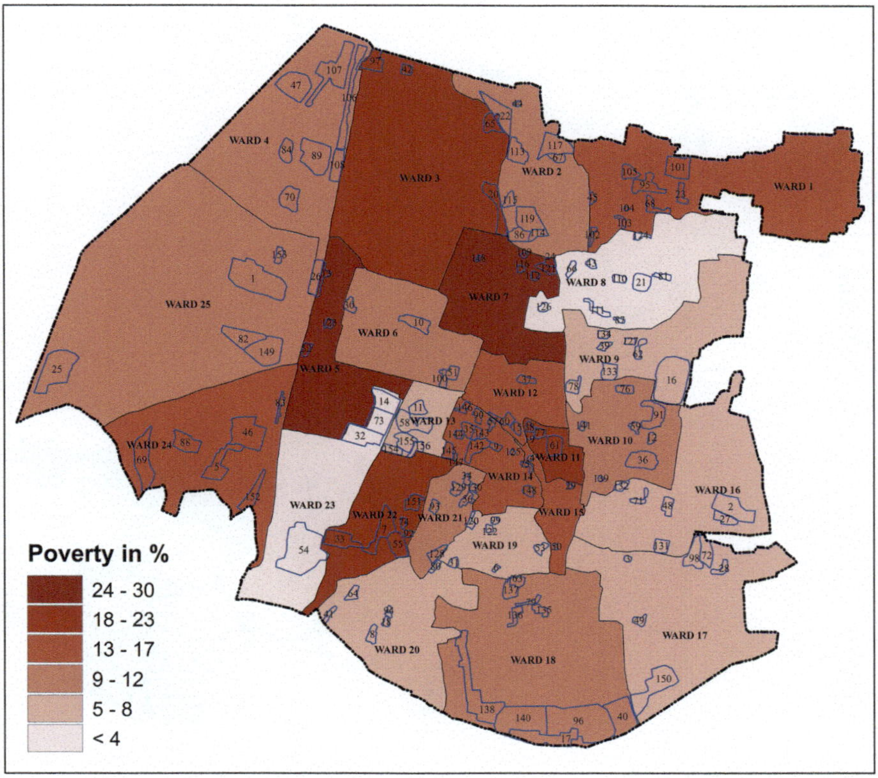

Fig. 4.24 Shows the poverty condition of each ward of the study area (Source: Author)

4.10.2.3 Elimination of Constraint Area

The land or areas of cities preoccupied with central and state government offices, universities and colleges, police headquarter areas, settlement areas, industries, recreational parks, etc. are known as constraint area. In the constraint areas, space is unavailable for the construction of any developmental project, such as a healthcare center. Therefore, these areas are not considered for site selection identified using the erase tool in ArcGIS v10.4 (Fig. 4.26).

4.10.2.4 Weight Calculation Based on OLS Technique

Ordinary least squares are applicable when many explanatory variables are in observations to determine the unknown parameters based on finite variances. The study has employed ordinary least squares (OLS) to explore the relationship between existing healthcare centers with other criteria, such as distance to slum areas, distance to poverty areas, distance to population density areas, distance to major roads,

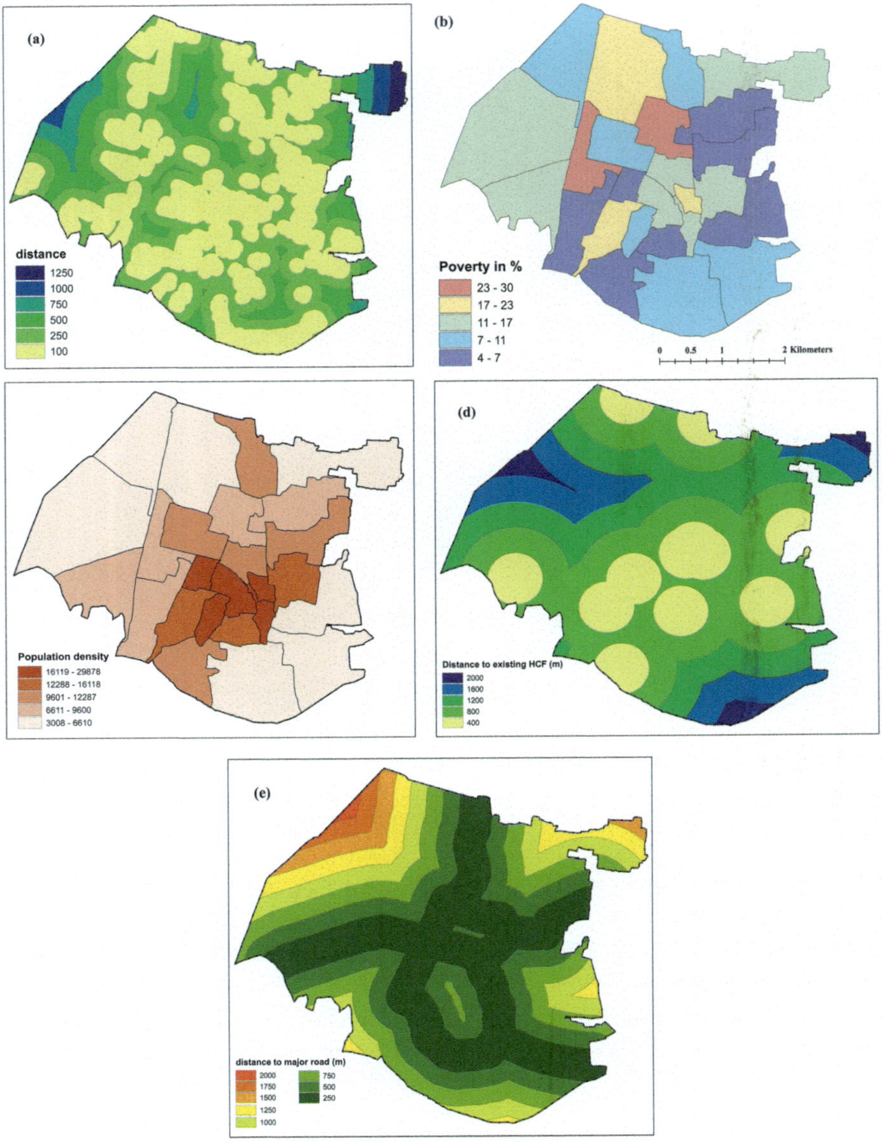

Fig. 4.25 Factors map for predicting suitable area and establish new HCF through AHP and OLS modelling (**a**) Distance to slum area (**b**) Distance to poverty affected area (**c**) Distance to population density areas (**d**) Distance to existing HCF (**e**) Distance to the major road network

and distance to existing HCF (Table 4.6). The coefficients of each criterion were estimated according to the mean distance between individual healthcare centers and other healthcare centers to find the effects of various criteria on existing healthcare centers (Table 4.7) (Abdullahi & Pradhan, 2012).

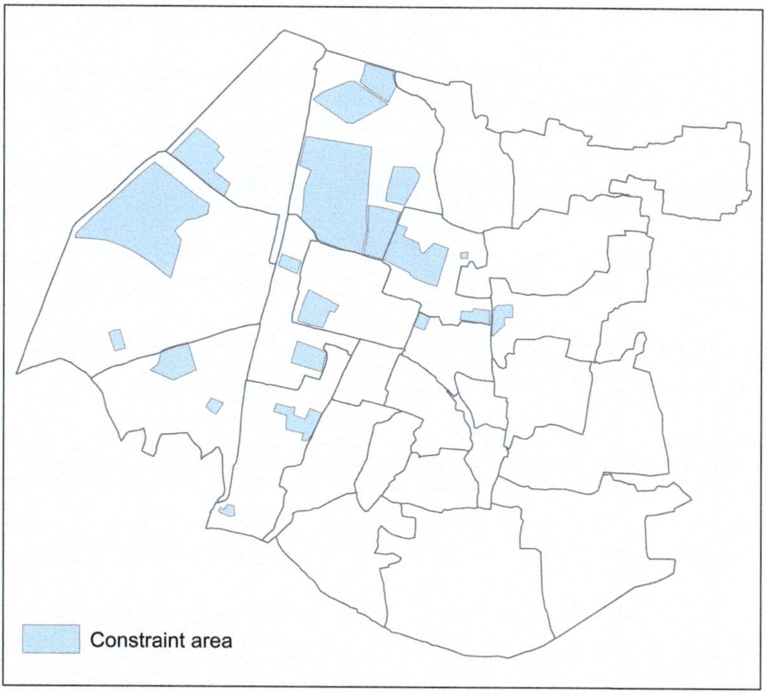

Fig. 4.26 Constraint zones in the study area

4.10.2.5 Estimation of Weights Using Analytical Hierarchy Process (AHP) Model

The analytical hierarchy process (AHP) was developed by Thomas Saaty in late 1970 through the framework of multi-criteria decision-making (MCDM) as a pair-wise comparison (Saaty, 1994) (Tables 4.8 and 4.9) technique by allocating various weights and scores based on his scale (Table 4.10) (Saaty, 1980; Abdullahi et al., 2013). Siddiqui et al. (1996) first attempted to use the AHP model with GIS in medical geography for site selection (Ahmed et al., 2016), including both subjective and objective facets of a decision. GIS, SPSS, and Microsoft Excel help to identify the different criteria, and analytical hierarchy processes (AHP) and ordinary least squares (OLS) generate weight used for mapping. The maps were developed in the GIS platform and classified from 1 (very poor, S1) to 4 (most suitable, S4) using two models (Fig. 4.27). Both maps reveal that only 2–3% of the area is most suitable (S4), 27% of the area is suitable (S3), and about 6% of the area is unsuitable (S1). However, about 65% area within the study area is not significant for the sites of PHCs due to the influence of existing healthcare centers. It is clear from the overall study that existing hospitals and PHCs are not properly distributed, as well as suitability maps indicating the northeastern and southeastern parts represent an area of high suitability (S4) and a narrow belt of the middle and northeastern part of the

Table 4.6 Average distance of each HCF to other HCFs and other criteria based on the road network

HCF	Distance to mean center	Distance to existing slum areas (a)	Distance to poverty-affected areas (b)	Distance to population density areas (c)	Distance to existing HCF (d)	Distance to major road (e)
Midnapore Medical College and Hospital	2000	2390	3150	2250	1900	650
Midnapore Homeopathic Medical College and Hospital	1950	2150	3146.6	2300	2050	700
Indian Red Cross Society	1780	2400	3000.6	2000.5	2180	500
Saratpally Urban Primary Health Center	2400	2000	2500	2100	2400	500
BerBallarpur Urban Primary Health Center	2800	2100	2100	2250	2130	500
Dharma Ramkrishnanagar Sub Health Center	2900	2200	2600	2400	2450	750
Udaypally Social Service Center	2600	2270.5	2400	2150	2500	900
Kuikota Urban Primary Health Center	2700	2390.8	2600	2500	2480	800
Vidyasagar Institute of Health	2050	2560.5	2900	2450	2070	700
BoxiBazar Saheb pukur Sub Health Center	1865.4	2000	2750.3	2150	2450	650
Palco Hero Pathanmohalla Urban Primary Health Center	1206.3	2000.6	2800	2280	1980	500
Najarganj Health Post	2800.5	2400	2950	2100	2450	600

Source: Author(s)

town representing moderately suitable (S3) areas for building new PHCs. Furthermore, there is no significant difference between AHP and OLS results (Table 4.11 and Fig. 4.28).

The town's settlement areas have grown rapidly since 1991 in different directions. Dinda et al. (2018) estimated the construction area of the town in 1991 was 15.29 km^2, which grew in 2001–20.10 km^2 with an advance rate of 31.48%, though it was high in the period between 2001 and 2017, which is about 35.39%. This

Table 4.7 Weight calculation based on the OLS technique

Criteria	Coefficients	Saaty's scale weight
Distance to existing HCF	0.472	5
Distance to existing slum areas	0.412	4
Distance to major road	−0.031	1
Distance to population density areas	0.106	1
Distance to poverty-affected areas	−0.486	1

Source: Author(s)

Table 4.8 Pair-wise comparison matrix

Criteria	Distance to existing HCF	Distance to existing slum areas	Distance to major road	Distance to population density areas	Distance to poverty-affected areas
Distance to existing HCF	1	3	3	3	2
Distance to existing slum areas		1	2	2	2
Distance to major roads			1	3	2
Distance to population density areas				1	2
Distance to poverty-affected areas					1

Source: Author(s)

Table 4.9 Weights are calculated by the authors

Criteria	Normalized value	Saaty's scale weight
Distance to existing HCF	0.387907	**4**
Distance to existing slum areas	0.209253	**2**
Distance to major road	0.176037	**2**
Distance to population density areas	0.12067	**1**
Distance to poverty-affected areas	0.106133	**1**

Source: Author(s)

Table 4.10 Saaty's scale (1980)

Weighted value	Description
1	Equal importance
2	Equal to moderate importance
3	Moderate importance
4	Moderate to strong importance
5	Strong importance
6	Strong to very strong importance
7	Very strong importance
8	Very strong to Extremely importance
9	Extremely importance

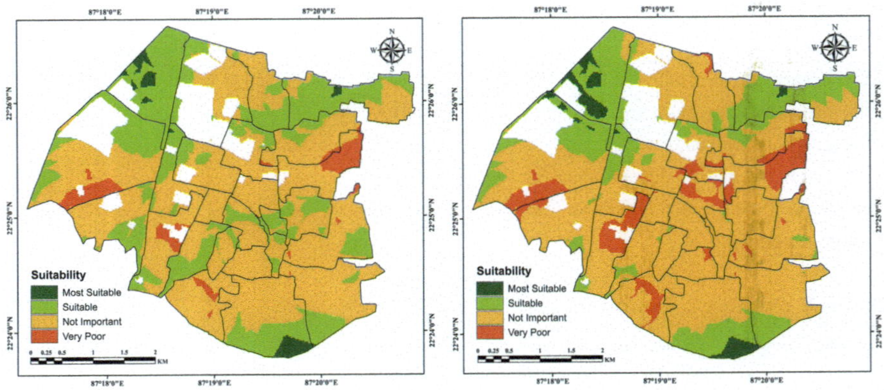

Fig. 4.27 Result of predicted suitable area for establishing new HCF by AHP and OLS model

Table 4.11 Spatial comparison of AHP and OLS results

Total area without constraint = 17.48 km^2					
Constraint area = 2.52 km^2					
AHP			OLS		
Suitability	Area (km^2)	%	Suitability	Area (km^2)	%
Poor (S1)	0.683696	4	Poor (S1)	1.451158	8
Not important (S2)	10.889701	62	Not important (S2)	11.721851	67
Suitable (S3)	5.626937	32	Suitable (S3)	3.855433	22
Most suitable (S4)	0.284803	2	Most suitable (S4)	0.454504	3

Source: Author(s)

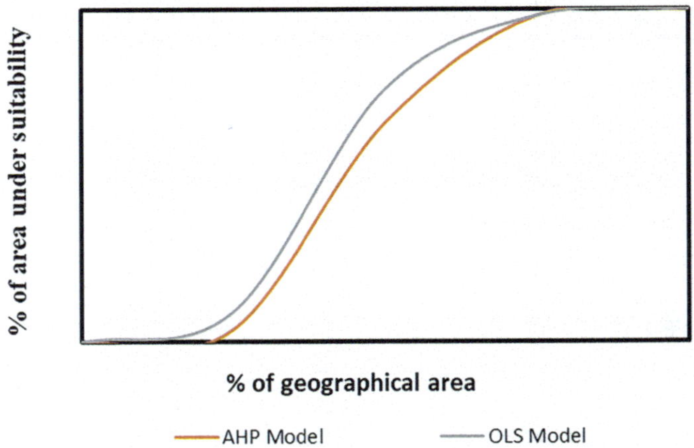

Fig. 4.28 Comparison of AHP and OLS results

growing trend is an indication of the growing population, especially in the new vacant areas, the so-called slum areas that demand healthcare services. Therefore, a suitability map can help the urban planner meet the basic healthcare needs of the urban poor in the study area and elsewhere.

4.11 Conclusion

Healthcare infrastructure and provision of the healthcare system are the basic amenities for sustainable development that can improve health and wealth. Additionally, the healthcare system can be enhanced through robust infrastructure, including both physical and nonmedical equipment, transportation, skilled personnel, and advanced technology. Furthermore, the availability of essential healthcare resources—such as funding, trained staff, medications, medical equipment, doctors, and nurses—is crucial for delivering effective healthcare interventions in accordance with WHO guidelines. Healthcare service varies from place to place depending on geographic accessibility, cost, facility availability, quality, etc. The slum neighborhoods of Midnapore municipality mostly choose government healthcare facilities due to healthcare expenditure. In the geospatial analysis of existing healthcare facilities and slum neighborhoods, the result shows that most of the slum neighborhoods in the periphery of Midnapore municipality are out of coverage from existing healthcare facilities. After finding the out-of-coverage slum neighborhoods, a site suitability analysis was performed to establish a new primary healthcare center regarding the hierarchy of the healthcare system, which can remove healthcare disparities and reduce health-related problems (Fig. 4.29).

Fig. 4.29 Major Primary Healthcare Centers (PHC) in Midnapore municipality. (**a**) Saratpally Urban Primary Health Centre (**b**) BerBallabpur Urban Primary Health Centre (**c**) Kuikota Urban Primary Health Centre (**d**) Sahebpukur Sub Health Centre (**e**) Najarganj Health Post (**f**) Ramkrishnanagar Sub-Health Centre (**g**) Pathanmohalla Urban Primary Health Centre (**h**) Udaypally Social Service Centre

References

Abdullahi, S., & Pradhan, B. (2012). Spatial modeling of site suitability assessment for hospitals using geographical information system-based multicriteria approach at Qazvin city, Iran. July 2015. https://doi.org/10.1080/10106049.2012.752531.

Abdullahi, S., & Ahmad, S. (2013). An analytical hierarchy process (AHP) based multi-criteria decision making approach for landfill site selection: A case study of Kuala Lumpur. *International Journal of Environmental Science and Development, 2*(1), 1–10.

Abujayyab, S., et al. (2016). GIS modelling for new landfill sites: Critical review of employed criteria and method of selection criteria. *IOP Conference Series Earth and Environmental Science, 5*(2), 101–118. https://doi.org/10.1088/1755-1315/37/1/012053

Ahmed, A. H., Mahmoud, H., & Aly, A. M. M. (2016). Site suitability evaluation for sustainable distribution of hospital using spatial information technologies and AHP: A case study of Upper Egypt, Aswan City. https://doi.org/10.4236/jgis.2016.85048.

Alabi. M. O. (2011). Towards a sustainable distribution of health centers using GIS. American Journals of Tropical Medicine and Public Health.

Alkire, S., et al. (2015). *Multidimensional poverty measurement*. Oxford University Press.

Al-Shalabi, M. A., Mansor, S. B., Ahmed, N. B., & Shiriff, R. (2006). GIS based multicriteria approaches to housing site suitability assessment. Paper presented at the XXIII FIG Congress, Shaping the Change, Munich, Germany.

Arcury, T. A., Preisser, J. S., Gesler, W. M., & Powers, J. M. (2005). Access to transportation and health care utilization in a rural region. *Journal of Rural Health, 21*(1), 31–38. https://doi.org/10.1111/j.1748-0361.2005.tb00059.x.

Cavapozzi, D., Han, W., & Miniach, R. (2015). Alternative weighting structures for multidimensional poverty assessment. *Journal of Economic Inequality, 13*(3), 425–447.

Census of India 1961. (1961, 2011).

Chahal, H., & Mehta, S. (2013). Modeling patient satisfaction construct in the Indian health care context. *International Journal of Pharmaceutical and Healthcare Marketing, 7*. https://doi.org/10.1108/17506121311315445

Christaller, W. (1966). *Central places in Southern Germany*. Prentice-Hall.

Danjuma, I. (2015). Locational analysis of healthcare facilities in Keffi local government area of Nasarawa State, Nigeria. Ph.D. Thesis. http://kubanni.abu.edu.ng/jspui/bitstream

Das, P., Paria, B. & Firdaush, S. (2020). Consumption poverty and multidimensional poverty: A study in Indian context, Social Indicators Research. https://doi.org/10.1007/s11205-020-02519-0.

Department of Health & Family Welfare (India). (2013). National health profile 2013. Central Bureau of Health Intelligence.

Dinda, S., Chaterjee, N. D., Das, K., & Ghosh, S. (2018). *Integration of GIS and statistical approach in the mapping of urban sprawl and predicting future growth in Midnapore town*. Springer.

Dutta, B., Das, M., Roy, U., & Rath, S. (2021). Spatial analysis and modelling for primary healthcare site selection in Midnapore town, West Bengal. *Geo Journal, 87*(6), 4807–4836. https://doi.org/10.1007/s10708-021-10528-w

Dutta, B., Roy, U., & Maity, I. (2018). Evaluating coping mechanism of slum dwellers in Midnapore Municipal Area, West Bengal. *Asian Journal of Multidimensional Research*.

Goel, S., Sharma, D., Bahuguna, P., Goel, S., & Singh, A. (2014). Predictors of patient satisfaction in three tiers of health care facilities of North India. *Journal of Community Medicine & Health Education*. https://doi.org/10.4172/2161-0711.S2-002

Jamal, I. A. (2016). Multi-criteria GIS analysis for school site selection in Gorno-Badakhshan Autonomous Oblast, Tajikistan. 55.

Kim, Y., Byon, Y., & Yeo, H. (2018). Enhancing healthcare accessibility measurements using GIS : A case study in Seoul, Korea. 1–19.

Kumar, S., Haque, A., & Tehrani, H. Y. (2012). High satisfaction rating by users of private-for-profit healthcare providers-evidence from a cross-sectional survey among inpatients of a private tertiary level hospital of North India. *North American Journal of Medical Sciences, 4*, 405–410.

Lawal, A. A., & Anyiam, F. E. (2019). Modelling geographic accessibility to primary health care facilities: Combining open data and geospatial analysis. *Geo-spatial Information Science, 22*(3), 174–184. https://doi.org/10.1080/10095020.2019.1645508

Lipton, R. B., Hamelsky, S. W., Kolodner, K. B., Steiner, T. J., & Stewart, W. F. (2000). Migraine, quality of life, and depression: A population-based case–control study. *Neurology, 55*(5), 629–636. https://doi.org/10.1212/WNL.55.5.629.

Malczewski, J. (2006). GIS-based multi-criteria decision analysis: A survey of the literature. *International Journal of Geographical Information Science, 20*(7), 703–726.
Mansour, S. (2016). Geo-spatial information science spatial analysis of public health facilities in Riyadh Governorate, Saudi Arabia : A GIS-based study to assess geographic variations of service provision and accessibility. *Geo-spatial Information Science, 5020*, 1–13. https://doi.org/10.1080/10095020.2016.1151205
Midnapore Municipality. (2022). Ward Services. https://midnaporemunicipality.com/.
Ministry of Health and Family Welfare, Government of India. (2013). National Urban Health Mission: Framework for implementation. https://nhm.gov.in/images/pdf/NUHM/Implementation_Framework_NUHM.pdf.
Ministry of Health and Family Welfare (India). (2011). India family welfare statistics 2011.
Murseli, R., & Isufi, F. (2014). Land suitability location analysis for housing development using GIS. Paper presented at the UBT International Conference, Durres, Albania. https://doi.org/10.33107/ubt-ic.2014.18.
Nemet, G. F., & Bailey, A. J. (2000). Distance and health care utilization among the rural elderly. *Social Science & Medicine, 50*(9), 1197–1208. https://doi.org/10.1016/S0277-9536(99)00365-2.
Parry, J. A., Ganaie, S. A., & Bhat, M. S. (2018). GIS-based land suitability analysis using AHP model for urban services planning in Srinagar and Jammu urban centers of J & K, India. *Journal of Urban Management, 7*(2), 46–56. https://doi.org/10.1016/j.jum.2018.05.002
Ravallion, M. (2011). On multidimensional indices of poverty. *The Journal of Economic Inequality, 9*(2), 235–248. https://doi.org/10.1007/s10888-011-9173-4.hdl:10986/3346
Saaty, T. (1980). *The analytical hierarchy process* (pp. 20–25). McGraw-Hill.
Saaty, T. (1994). Highlights and critical points in the theory and application of the analytic hierarchy process. *European Journal of Operational Research, 74*(3), 426–447.
Sandipan, D., Anirban, B., & Sagar, M. (2013). Study on urban land suitability assessment using remote sensing and GIS: A case study of Khairagarh, in Chhattisgarh. *International Journal of Computer Applications, 74*, 10.
Sedenu et al. (2016). Spatial analysis of the distribution pattern of primary healthcare facilities in spatial analysis of the distribution pattern of primary healthcare facilities in Ile—Ife metropolis using geographic information system. June. https://doi.org/10.9734/JSRR/2016/22339.
Shukla, A., Kumar, V., and Jain, K. (2017). Site suitability evaluation for urban development using remote sensing, GIS, and analytic hierarchy process (AHP). Proceedings of International Conference on Computer Vision and Image Processing. https://doi.org/10.1007/978-981-10-2107-7_34.
Siddiqui, M. Z., Everett, J. W., & Vieux, B. E. (1996). Landfill siting using geographic information systems: A demonstration. *Journal of Environmental Engineering, 122*(6), 515–523.
Wang, F., & Luo, W. (2005). Assessing spatial and nonspatial factors for healthcare access : towards an integrated approach to defining health professional shortage areas. 11, 131–146. https://doi.org/10.1016/j.healthplace.2004.02.003.
Wang, L., Fan, H., & Wang, Y. (2018). Site selection of retail shops based on spatial accessibility and Hybrid BP Neural Network. https://doi.org/10.3390/ijgi7060202.
World Health Organization. (2004). The world health report 2004: Changing history. https://www.who.int/publications/i/item/924156265X.
World Health Organization. (2010). The world health report 2010: Health systems financing—The path to universal coverage. https://www.who.int/publications/i/item/9789241564021.
World Health Organization. (2001). The world health report 2001: Mental health: New understanding, new hope. https://iris.who.int/handle/10665/42390.
World Health Organization. (2016). World health statistics 2016: Monitoring health for the SDGs, sustainable development goals. https://www.who.int/publications/i/item/9789241565264.
Zhou, J., & Daniel, I. (1996). Using genetic learning neural networks for spatial decision making in GIs. *Photogrammetric Engineering & Remote Sensing, 62*(11), 1287–1295.
Zhou, L., & Wu, J. (2012). GIS-based multi-criteria analysis for hospital site selection in Haidian District of Beijing.

Chapter 5
Utilization and Accessibility of Healthcare Services

5.1 Introduction

Health is an indispensable requisite for the standard of living and a better quality of life. Decent health and well-being reduces health expenditure and thereby maintains the standard of living. In this context, the utilization and accessibility of healthcare services are significant for those living in rural and urban slums with poor conditions. The utilization of healthcare services refers to households using healthcare services through the accessibility and affordability of those services to prevent health risks, care for health problems, and promote and maintain health, well-being, and prognosis. The utilization of healthcare services depends upon the frequency of use, timing, distance, cost, needs, types of illness and disability, availability of resources, behaviors, cultural beliefs, and socioeconomic status with the quality of the services (Carrasquillo, 2013). The utilization of healthcare services can be measured by the rate and percentage of utilization of healthcare services over time. The utilization of healthcare services is primarily determined by accessibility and affordability of the available services; thus, accessibility of healthcare is defined as the extent to which healthcare services are available and accessible to households. It essentially denotes the actual utilization of the available existing services. The accessibility of healthcare services also depends on the available transportation, time and distance, cultural taboos, available medical services, cost and willingness, etc. The accessibility of healthcare services may be geographical or spatial, financial, cultural, and potential, depending on the space-time matrix. The existing healthcare system in India fails to cover its population, particularly in rural and urban slum areas, and net underutilization. In urban slums, health infrastructure is inadequate, poorly distributed, and neglected from available healthcare services. Besides, due to a lack of awareness and poverty, the urban poor have the least access to available healthcare services, which varies in space and across the country. By examining the status of healthcare utilization and accessibility among slum

© The Author(s), under exclusive license to Springer Nature Switzerland AG 2025

B. Dutta et al., *Using GIS to Identify Opportunities for Improved Health in Urban Slums*, Global Perspectives on Health Geography,
https://doi.org/10.1007/978-3-031-89617-0_5

dwellers, we can see the demand for healthcare services in slum neighborhoods. Accordingly, this chapter aims to assess the status of healthcare utilization and accessibility among slum dwellers of Midnapore town and to identify its influential factors to determine the level of healthcare utilization and accessibility among these groups of people and improve healthcare delivery systems' performance.

5.2 Utilization of Healthcare Services (Medical Tests) and Supplements (Medicine)

Heath checkups and medical tests are common for every human body during illness and disability. Regular health check-ups during childhood, before marriage, and old age are crucial for preventive and promotive healthcare. Early screening for genetic, infectious, and non-communicable diseases supports timely intervention and improved prognosis. However, the place of health checkups or necessary medical tests varies, including accessibility to test facilities, distance to the adjacent laboratory, socioeconomic status, cost, cultural beliefs, drive-by doctors, etc. However, health checkups and medical tests are common in rich people, while poor people are not interested due to poverty, lack of awareness, superstition, and so many causes. In the study area, most of the slum dwellers chose blood tests (50.45%) (Fig. 5.1) and medicine collection (Tables 5.1 and 5.2) from the government hospital or healthcare center (50.24%) (Fig. 5.2) and also for their all medical test and medicine collection. Due to the cost of the test (49.62%) and lack of family awareness (32.12%) (Fig. 5.3 and Table 5.1), households are not willing to take medical tests before marriage and during illness.

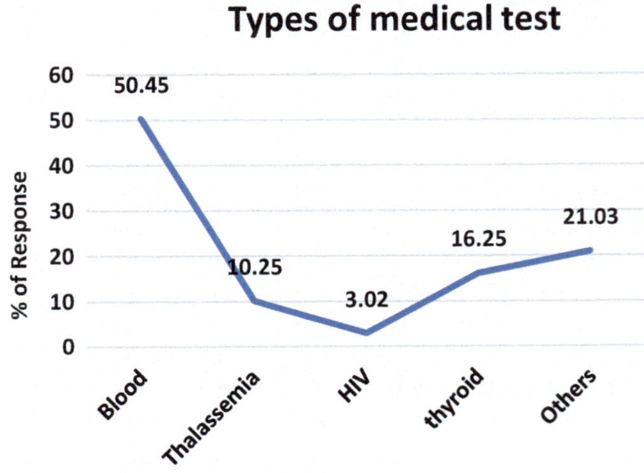

Fig. 5.1 Types of medical test (Field survey, 2022)

Table 5.1 Place of antenatal care, delivery care, postnatal care, immunization, early childhood care, birth certificate collection, sterilization, minor illness checkup, treatment for sick children and aged persons, treatment for other family members and chronic health problems, and collection of medicine

Place	Antenatal care	Delivery care	Postnatal care	Immunization	Early childhood care	Birth certificate collection	Sterilization	Minor illness checkup	Treatment for sick children	Treatment for aged person	Treatment for other family members	Chronic health problem	Collection of medicine
	Percentage of households												
Anganwadi/ICDS center	5.32	–	–	–	5.38	–	–	–	–	–	–	–	–
Pharmacy/drug stores/diagnostic center	–	–	–	–	–	–	–	15.38	–	–	2.38	–	14.68
Govt. hospital/trust hospital/urban health center/govt. mobile clinic/govt. dispensary/RCHCs	62.63	76.15	51.13	63.13	45.90	83.21	63.27	32.05	40.56	40.58	40.80	64.05	40.79
Pvt. clinic or nursing home/quack doctor dispensary	15.5	9.38	24.06	9.01	26.29	16.89	6.67	35.07	51.18	40.41	46.03	29.87	39.17
Vaidya/hakim/homeopath dispensary	–	–	–	–	4.25	–	–	10.24	5.13	4.65	5.52	2.70	5.36
Spiritual treatment center	–	–	4.69	1.25	1.25	–	4.88	2.38	–	2.23	–	–	–
Home treatment/parents' home	6.75	14.47	20.12	4.13	9.75	–	7.72	4.88	3.13	5	5.27	3.38	–
Not taken	–	–	–	–	5.18	–	17.46	–	–	7.13	–	–	–
Others (specify)	9.80	–	–	–	2.00	–	–	–	–	–	–	–	–

Source: Field survey (2022)

Table 5.2 Reason for choosing the place of antenatal care, delivery care, postnatal care, sterilization, and medicine collection

	Antenatal care	Delivery care	Postnatal care	Sterilization	Medicine collection
Reason	Percentage of households				
No charge	23.63	26.28	40.64	20.55	13.33
Low cost	40.64	26.92	23.63	20.45	20
Nearby	–	11.73	5.76	–	26.67
Standard facilities	23.97	22.12	23.97	25.26	26.67
Get JSY money (Rs. 6000/)/ sterilization money facility	–	–	-	13.62	–
Others	11.76	12.95	6.0	20.12	13.33

Source: Field survey (2022)

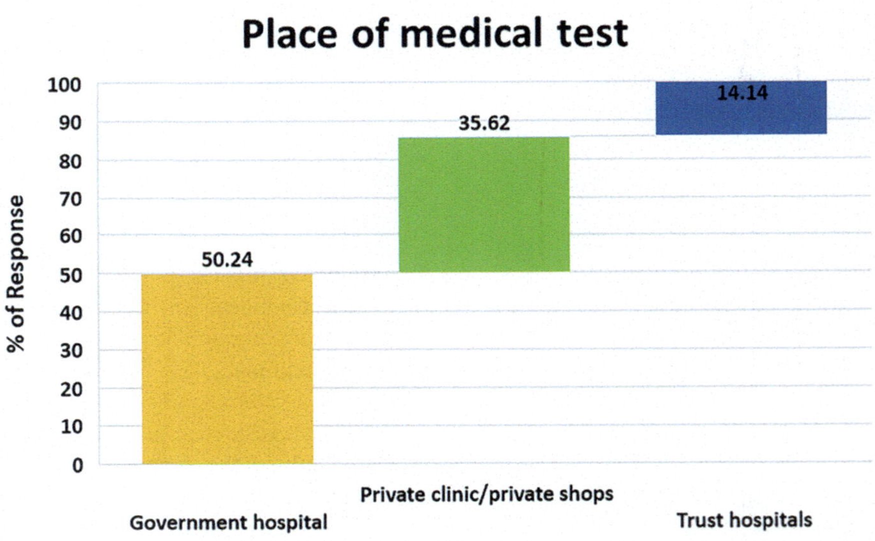

Fig. 5.2 Place of medical test (Field survey, 2022)

5.3 Women's Healthcare and Services

Women's health is an essential and foremost part of any country for their overall societal development from the perspective of demography and medical care. Based on the evidence, several possible relationships exist between women's status in the family and society and different dimensions of population dynamics, particularly fertility and mortality transition (Chaurasia & Gulati, 2013). Women's health broadly refers to the health during the menstruation period, contraceptive use, HIV

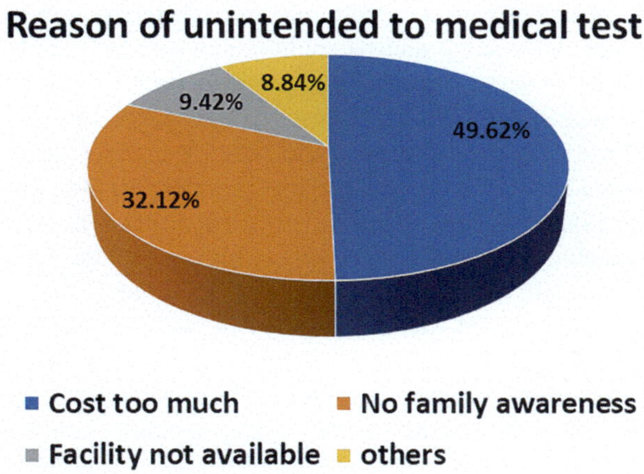

Fig. 5.3 Reason of unintended to medical test (Field survey, 2022)

care, pregnancy, delivery, and the postpartum period. Thus, women are associated with suffering from different diseases, illnesses, and even mortality from the menstruation stage to the postpartum period. Therefore, women's health is a prime concern of any nation. Hence, menstruation care, age at marriage, contraceptive care, family planning, antenatal care, place and assistance of delivery care, postnatal care, birth interval, socioeconomic status, occupation, literacy, etc. are the significant issues for assessing women's health status. Poor women are mostly suffering about their health and healthcare due to poor socioeconomic conditions, lack of awareness, cultural taboos, low level of literacy, lack of knowledge and family care, etc. Every government has taken many initiatives to improve healthcare for the urban poor. Nonetheless, urban poor women face many problems in terms of their health and healthcare. Women in poor health are more likely to give birth to low-weight infants. They also are less likely to be able to provide food (breastfeeding) and adequate care for their children. Moreover, a woman's health affects the household's economic well-being, as women in poor health will be less productive in the labor force. Therefore, the study was undertaken to assess the women's health status in poor neighborhoods of the town, considering the existing healthcare facilities.

5.3.1 Menstruation Care

Menstruation care is important for women's health because a lack of menstruation care may cause many gynecological problems and diseases, which may further create problems for their reproductive health, cancer, and even death (Kirk & Sommer, 2006). Therefore, special care is given to girls at their stage of puberty. However, menstruation is associated with sociocultural taboos in India. Sociocultural taboos and beliefs, literacy, knowledge and perception about puberty, and menstruation, economic status, awareness programs, etc. are also used to assess menstruation care. However, perceptions of menstruation care differ across rural and urban areas. For example, 57.6% of girls and women use sanitary napkins in India (NFHS-4, 2015–2016), while almost 77% of rural girls and women use old clothes and sometimes reuse them (SOS, 2015). Urban poor girls and women are also neglected in their menstruation care in India as they belong to the poor socioeconomic condition where meeting their daily livelihood is a prime concern, and menstruation care is less important. In this regard, overall pictures of the study area are shown in Figs. 5.4, 5.5 and 5.6. Still, 26.28% of girls and women used homemade cloth during their menstruation period living in slum neighborhoods due to the high cost of modern sanitary napkins. Those who used sanitary napkins mainly collected from private shops, private clinics (61.09% and 15.38%), and sometimes nearby PHCs (18.65%)

5.3.2 *Contraceptive Care*

Contraceptive care is most important in recent times because the lack of contraceptive care brings unwanted pregnancies, sexually transmitted infections (STIs) such as human immunodeficiency virus (HIV) or acquired immunodeficiency syndrome (AIDS), abortion, and other gynecological problems. From the perspective of demography, contraceptive care may play an essential role in the population composition of any country. Contraceptive care can protect against unwanted pregnancies and births by giving sexual pleasure. Therefore, any country has now given the focus on birth control through the awareness of contraceptive care. However, both

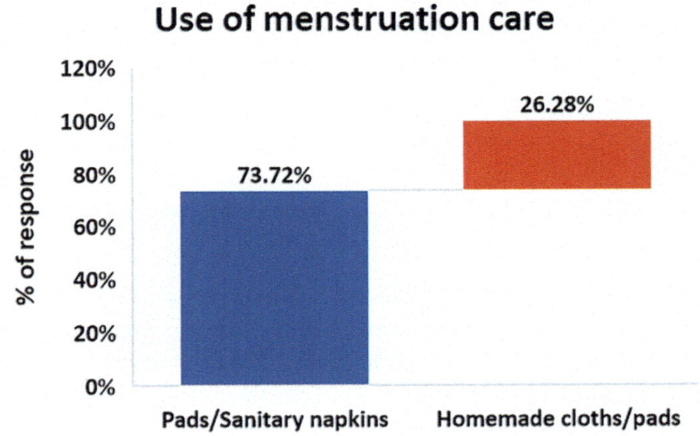

Fig. 5.4 Types of menstruation care used (Field survey, 2022)

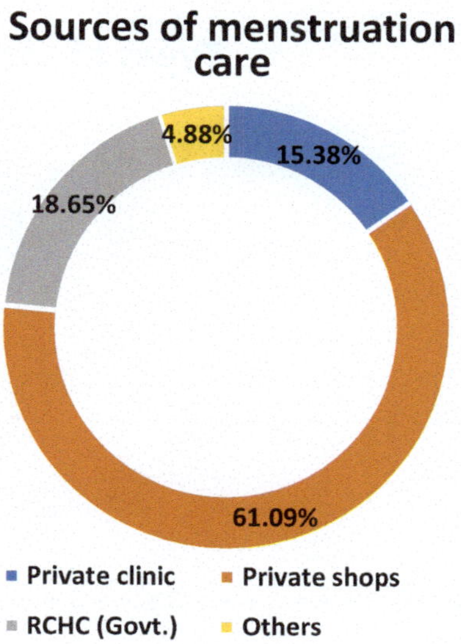

Fig. 5.5 Sources of menstruation care (Field survey, 2022)

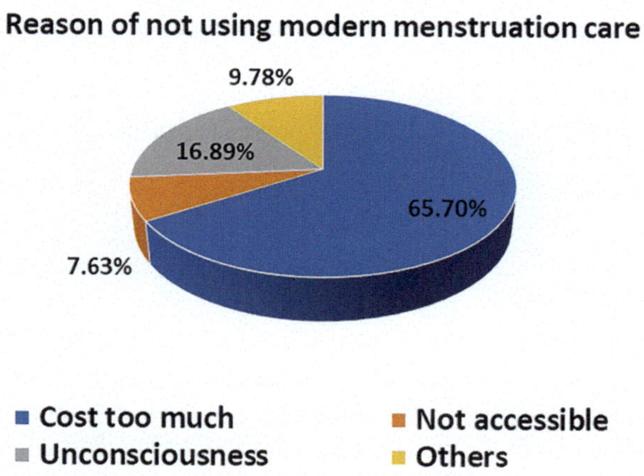

Fig. 5.6 Reason of unintended to use modern menstruation care (Field survey, 2022)

men and women are responsible for contraceptive care. Thus, socioeconomic status, knowledge and awareness, religion, ethnicity, cultural belief, education, and available contraceptive resources have an impact on the contraceptive care of a couple of households across rural and urban areas in India. It is also essential to understand that contraceptive care is not popular in rural and urban areas due to their socioeconomic and cultural backwardness. In India, 4.4% of women used pills, and 36.3% of females have been sterilized, which is higher in rural areas than in urban areas (NFHS-5, 2019–2020). The present study attempted to explore the nature of contraceptive care among slum dwellers in Midnapore. The study reported that dwellers mostly used pills (49%) collected from a private clinic or shop; sometimes, they

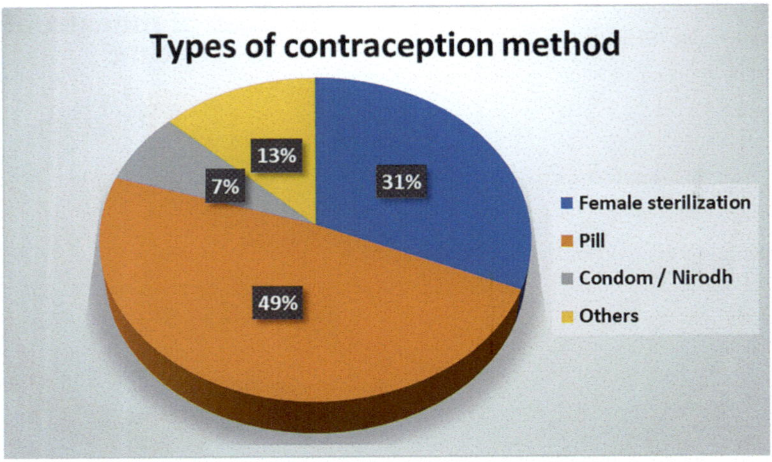

Fig. 5.7 Sources of collection for contraception (Field survey, 2022)

used unwanted pills collected from a private shop (Figs. 5.7 and 5.8) and sterilization (31%) for the prevention of unwanted pregnancies or births abundantly (Fig. 5.9). The study reveals that a few slum dwellers used condoms/Nirodh due to the high cost and shyness to collect from private shops. They also reported that they sometimes feel gynecological problems or pregnancy difficulties due to the use of the unwanted pill.

5.3.3 Antenatal Care

Antenatal care (ANC) is the treatment of married women during the whole pregnancy period up to the delivery. The importance of ANC is to prevent fetal death, miscarriage, difficulties in the pregnancy period and during delivery, as well as premature birth, stunted baby birth, and, finally, infant and maternal mortality during delivery. Therefore, the government has taken the initiative to develop ANC by developing PHCs and employing ASHA and other health workers to observe pregnant women and their problems up to delivery regularly. Nonetheless, people experiencing poverty are less aware and not given importance in the utilization of the ANC. Sometimes, they avoid or do not have information about ANC, rather entirely relying on local or quack doctors' treatment. Thus, maternal education, men participation, higher birth order, health insurance, teenage and intentional pregnancies, place of residence, caste, religion, ethnicity, cultural belief and superstition, as well as the use of ICDS services, etc., assessing the utilization of ANC across states of India. NFHS-5 data explore mothers who had at least four antenatal care visits in urban areas (68.1%) and mothers who had an antenatal checkup in the first trimester (75.5%) (NFHS-5, 2019–2020; Gunjan Kumar et al., 2019). In the study area, 60%

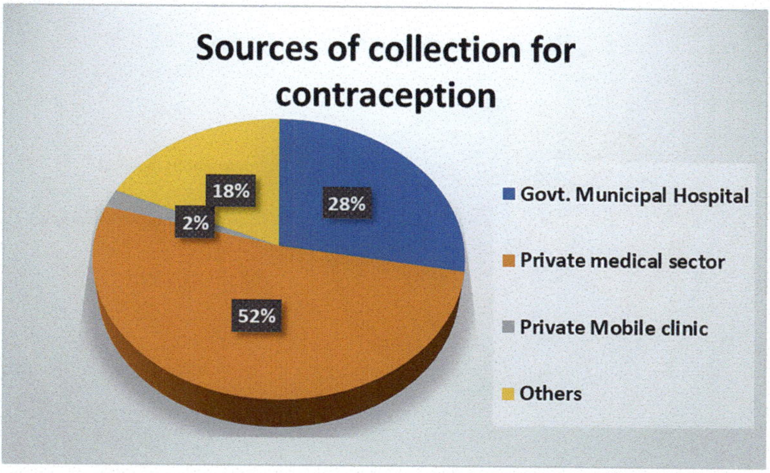

Fig. 5.8 Use of emergency contraception (Field survey, 2022)

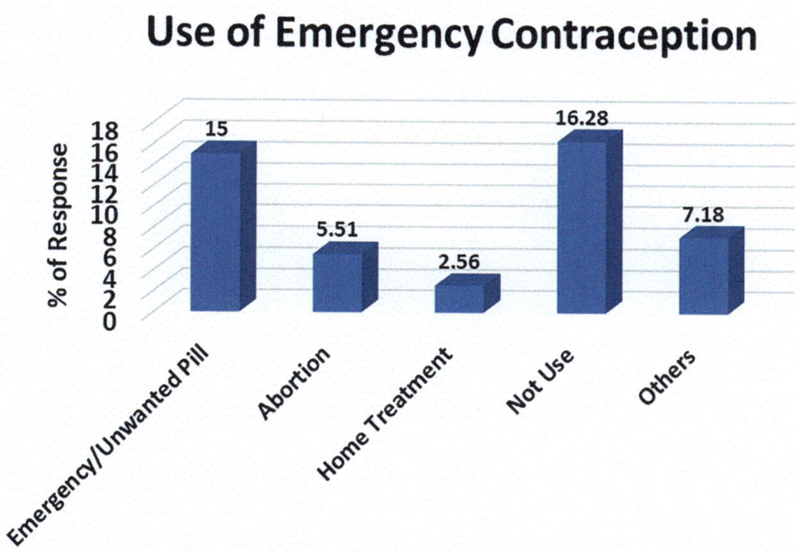

Fig. 5.9 Types of contraceptive method used (Field survey, 2022)

of women utilized ICDS services (Table 5.1), and 80% (Table 5.2 and Fig. 5.23) of women tested their pregnancy by Pregnancy Strip at Govt. hospital/Sub-center, and they mostly took ANC from Govt. Hospital/Trust Hospital/Urban Health Center/ Govt. Mobile Clinic/Govt. Dispensary/RCHCs (Fig. 5.10) due to low cost (Fig. 5.11). About 21.08% of women taking iron tablets, folic acid, combined iron and folic acid, weight, BP measured, urine sample, blood sample, abdomen check up, delivery date, place of delivery, nutrition chart and others facilities (Fig. 5.12).

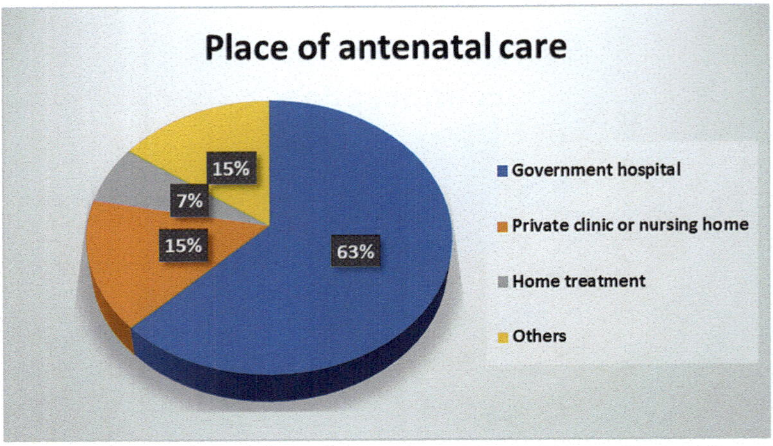

Fig. 5.10 Place of antenatal care (Field survey, 2022)

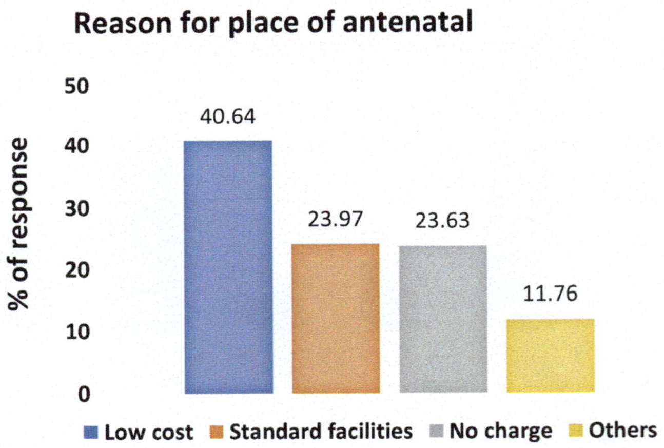

Fig. 5.11 Reason for place of antenatal care (Field survey, 2022)

Therefore, ANC is somewhat better in Midnapore town than the average ANC in India.

5.3.4 Delivery Care

Globally, the maternal and child death rate is high in India. Delivery care is an essential prime concern for reproductive and child health. The reproductive and child health program aims to encourage deliveries in proper hygienic conditions under the supervision of trained healthcare professionals. Therefore, if we want to

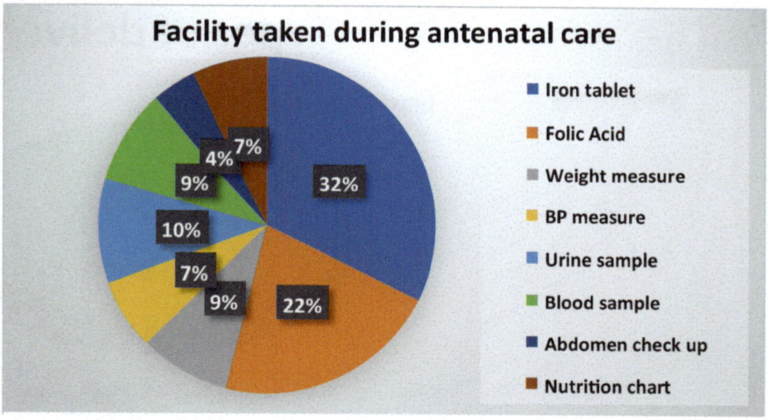

Fig. 5.12 Facility taken during antenatal care (Field survey, 2022)

reduce maternal and child deaths, then institutional delivery should be promoted. Selection of places of delivery, maternal and family educational level, and information regarding delivery care are the most critical factors for delivery care. The NFHS-5 report reveals that institutional delivery in urban India is 93.8%, with cesarean cases high in private healthcare centers. Still, 14.47% of births to the total take place at home.

The institutional delivery rate is relatively high (85.53%) in Midnapore urban areas. In the study area, 26.48% of women with normal delivery (Fig. 5.13) by the doctor (Fig. 5.14) at government hospitals/trust hospitals/urban health centers/government mobile clinics/government dispensaries/RCHCs (76.15%) (Fig. 5.15) due to no charge, i.e., free of cost (Fig. 5.16). Moreover, women have received Rs. 6000 under the JSY scheme; however, this amount has been considered sufficient by many, resulting in 60% of them opting not to use modern delivery care facilities (Tables 5.1 and 5.2). Various health institutes have emerged to offer delivery care to pregnant women. Assistance in delivery from trained personnel is available at a low cost in those institutes. In this respect, the Midnapore District Hospital, Red Cross Society's Hospital, urban community health center, Integrated Child Development Services (ICDS), public health center (PHC), paramedical institute, and low-cost maternity clinics play a significant role.

5.3.5 *Postnatal Care*

The health of the mother and her newborn child depends not only on the healthcare received during her pregnancy and delivery but also on the care she and the infant receive during the first few weeks after delivery. On the importance of postnatal checkups, the reproductive and child health (RCH) program recommends three postnatal visits at 2, 4, and 6 weeks after delivery for both the mother and her child

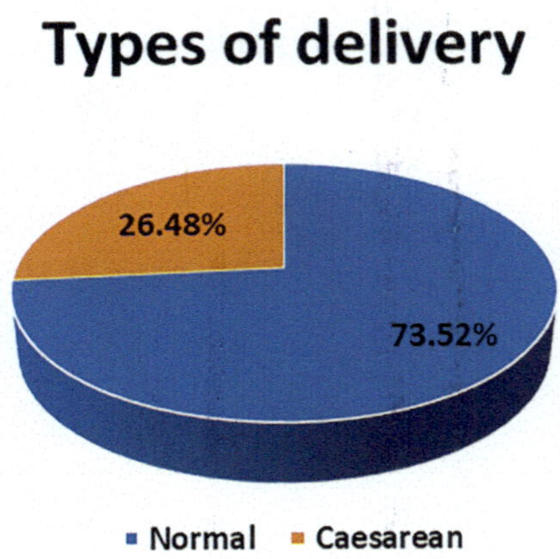

Fig. 5.13 Types of delivery for birth (Field survey, 2022)

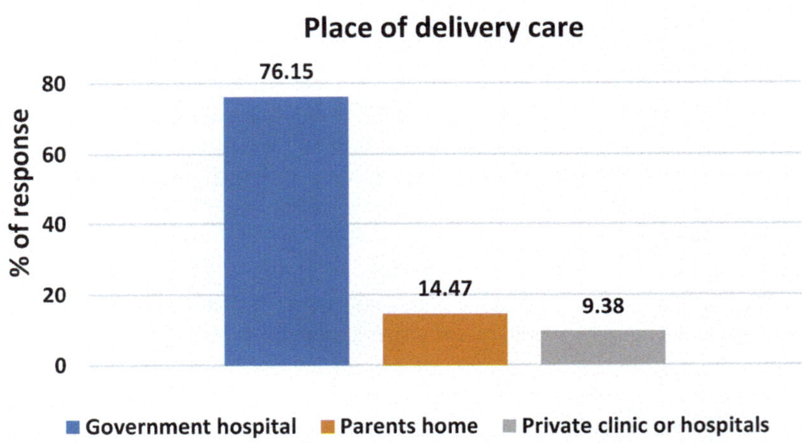

Fig. 5.14 Health personnel for delivery (Field survey, 2022)

for weight checks and immunizations (Ministry of Health and Family Welfare, 2020). After delivery, maternal health complications arise in most cases, as well as children who are getting problems due to complications of maternal health, premature birth, respiratory problems, and other genetic problems, which need to be taken care of at an institutional healthcare center with a gynecologist and pediatrician. Postnatal checkups are essential in reducing maternal health and child problems, morbidity, and mortality. Regular postnatal checkups also reduce the extra burden of further healthcare expenditure, especially for the poor neighborhoods. Therefore, an institutional healthcare center is a significant provision in this context. It is found that 54.76% of households use a private clinic or nursing home/quack doctor dispensary and only 38.15% of households use a government hospital/trust hospital/urban health center/government mobile clinic/government dispensary/RCHCs for postnatal care (Table 5.1 and Fig. 5.17). The reasons behind the choice of place of postnatal care, whether private or government institution, reported by the slum

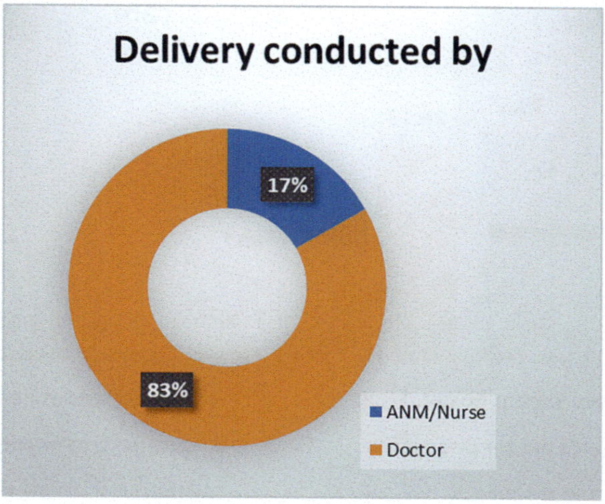

Fig. 5.15 Place of delivery care (Field survey, 2022)

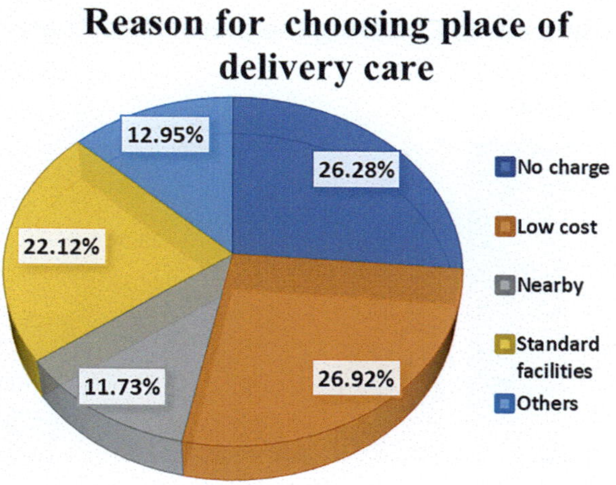

Fig. 5.16 Reason behind the place of delivery care (Field survey, 2022)

dwellers, are mainly free healthcare services, low cost, and standard facilities given by the institutions (Figs. 5.18 and 5.23 and Table 5.2).

5.4 Child Healthcare and Services

Despite the growth of the urban population and economy in India, poverty, malnutrition, poor sanitation, and urban slum development are significant constraints to the development of proper healthcare services, especially for the growing child

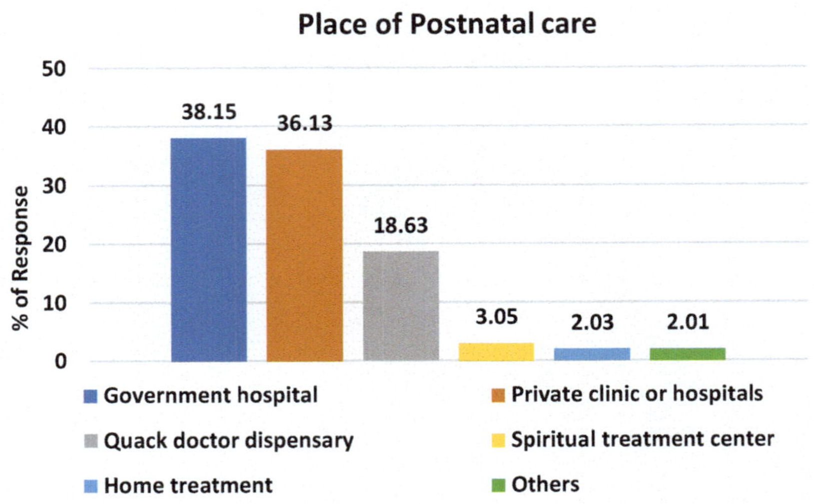

Fig. 5.17 Place of postnatal care (Field survey, 2022)

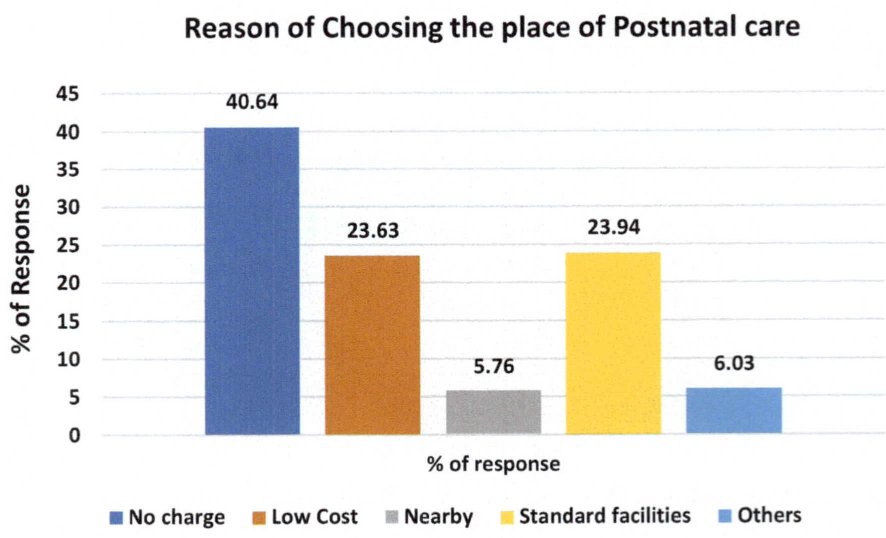

Fig. 5.18 Reason behind the choosing of place of postnatal care (Field survey, 2022)

population. The inadequate healthcare accessibility led to health risks to the newborn and her/his mother. Child immunization and proper nutrition are most important for newborn babies after a few weeks, which may reduce the child mortality rate and the mother's health problems. The government of India has introduced a reproductive and child health (RCH) program launched in 1996 to provide essential services to pregnant women and their children. However, still, now child mortality, malnutrition, and stunted children are high in India, between 30% and 40% (NFHS-5, 2019–2020). Therefore, it is most important to focus on child healthcare services and immunization received and the place or institution from which they access these services. The health personnel and pediatrician have recommended the following immunizations after 6 weeks of newborn babies:

- DPT (first dose).
- HIB (first dose).
- Oral polio vaccine (second dose).
- Hepatitis B (second dose).
- Pneumococcal vaccine (first dose).
- Rotavirus (first dose).

After 10 weeks, babies should have the following immunizations:

- DPT (second dose).
- HIB (second dose).
- Oral polio vaccine (third dose).
- Rotavirus (second dose).

After 14 weeks, babies should have the following immunizations:

- DPT (third dose).
- HIB (third dose).
- Oral polio vaccine (fourth dose).
- Rotavirus (third dose).
- Pneumococcal vaccine (second dose).

After 20 weeks, babies should have the following immunization:

- Pneumococcal vaccine (third dose).

After 6 months, babies should have the following immunization:

- Influenza vaccine.

The study collected data on common child immunizations known to common people. All vaccination received by children in the slum of Midnapore town is only 21.4% in which polio, BCG, and DPT are common (Fig. 5.19) from government hospitals/trust hospitals/urban health centers/government mobile clinics/government dispensaries/RCHCs (81.44%) as shown in Fig. 5.20. Table 5.1 and Figs. 5.21 and 5.22 show the overall childcare pictures of the study area (Fig. 5.23). The Department of Women and Child Development within the Ministry of Human Resource Development initiated the Integrated Child Development Services (ICDS) in 1976. Under the ICDS program, Anganwadi centers provide children with health, nutrition, and education services from birth to 6 years of age, as well as nutritional and health services for pregnant and breastfeeding mothers. The town of Midnapore has 102 ICDS (see Fig. 4.5) centers, and 4–5 ICDS centers have been established within 1 municipal ward. It reveals quite a better situation across the neighborhoods in the study area. At the same time, the immunization rate did not reach the mark due to a lack of awareness, childcare knowledge, etc.

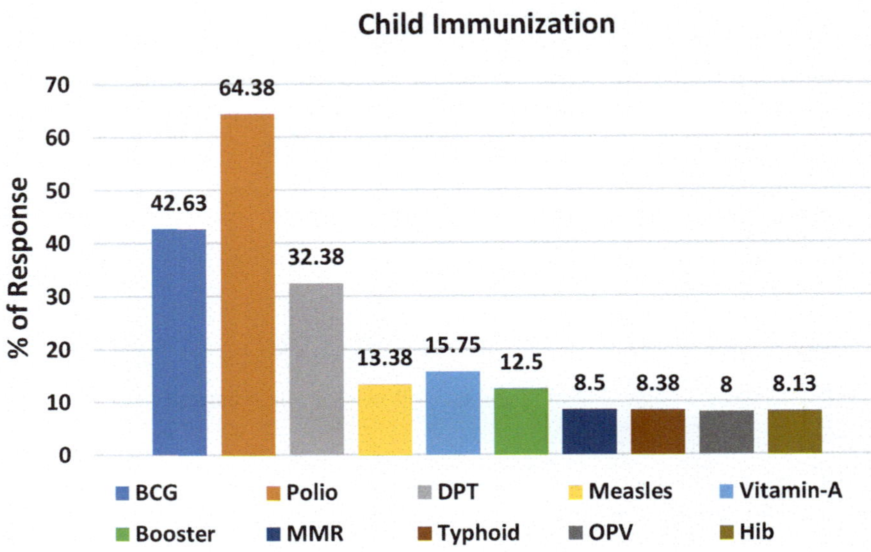

Fig. 5.19 Child immunization taken (Field survey, 2022)

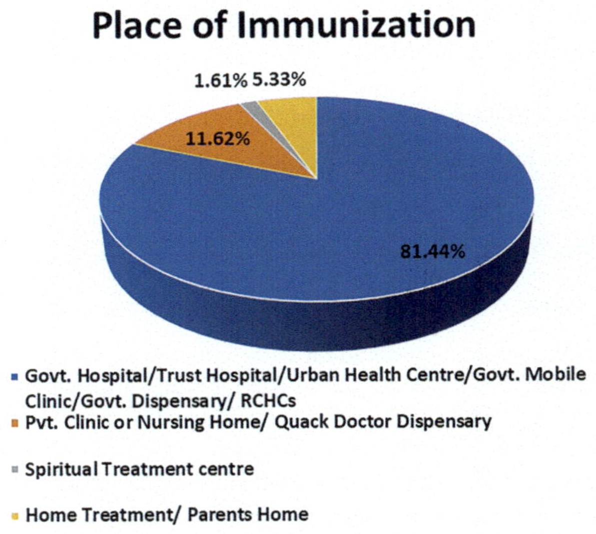

Fig. 5.20 Places of taken for overall family healthcare facilities (Field survey, 2022)

5.5 Family Planning and Family Planning Participation

Family planning means couples want to anticipate and achieve their desired number of children with the spacing and control of their births through the practices of contraceptive methods. Moreover, family planning depends on couples' decision and their voluntary participation. In India, awareness about contraception and family

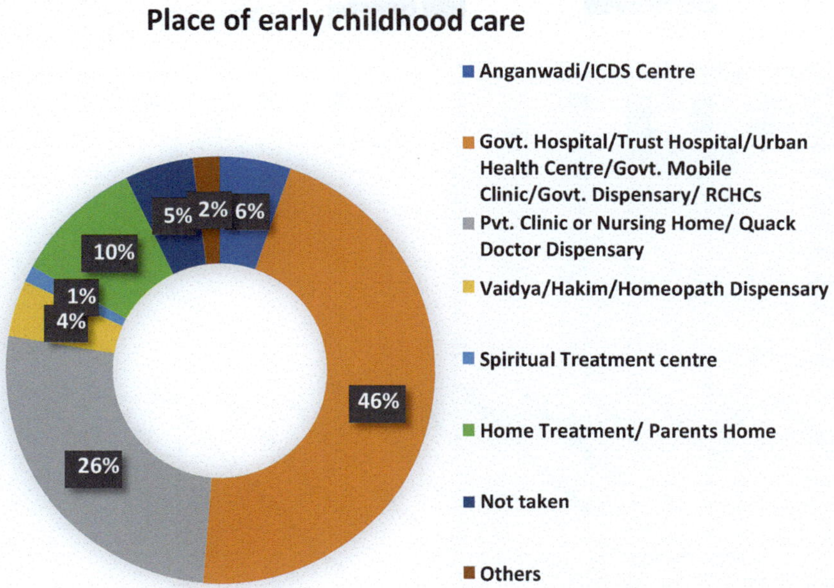

Fig. 5.21 Places of early childhood care (Field survey, 2022)

planning is not at a significant level. The NFHS-5 (2019–2020) report reveals that married women have mostly preferred sterilization (36.30%) in urban areas for the birth limits over birth planning. Family planning decision with birth spacing has a direct impact on the health of married women for their pregnancy outcome and child.

In the study area, the dwellers have positive intentions toward family planning and birth control (Fig. 5.24) but lack of desire to use family planning methods. For example, most slum married women have used contraceptive pills (49%), and sterilization (around 31%) is mainly preferred in governmental institutions (63.13%; see Table 5.1) voluntarily or sometimes involuntarily due to low socioeconomic status and related problems. Birth spacing or interval is satisfactory among the slum married couples (Fig. 5.25). More than 90% of married couples have articulated their views on the ideal small family size ranging from two to four members (Fig. 5.26) in the family, including both couples. The study also adopted the fertility transition model after Chaurasia and Gulati (2013) estimated the value as 0.65, which indicates that the sample slum dwellers are going through the middle stage of fertility transition by trying to adopt small family norms, perhaps due to their low socioeconomic status and related deprivations. Family planning services and supplies in this urban area are provided primarily through a network of government hospitals, urban community health centers, sub-centers, private hospitals, and private healthcare institutes. From time to time, sterilization camps are organized to provide sterilization services by the hospital, urban community health center, and Red Cross Society hospital in the study area. Regarding the dissemination of family

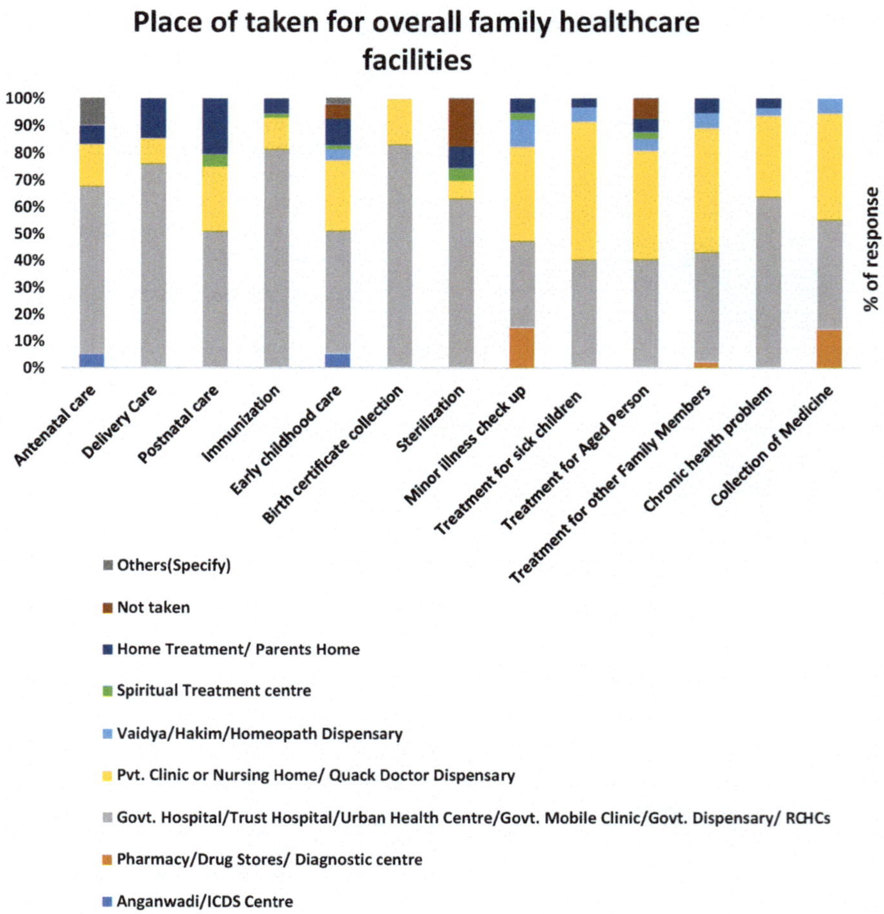

Fig. 5.22 Place of taken for overall family healthcare facilities (Field survey, 2022)

planning information, the functioning of the government-run healthcare system needs to be more emphasized in the study area. Only one-fourth (25%) of the respondents have been found to become aware of family planning practices through the state-run mechanism.

5.6 Spatial Accessibility of Healthcare Center

In developing countries, geographic access to healthcare facilities (Evans & Stoddart, 1990; Rekha et al., 2017) is one of the key challenges in ensuring healthy lives and promoting well-being for all at all ages in achieving the third sustainable development goal due to distance to facilities, ease of travel cost, and travel time

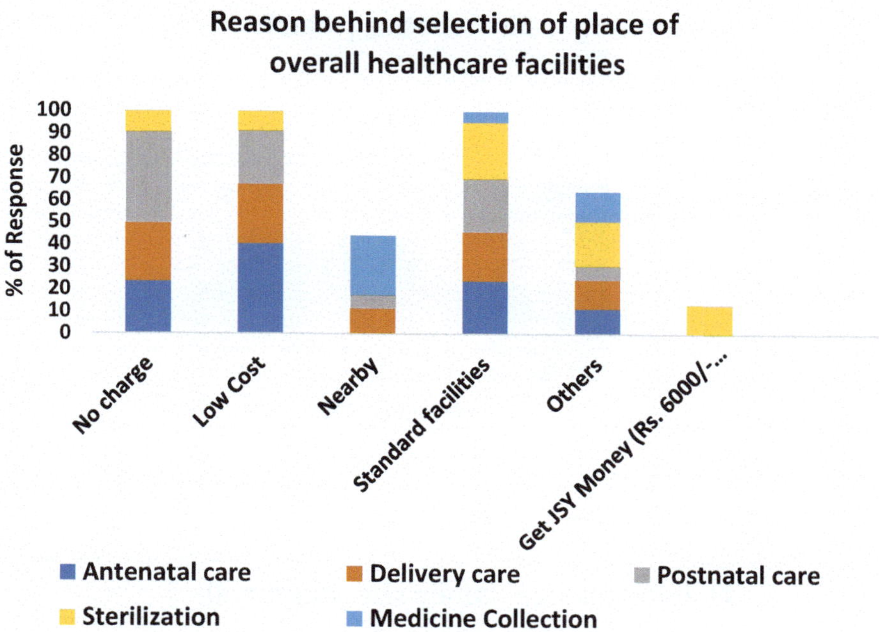

Fig. 5.23 Reason behind the selection of place of the overall healthcare facilities (Field survey, 2022)

Fig. 5.24 Family planning of the households (Field survey, 2022)

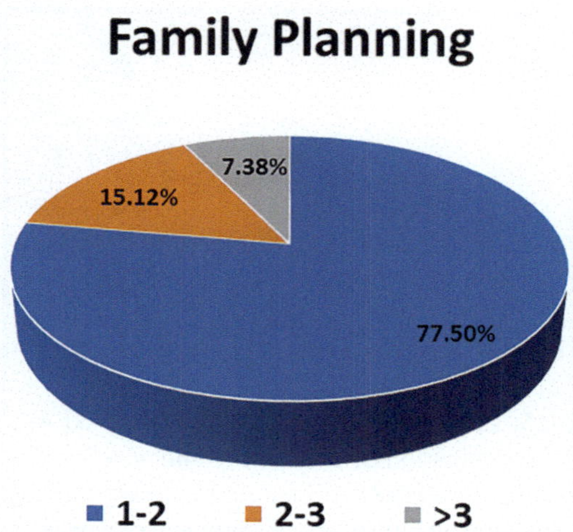

(Pyrialakou et al., 2016; Lawal & Anyiam, 2019; Rekha et al., 2017). Geographic or spatial accessibility (Basu et al., 2018; Bennett, 1981; Hossain & Laditka, 2009; Huerta Munoz & Källestål, 2012; Ye & Kim, 2014) refers to physical access of people to a location, e.g., healthcare facility centers (McGrail & Humphreys, 2014; Owen et al., 2010). Spatial accessibility (Khan, 1992; Wang et al., 2018; Polo et al.,

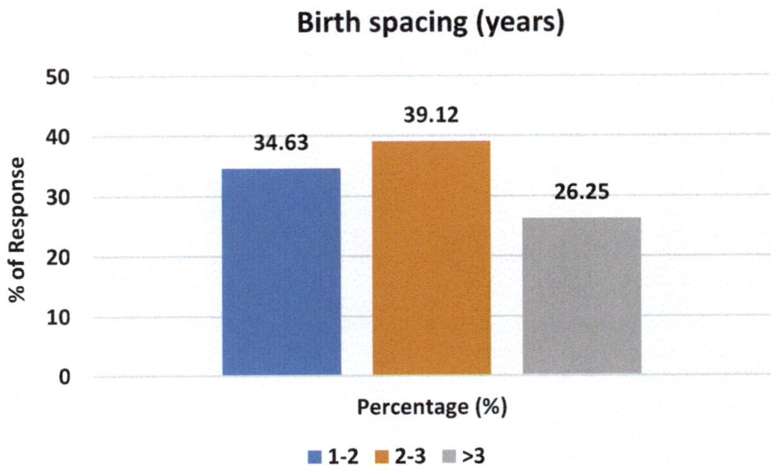

Fig. 5.25 Birth interval of the slum married couple (Field survey, 2022)

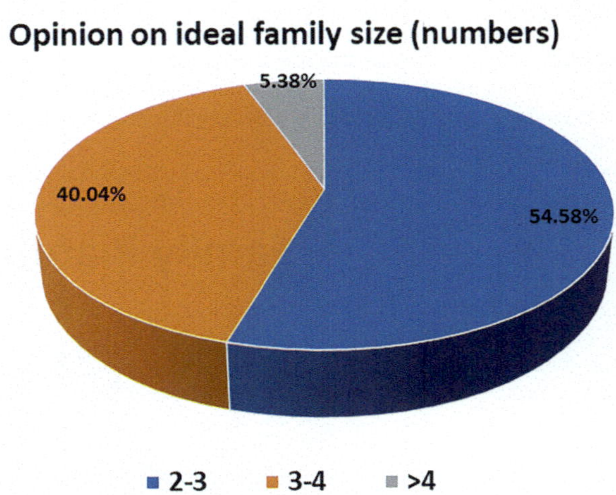

Fig. 5.26 Opinion on ideal family size among the slum married couple (Field survey, 2022)

2015; Langford & Higgs, 2006; Murawski & Church, 2009; McLafferty, 2003) is the combination of availability and accessibility dimension as stated by Penchasky and Thomas (1981) (Guagliardo, 2004; Wang & Luo, 2005; Luo, 2004) that can be measured by a time-space framework such as vehicular travel time, distance (Owen et al., 2010; Basu et al., 2018; Huerta Munoz & Källestål, 2012; Jalil et al., 2018; Ye & Kim, 2014; Guagliardo, 2004), and Euclidean distance, which is influenced by the topology and structure of the road network (Geurs & Wee, 2004; Maleki Rastaghi et al., 2018; Jalil et al., 2018; Huerta Munoz & Källestål, 2012) for identifying deprived areas and people (Lawal & Anyiam, 2019). Geographic access to healthcare facilities is a multidimensional process (Ye & Kim, 2014; Rekha et al.,

2017) and can be described by five key dimensions—availability, accessibility, accommodation, affordability, and acceptability (Penchasky & Thomas, 1981; Shah et al., 2016; Rekha et al., 2017)—which can be more explicitly described as the supply of healthcare services with physical proximity (travel distance and time) (Huerta Munoz & Källestål, 2012; Jalil et al., 2018) on a particular geographical location at a reasonable price concerning the population in need (Shaw & Sahoo, 2020). The location of healthcare centers near homes can lead to more recurrent visits (Ye & Kim, 2014; Sedenu et al., 2016), resulting in better family health conditions.

The "Health for all" doctrine indicates that everyone should have access to appropriate healthcare facilities in an acceptable and affordable time-space framework (Lawal & Anyiam, 2019; Ye & Kim, 2014). Affordable and accessible healthcare can reduce potential health risks and the rate of hospitalization. Dependency and a higher rate of hospital admissions in an area may provide evidence of the inaccessibility of people to primary healthcare centers (Hossain & Laditka, 2009). The steady expansion of urban areas with population growth requires regular assessment of the accessibility of healthcare facilities across metropolitan areas (Balasubramani et al., 2016; Lawal & Anyiam, 2019; Abdelkarim, 2019) because limited access to facilities may cause an epidemic outburst such as the COVID-19 virus (Lawal & Anyiam, 2019). Higher population density needs the provision of three tiers of healthcare facilities to satisfy its threshold population implicit in the central place theory (Sedenu et al., 2016), which is relevant to the analysis and planning of healthcare facilities in urban areas of India, thereby reducing the spatial inequality (Bindu & Joshi, 2013; Shaw & Sahoo, 2020) of healthcare facilities.

In developing countries, locations of primary healthcare are generally decided by local government officers and often based on political or pragmatic considerations; as a result, they are far from optimal locations (El Karim & Awawdeh, 2020; Rahman & Smith, 2000; Sedenu et al., 2016). There is a close relationship between infant as well as maternal mortality and distance to healthcare facilities, as PHCs act as the distributors of polio and other significant vaccines for infants and pregnant women (Rahman & Smith, 2000). Recently, it has been necessary to make locational data of healthcare facility centers available to everyone to improve healthcare services. Furthermore, analysis of these data can help to find a spatial pattern of healthcare facilities, thereby preventing disease progression and loss of lives, improving the quality of service facilities, and reducing the higher rate of hospital admissions (Basu et al., 2018; Muggah et al., 2013) as well as the cost of healthcare services (Raghupathi & Raghupathi, 2014; Lawal & Anyiam, 2019). Lack of accessibility to healthcare facilities as the result of low public investment in health sectors encourages entrepreneurs of private healthcare facilities, forcing people to access these facilities (MoSPI, 2015), which often leads to higher healthcare expenditures, health shocks, and medical poverty trap. Private facilities do not provide affordable healthcare facilities to the economically weaker sections and lower-income deprived groups (Patel et al., 2015; WHO, 1986; Iyer et al., 2008; Basu et al., 2018). Therefore, the allocation of public healthcare facilities should be based on accessibility and affordability gaps as per the guidelines of the National Health Policy (NHP) (2002)

(MoHFW, 2002) and the National Urban Health Mission (NUHM) (2013) (MoHFW, 2012) as to develop smart and healthy city (WHO, 1986) as well as provide affordable and essential primary healthcare services to the urban poor so that the needs of the healthcare are met. Suitable locations (Ye & Kim, 2014; Sedenu et al., 2016) of healthcare facilities reduce the travel time cost of healthcare, in addition to mortality and morbidity rates (Maleki Rastaghi et al., 2018; Rahman & Smith, 2000; Shaw & Sahoo, 2020; Buchmueller et al., 2006; Ye & Kim, 2014).

As per the NFHS-4 report (2015–2016), antenatal care visits and maternal care are much lower in Paschim Medinipur district in West Bengal. Hence, child and maternal morbidity remain a significant concern (Shaw & Sahoo, 2020; Baru et al., 2010; Horton & Das, 2011; Rekha et al., 2017). The study area is the eyes of the Paschim Medinipur district, where only the tertiary categories of healthcare services are available with the most specialized doctors, and facilities attract vast numbers of people from rural and slum areas. In this study, the lack of healthcare facilities and accessibility areas are analyzed and help to enhance the distribution of new PHCFs.

5.6.1 Data and Method

The study uses different data required for the spatial accessibility analysis of healthcare facility centers (HCFs) from other sources, such as the location of HCF, municipal and slum boundary, road network, and slope, summarized in Table 5.3. After the data collection, all data are projected into the GCS WGS-1984 coordinate system to create a GIS database. The methodology has been developed in the ArcGIS v10.4 algorithms through the network analysis tools (e.g., service area and location-allocation model) and predictive tools (e.g., speed model builder, distance from and time from the tool). A predictive tool is a raster-based proximity analytical tool that

Table 5.3 Type of data used for spatial accessibility analysis

Data type	Sources	Data processing	Types of data	Types of data format	Scale	Coordinate system
Midnapore municipality and ward boundary	Midnapore municipality (.tiff)	Digitize	Polygon	Vector	1:100,000	UTM WGS 1984
Slum boundary	Midnapore municipality	Digitize	Polygon	Vector	1:100,000	UTM WGS 1984
HCF	GPS point data collection	Join and relate	Point	Vector	1:100,000	UTM WGS 1984
Road network	OSM	Clipping	Line	Vector	1:100,000	UTM WGS 1984
ASTER DEM	USGS	Extract by mask	Pixel	Raster	30 × 30 m spatial resolution	UTM WGS 1984

determines the travel time by considering the slope of available roads (Lawal & Anyiam, 2019). Network analysis is used to assess the service area and proposed location of the new healthcare facility center based on the road network (Abdelkarim, 2019; Yasenovskiy & Hodgson, 2007). To set the facilities of the study area, 12 HCFs were selected for the required point (Table 5.4), and demand points are 156 slum neighborhoods. The purpose of the methodological development of geographic accessibility is to reduce the inaccessibility to healthcare facility centers (Rekha et al., 2017), as shown in Fig. 5.27.

5.6.1.1 Near Analysis

Near analysis in GIS algorithm measures the distance of nearest points (HCFs) to the input points (slum neighborhoods) through using Euclidean distance (meter or km) with specified distance or radius within a particular geographical region (Zhou & Wu, 2012; Kim et al., 2018; Wang et al., 2018; Danjuma, 2015; Wang & Luo, 2005). The analysis reveals (see Fig. 4.16) that 16.5% of slum neighborhoods avail themselves of healthcare facilities within 200 m of their residence. In comparison, about 77% of slum neighborhoods have healthcare facilities that range from 200 m to 1 km. In contrast, only 6.50% of slum neighborhoods avail healthcare services further from proximity to healthcare centers (HCFs) outside 1 km.

Table 5.4 Major healthcare facility (HCF) center

Sl no.	Name of the HCF	Location		Ward	Types of HCF
		Latitude	Longitude		
HCF 1	Vidyasagar Institute of Health	22°25′1.2″	87°17′52.8″	25	Hospital
HCF 2	Saratpally Urban Primary Health Center	22°24′47.31″	87°18′39.3″	23	PHC
HCF 3	Pathan Mohalla Urban Primary Health Center	22°25′1.5″	87°18′53.09″	13	PHC
HCF 4	Midnapore Medical College and Hospital	22°25′12.36″	87*19′19.92	12	Hospital
HCF 5	Indian Red Cross Society	22°25′15.6″	87°19′22.8″	12	Hospital
HCF 6	Midnapore Homeopathic Medical College	22°25′12.36″	87*19′30	12	Hospital
HCF 7	Saheb Pukur Sub Health Center	22°24′45.04″	87°19′16.85″	14	PHC
HCF 8	Najarganj Health Post	22°24′15.52″	87°18′46.59″	20	PHC
HCF 9	Berballavpur Urban Primary Health Center	22°24′49.47″	87*20′0.43	16	PHC
HCF 10	Ramkrishnanagar Sub Health Center	22°25′23.32″	87°20′13.81″	9	PHC
HCF 11	Kuikota Urban Primary Health Center	22°26′21.87″	87°19′32.78″	2	PHC
HCF 12	Udaypally Social Service Center	22°26′33.18″	87°18′0.51″	3	PHC

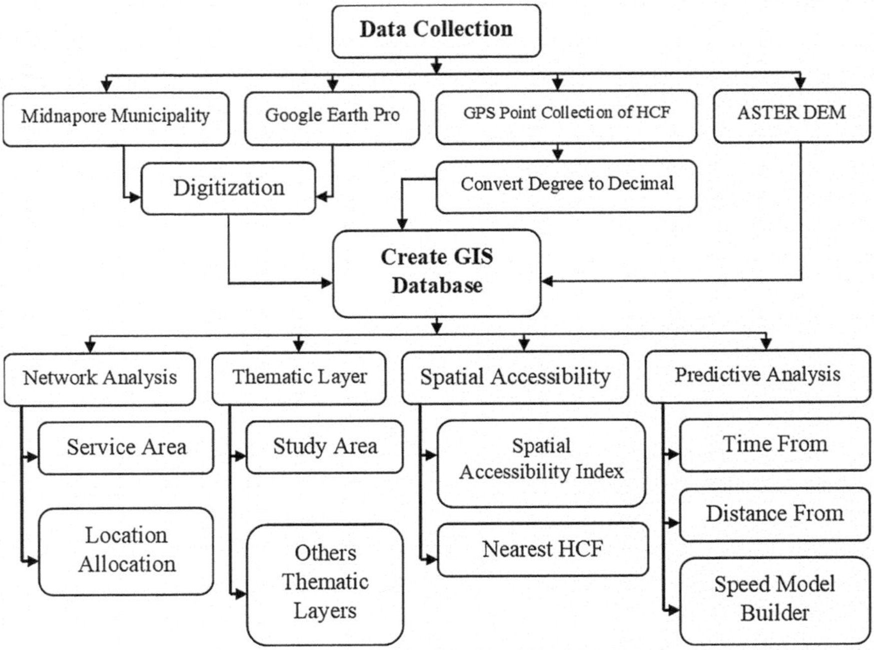

Fig. 5.27 Methodology of spatial accessibility analysis (Source: Das et al., 2023)

5.6.1.2 Network Analysis

Network analysis in the extension of ArcGIS v10.4 is used to recognize the most effective and functioning road for distributing services. The network dataset (e.g., network-based impedance such as travel time or distance) with demand site and requisite facilities is the input for the analysis. Various tools and models operate in network analysis, such as service area, OD cost matrix, location-allocation, shortest path, etc. In this study, the service area tool and location-allocation model have been used based on the available network dataset of the catchment area with the demand side (HCF).

5.6.1.3 Service Area

A service area in a GIS Network Analyst extension is an area that provides services with distance and time from the slum neighborhoods to the nearest public health-care center, including all access roads, expressed as (Eq. 5.1)

$$\text{Time} = \frac{\text{Distance}}{\text{Speed}} \tag{5.1}$$

The service area map (Fig. 5.32) shows the most accessible or the shortest or least-cost routes or roads or streets to the nearest healthcare center at a constant speed of 40 km/h (Sedenu et al., 2016) within a given time of 2, 4, 6, and 8 min. The network also estimated the total served and unserved areas of healthcare services are 12.28 and 7.72 km^2, respectively (see Fig. 4.23). The map (see Fig. 4.23) shows that 90% of areas remain unserved in nine municipal wards, while four municipal wards have between 30% and 50% unserved areas, and three municipal wards have approximately 20% of their areas unserved. At the same time, public healthcare facilities with high accessibility completely enclose eight municipal wards. Therefore, it is suggested that healthcare services be appropriately allocated to unserved areas.

5.6.1.4 Location-Allocation Model

A location-allocation algorithm is used to identify the optimum demand location (Awasthi et al., 2016). Also, it is employed to find the most negligible travel costs and scarcity of facilities, suggesting new demand locations and coverage capacity of a facility center. The location-allocation model is available in the network analysis extension of the ArcGIS environment. The location-allocation model is used in different aspects. However, the three main components of this model are as follows: supporting road network is the existing road network, demand site, and facility center. Accessibility between demand site and facility center: distance has a crucial role in quantification, e.g., Euclidean distance, Manhattan distance, network distance, and time distance. Most researchers used network distance from those distances and impedance distances. A user is willingly traveling a maximum acceptable distance to access a specific facility that is impedance distance of accessibility. In this study, the location-allocation model is employed to determine the accessibility and inaccessibility of health shocks between slum neighborhoods and healthcare facility centers (HCF), i.e., coverage and non-coverage slums by the facilities. Here, the impendence cutoff with maximum coverage is 2 km.

5.6.1.5 Predictive Analysis Tool

Predictive analysis is an add-in tool in the ArcGIS environment that predicts multiple raster datasets. This tool also carried out complex queries on multiple or single raster bands. The capability of this tool is to extract information from raster bands using different queries to take into different obstacles such as buildings, water bodies, etc. This tool also contains the speed model and distance from and time from those identified as travel times from the raster band.

Distance from

The distance from the tool can find how far an unknown location of an area of interest is from a known location (e.g., point, line, or polygons). This tool finds the distance based on the shortest linear distance per set of locations. The study uses this tool to identify measuring distance from healthcare facility centers over the study area.

Time from

The time from the tool in predictive analysis is used to find the least collective travel time from the adjoining source location or facility center to each cell of the raster dataset. This tool also finds distance in terms of travel time in a constant velocity using the algorithm of the raster calculator. Also, this tool lets speed model as an input, where multiple effects can control travel time and speed. This study uses this tool to identify accessibility in terms of traveling time from healthcare facility centers to slum neighborhoods over the study area.

Speed Model Builder

The speed model is a complex query-based building model available in predictive analysis tools within ArcGIS (ESRI, 2014). This tool supports an analyst-specified condition building model, which predicts moving or fixed objects and facilities' locations. The speed model functions as proximity analysis over a raster band, but objects or facilities can be specified as constraints to movement computed travel time and distance to a destination. The specified speed depends on different surfaces and terrain, such as a conditioning surface created by the speed model. For example, the high terrain slope affected vehicles' speed and traffic jams, and road environments affected vehicle speed. Also, the speed model depends on local knowledge and understanding. During health shocks, the achievable speed for access to healthcare facility centers ranges between 10 km/h and 40 km/h, estimated by slope and distance parameters based on terrain conditions and road network while considering obstacles of the study area (see Table 1.3). The accommodated scenarios are conditions of the speed model using time from the tool to compute traveling time to the nearest health center facilities.

5.6.2 Result and Discussion

5.6.2.1 Catchment Area of HCFs and Accessibility

In general, slum neighborhoods are confused about choosing the nearest healthcare facility center during an emergency illness. Therefore, during health shocks, they generally travel too much distance, which increases a patient's mortality risk. A vector-based near tool can be applied to identify the closest healthcare facility center to make the right choice during an emergency. The attributes (1 to 12) of slum neighborhoods shown in the map (Fig. 5.28) indicate which healthcare facility center is much nearer. Also, this analysis shows that slum people travel a minimum distance to a minimum healthcare facility center. However, from Fig. 5.29, the output map represents the accessibility of each slum neighborhood to the healthcare facility center. Most of the slum neighborhoods (such as Raja Bazar, Sepoy Bazar,

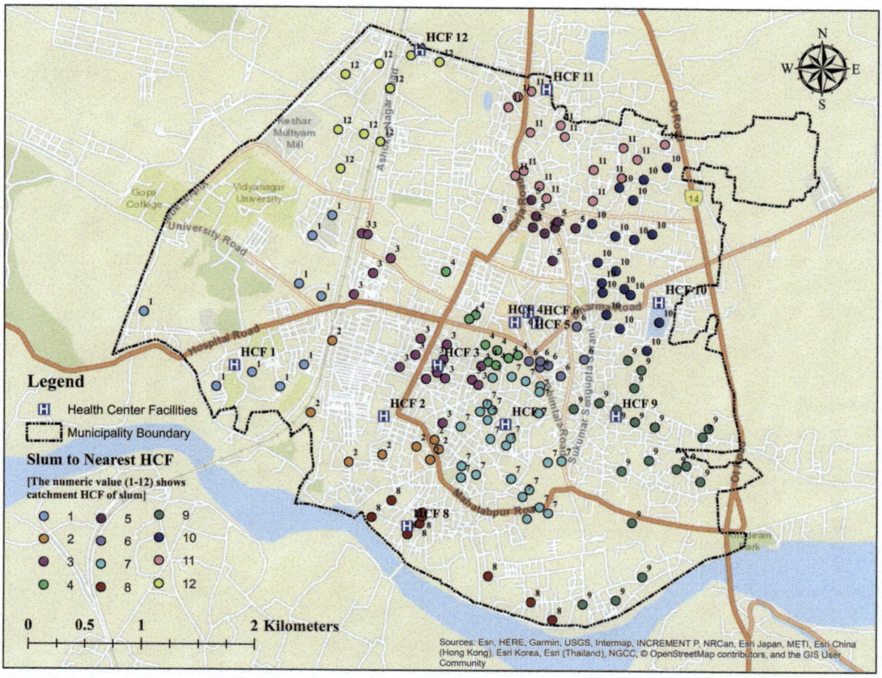

Fig. 5.28 Map shows slum to nearest HCF in the study area

Keranitola, Barabazar, and others) have much better accessibility (70%), and about 23% of slum neighborhoods (such as station road, Bhuniapara, Ashok Nagar, and others) have moderate accessibility to a healthcare facility center across the study area. Only 8% of the slum neighborhoods are in low accessibility zones (such as Gandhi Ghat, Kabar Danga, Tantigeria, Baro Astana). They must travel to a single facility or, during emergencies, to the existing healthcare facility centers. Due to the uneven distribution of healthcare facility centers, these slums are not facilitated as per their requirements.

5.6.2.2 Spatial Accessibility by Predictive Analysis

The complex query and raster-based proximity analysis tool, the so-called predictive analysis, has been used to find the traveling distance and time from the healthcare facility center (HCF) to slum neighborhoods across the study area. Distance from the predictive analysis tool was deployed to find the distance from the road to slum neighborhoods and every pixel across the study area. The key point from this analysis is that the periphery regions of the study area are around 1 km from the major road network, which is considered off-road in speed model building (Fig. 5.30).

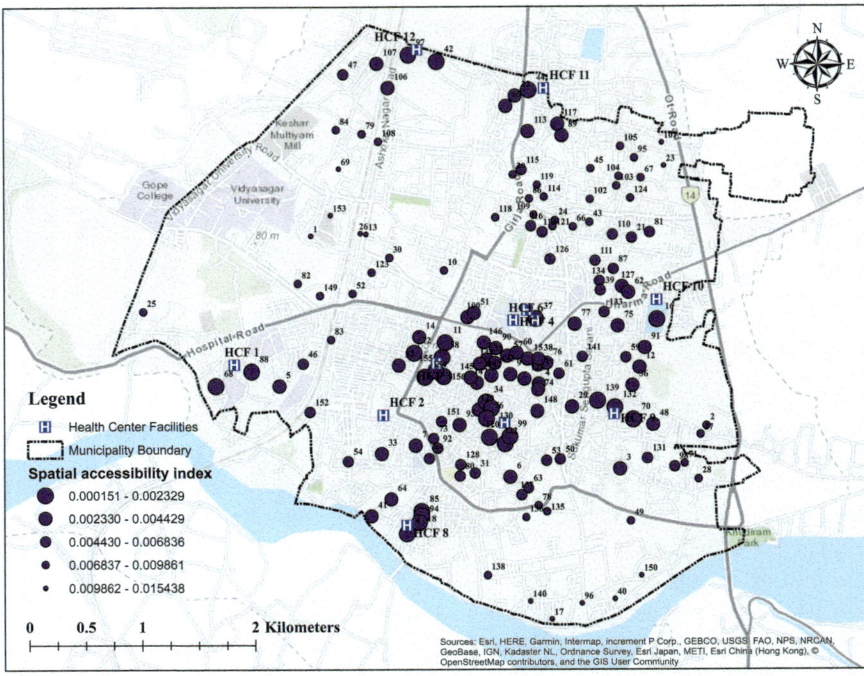

Fig. 5.29 Map shows spatial accessibility of slum to nearest HCF in the study area

However, the speed model builder was created by combining the road environment, distance from the road network, terrain attributes, and healthcare facility centers (HCF). These conditions are combined to find the travel time from each healthcare facility center (HCF) to slum neighborhoods across the study area based on conditions of the speed model (see Table 1.3). The analysis shows that accessibility to HCF in terms of travel time is a maximum of ±31 min, while the average travel time is around ±8 min over the study area.

Most of the slum neighborhoods under the study area have high spatial accessibility to healthcare facility centers except periphery regions such as Golapi Chak, Baishakhi Palli, Naram Pur, Bara Astana, Rajar Pukur, and others (Fig. 5.31). Most slum neighborhoods are in a virtuous position to access emergency care. From Fig. 5.31, the average traveling time from slum neighborhoods to healthcare facility centers ranges between 5 and 11 min. The peripheral regions of the study area face problems during health shocks in accessing medical care within half an hour. Following the Lawal and Anyiam (2019) works, the level of accessibility in terms of traveling time of over 20 min can increase the mortality risk during health shocks such as vehicle accidents, trauma, gynecological problems, etc. However, the analysis could not include substantial traffic jams (in Barabazar, Keranitola, Panchur Chak, Raja Bazar, Sepoi Bazar, LIC More, Midnapore Central bus Stand, Ashok

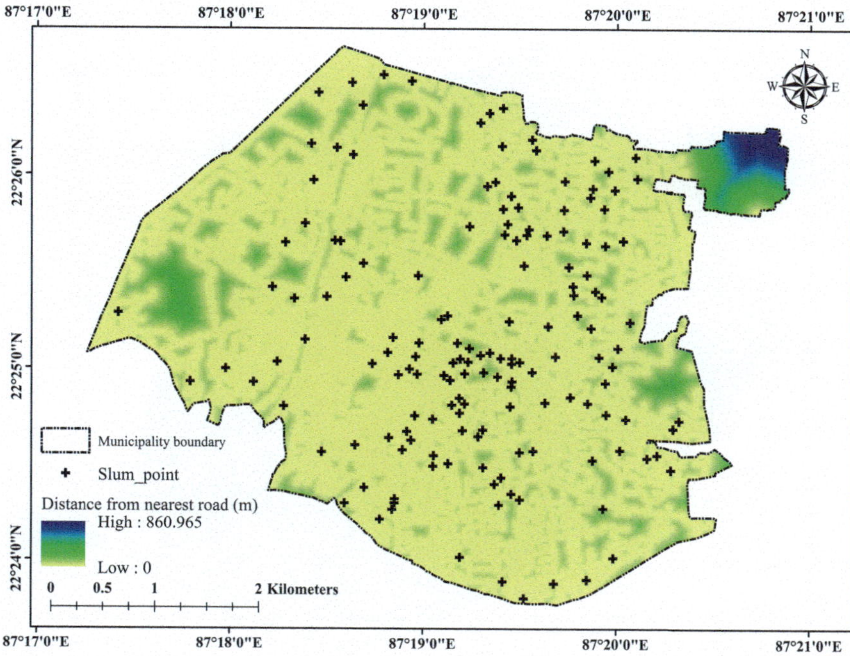

Fig. 5.30 Map shows distance from nearest road in the study area

Nagar, Midnapore railway station road areas) and lack of transport availability and ambulance services, which are an essential aspect of predicting accessibility.

The results also identified characteristics of road environments and cultural and environmental obstacles, such as off-road, buildings, ponds, parks, etc., as the reasons for the worst access to a healthcare facility, which produces difficulty in navigation during health shocks. For the foregone care, high spatial accessibility has been found in the central part, while low pocket accessibility areas are in the study area's northeastern, southern, and western parts. After estimating the accessibility in the study area, about 84% of the area has high spatial accessibility, and 16% has very poor to poor accessibility. Moreover, the analysis indicates that most of the HCFs are clustered in the center and distributed northeast to southwest directionally (Dutta et al., 2022), which hinders the accessibility to healthcare facilities toward the periphery. Despite this, two phenomena are responsible for the inaccessibility of healthcare facilities. Firstly, slum neighborhoods are not facilitated easily by healthcare facility centers (HCF) because, most of the time, primary health centers cannot treat health shocks. Secondly, the questionnaire survey on the study area shows that poverty is more responsible than the distance to the healthcare facility center. As a result, they are forced and pressured to get healthcare services to the district medical college and hospital. Therefore, the rapidly increasing population and low population to HCF ratio demand for launching new healthcare centers in the deficient and demand zone across the study area.

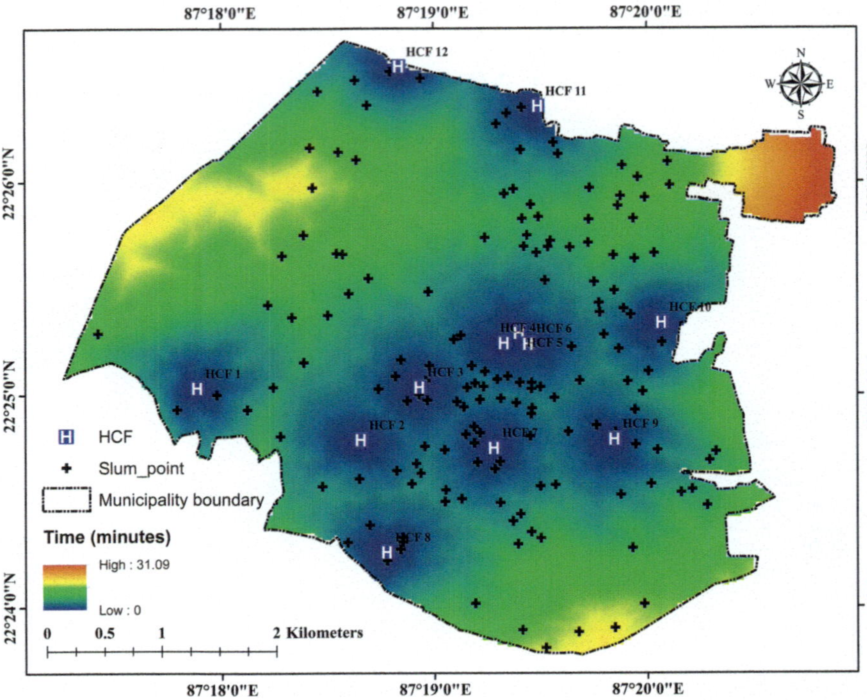

Fig. 5.31 Map shows time to travel nearest HCF using speed model builder in the study area

5.6.2.3 Spatial Accessibility Through Network Analysis

The service area analysis (network tool) evaluates geographic accessibility in terms of coverage areas through the traveling time from existing healthcare facility centers (HCFs) across the study area by road network during health shocks (Fig. 5.32). The results of service area analysis show that there are approximately 31 slum neighborhoods (such as Kabar Danga, Golapi Chawk, Tantigeria, nibeditapalli, Bara Astana, and others) that are not under the coverage area of the existing healthcare facility center (HCF); these are representing about 19.8% of total slums and 16,000 populations of total slum population approximately (Fig. 5.32). And, the other 125 slum neighborhoods (80.2%), as well as 34,000 populations of the total slum population, are under the coverage area of the existing healthcare facility center accordingly. Also, the results indicate that the uncovered area (around 8 km^2 area) traveled over 20 min during an emergency for a healthcare facility. Approximately 27 slums (around 2.8 km^2) (such as Momin Mahalla, Pakija bastee, Surja Nagar, Berballavpur, Narampur, and others) are under the coverage area of 0–2 min traveling time; also, 40 slums (around 4.7 km^2) (such as Jelepara, Daspara, Dewan Nagar, and others) are under the coverage area of 2–4 minutes traveling time; then, 28 slums (around 5.2 km^2) (such as Halugali baste, Sekhpura, Bhunia para, and others) are under the coverage area of 4–6 min traveling time; and lastly, 30 slums (around 7.3 km^2) (such as Rajarpukur, station road, Jelepara baste, and others) are under the coverage area of

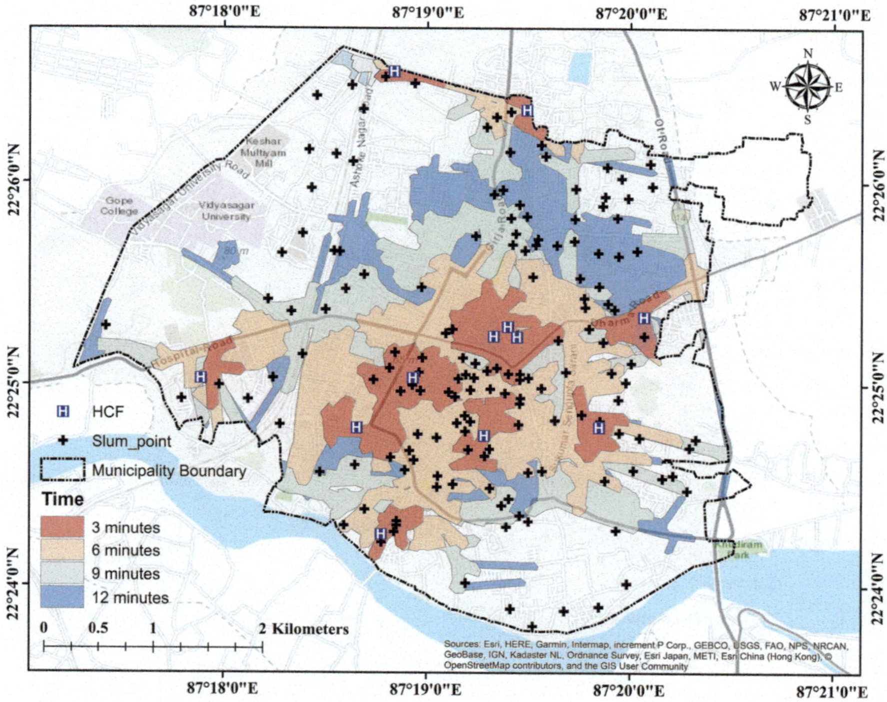

Fig. 5.32 Map shows service area of HCF in the study area

6–8 min traveling time simultaneously (Fig. 5.32). Most of the slum neighborhoods of the northwestern part and a few of the northeastern, southwestern, and southern parts are considered to have low accessibility regarding traveling time during emergencies. The analysis recommended to add a provisional healthcare facility center in the deprived healthcare service areas with the help of geospatial analysis.

The location-allocation model is useful in this context to measure and authenticate the level of accessibility from HCF to slum neighborhoods of the Midnapore municipality. The location-allocation model for this study is to set an impedance distance or threshold distance of 2 km during health shocks. The result (Fig. 5.33) indicates that the catchment area covering slum neighborhoods of existing healthcare facility centers (HCFs) and distance to access from HCF represents the pockets of inaccessibility or deprived healthcare facilities of slum neighborhoods. Most of the slum neighborhoods in the central part (such as Barabazar, Nimtala Chawk, Choto Bazar, Berballavpur, Raja Bazar, Burdge Town, Keranitola, Narampur, and others) of the study areas have decent spatial accessibility through road network, but slum neighborhoods in the northwestern part (such as Golapi Chawk, Nepali Para, and others) and southern part (such as Gandhi Ghat, Mahatabpur, and others) have spatial inaccessibility. Also, the results indicated that about 15% and 85% of slum neighborhoods have spatial inaccessibility and high spatial accessibility, respectively. However, there are some slum neighborhoods under the catchment area of HCF in this model that have too much travel time during emergencies.

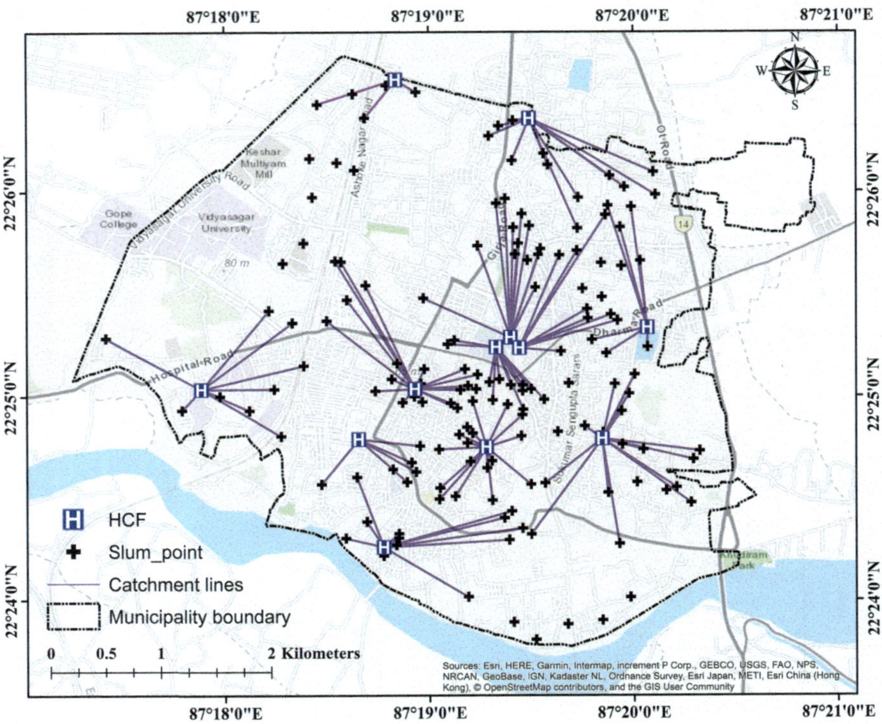

Fig. 5.33 Map shows location-allocation of existing HCF in study area

The location-allocation model has identified areas with poor to very poor accessibility or high levels of deprivation as priority zones for the demand of new healthcare facility centers (HCFs). The general law is that a facility with a suitable location plays a vital role in socioeconomic aspects across a region (Mishra et al., 2021). The inappropriate location of the facilities, like the existing HCF in the study area, severely affects the service quality and other costs. However, adding healthcare facility centers is not a solution whenever healthcare facility centers are not in a proper optimal location. The proposed HCF is the location-allocation model reemployed to find the recommended potential optimal location for a newly established HCF (Fig. 5.34). There are four proposed HCFs, and existing HCFs can serve most of the slum neighborhoods demanded in the study area (Fig. 5.34). However, a few slum neighborhoods are very close to the proposed and existing HCF, but still not under the catchment area due to the road network. The result (Fig. 5.34) was also validated by a survey (field survey in 2022), where people in deprived areas reported that they had been demanding healthcare facilities for a long time with the development of road networks from the local authority.

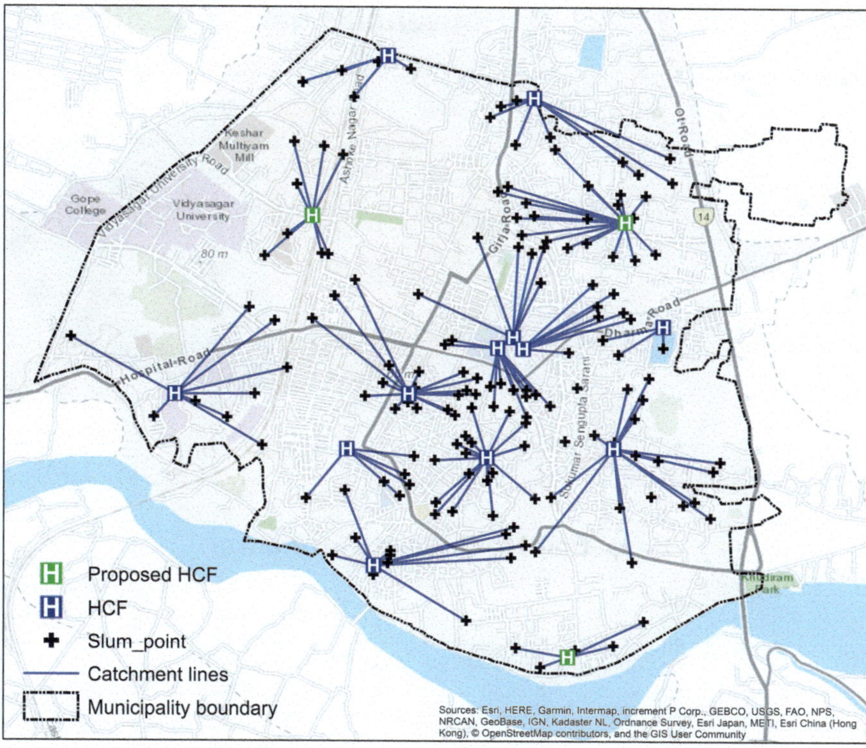

Fig. 5.34 Map shows location-allocation after proposing new HCF in the study area

5.7 Use of Government E-Healthcare Facilities: Health Card, Insurance, and Schemes

Present-day expenditure on healthcare treatment is a vital issue worldwide. Environmental degradation, lifestyle changes, frequent natural disasters, and overpopulation in limited spaces have contributed to the emergence and spread of various diseases, viruses (such as COVID-19), illnesses, and increased morbidity. Therefore, healthcare costs are increasing daily for an individual, even family members. For example, due to COVID-19 spreading over the world, we have experienced expenses on personal hygiene maintenance, hospitalization, frequent health checkups, and illness, even loss of life. A close relationship exists between an unhygienic living environment and disease burden or health shocks. Urban poor are faced with disease burden and health shock than other groups due to their dilapidated livelihood status, which need ensuing health insurance and schemes so that they can reduce their healthcare expenditure and flee from the "medical poverty trap." It is also found that among the urban rich, 90% have either health insurance or healthcare planning to meet the future healthcare cost. However, India's state and central governments have initiated different healthcare insurance schemes for poor people. For example,

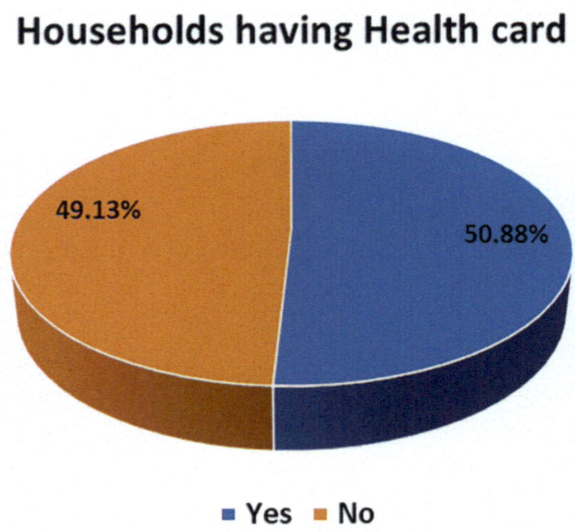

Fig. 5.35 Households having health card (Source: Field survey, 2022)

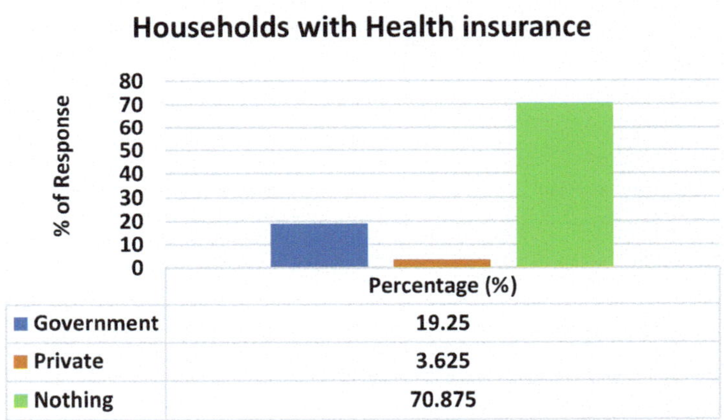

Fig. 5.36 Households having health insurance (Source: Field survey, 2022)

"Ayushman Bharat" was launched by the government of India, and "Swasthya Sathi" was launched by the government of West Bengal, especially for the poor.

The study investigated the fact that most slum households do not have health insurance and health schemes. Few people have government and private health insurance. Roughly 51% of households have health cards such as "Swasthya Sathi" or "Ayushman Bharat," and 49% do not have health insurance cards until the end of 2022 (Figs. 5.35 and 5.36). Moreover, the government of India has initiated different health programs and schemes under the Ministry of Health and Family Welfare's policy such as National Urban Health Mission (NUHM), Rashtriya Swasthya Bima Yojana (RSBY), Urban Health Post (UHP), Urban Family Welfare Center (UFWC), Adolescent Reproductive and Sexual Health (ARSH), Integrated Child Development Services (ICDS), Accredited Social Health Activist (ASHA), Auxiliary Nurse

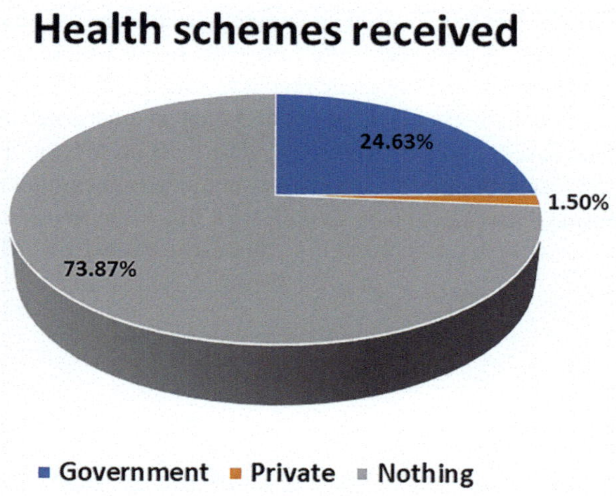

Fig. 5.37 Sources of health schemes received (Source: Field survey, 2022)

Midwife (ANM) and Anganwadi Worker (AWW) services, etc. especially for the urban poor with the aim of reducing hospitalization and cost of health checkup, infant malnutrition and mortality, maternal mortality and health and well-being, etc. These schemes are funded by central and state governments up to 5 lakhs per individual in the family and free of cost for the maximum cases. Nonetheless, 20–25% of the urban poor either lack access to these services, are socially excluded from them, or are unaware of the health schemes available to them. The study revealed that roughly 74% of slum households have received mostly government healthcare schemes, and still 25% (up to December 2022) households did not receive health schemes from any of the organizations (Fig. 5.37). The study investigates the underlying reasons for this phenomenon through face-to-face interviews, revealing that many individuals are unaware of these government schemes, and even when they are aware, they often face challenges in accessing them due to lengthy implementation processes and extended waiting times. Therefore, "Duare Sarkar" (government services at home) launched by the government of West Bengal may be the most effective effort in this context.

5.8 Conclusion

Health is required in daily human life and for the standard of living and better quality of life. Maintaining decent health increases the standard of living and quality of life and reduces health expenditure. However, slum neighborhoods of the study area are facing health-related problems like malnutrition, bad hygienic practices, and a deprived housing environment, which can pose health risks. During minor health problems, they can mostly prefer quack doctors and otherwise government healthcare facilities. They mostly used government healthcare facilities due to free-of-cost facilities. Lack of awareness often prevents many women from accessing sanitary napkins, leading to poor hygiene practices. Also, accessibility to healthcare facilities from slum neighborhoods is essential during an emergency. After employing different GIS-based analyses, we can find the different spatial accessibility maps that can represent the lack of healthcare facilities in the municipality, which refers to establishing new healthcare facilities. In this scenario, the awareness of health consciousness is very important in slum neighborhoods, which can prevent health risks.

References

Abdelkarim, A. (2019). Integration of location-allocation and accessibility models in GIS to improve urban planning for health Services in Al-Madinah Al-Munawwarah, Saudi Arabia. *Journal of Geographic Information System, 11*(06), 633–662. https://doi.org/10.4236/jgis.2019.116039

Awasthi, A., Mishra, N., & Pandey, C. (2016). A study to assess patient satisfaction in outpatient department of a tertiary care hospital in North India. *International Journal of Community Medicine and Public Health.*, 328, 10.18203/2394-6040.ijcmph20151586–334.

Balasubramani, K., Gomathi, M., & Prasad, S. (2016). GIS-Based Service area analysis for optimal planning strategies: A case study of fire service station in Madurai city. *Geographic Analysis of Union Geographic Information Technologists, 5*(2), 11–18.

Baru, R. V., Acharya, A., Acharya, S., Shiva Kumar, A., & Nagaraj, K. (2010). Inequities in access to health services in India. *Economic & Political Weekly, xlv*(38), 49–58. http://www.mendeley.com/research/inequities-access-health-services-india/

Basu, R., Jana, A., & Bardhan, R. (2018). A health care facility allocation model for expanding cities in developing nations: Strategizing urban health policy implementation. *Applied Spatial Analysis and Policy, 11*(1), 21–36. https://doi.org/10.1007/s12061-016-9208-0

Bennett, W. D. (1981). A location-allocation approach to health care facility location: A study of the undoctored population in Lansing, Michigan. *Social Science and Medicine. Part C Medical Geography, 15*(2), 305–312. https://doi.org/10.1016/0160-8002(81)90006-X

Bindu, B., & Joshi, J. P. (2013). A geospatial approach for assessing and modeling spatial accessibility of the primary health centers in the tribal Talukas of the Vadodara district. *International Journal of Geomatics and Geosciences, 3*(3), 582–591. http://www.ipublishing.co.in/jggsvol1no12010/volthree/EIJGGS3148.pdf%5Cn, http://ovidsp.ovid.com/ovidweb.cgi?T=JS&CSC=Y&NEWS=N&PAGE=fulltext&D=cagh&AN=20133154866%5Cn, http://lshtmsfx.hosted.exlibrisgroup.com/lshtm?sid=OVID:caghdb&id=pmid:&id=doi:&issn=0976-4

Buchmueller, T. C., Jacobson, M., & Wold, C. (2006). How far to the hospital?. The effect of hospital closures on access to care. *Journal of Health Economics, 25*(4), 740–761. https://doi.org/10.1016/j.jhealeco.2005.10.006

Carrasquillo, O. (2013). Health care utilization. In M. D. Gellman & J. R. Turner (Eds.), *Encyclopedia of behavioral medicine*. Springer. https://doi.org/10.1007/978-1-4419-1005-9_885

Chaurasia, A., & Gulati, S. (2013). Fertility, mortality and age composition effects of population transition in China and India: 1950-2015. *Comparative Population Studies, 42*, 149–186.

Danjuma, I. (2015). Locational analysis of healthcare facilities in Keffi Local Government Area of Nasarawa State, Nigeria. Ph.D. Thesis. http://kubanni.abu.edu.ng/jspui/bitstream

Das, M., Dutta, B., Roy, U., & Rath, S. (2023). Spatial accessibility modeling to healthcare facilities in the case of health shocks of Midnapore municipality. *India. Geo Journal, 88*(4), 3749–3772. https://doi.org/10.1007/s10708-023-00927-4

Dutta, B., Das, M., Roy, U., Das, S., & Rath, S. (2022). Spatial analysis and modelling for primary healthcare site selection in Midnapore town, West Bengal. *GeoJournal, 87*(6), 4807–4836. https://doi.org/10.1007/s10708-021-10528-w

El Karim, A. A., & Awawdeh, M. M. (2020). Integrating GIS accessibility and location-allocation models with multicriteria decision analysis for evaluating the quality of life in Buraidah city, KSA. *Sustainability (Switzerland), 12*, 4. https://doi.org/10.3390/su12041412

ESRI. (2014). Getting started with the ArcGIS Predictive Analysis. 45.

Evans, R. G., & Stoddart, G. L. (1990). Producing health, consuming health care. *Social Science and Medicine, 31*(12), 1347–1363. https://doi.org/10.1016/0277-9536(90)90074-3

Geurs, K. T., & Wee, B. V. (2004). Accessibility evaluation of land-use and transport strategies: Review and research directions. *Journal of Transport Geography, 12*(2), 127–140. https://doi.org/10.1016/j.jtrangeo.2003.10.005

Guagliardo, M. F. (2004). Spatial accessibility of primary care : Concepts, methods. *and challenges., 13*, 1–13.

Horton, R., & Das, P. (2011). Indian health: The path from crisis to progress. *The Lancet, 377*(9761), 181–183. https://doi.org/10.1016/S0140-6736(10)62179-4

Hossain, M. M., & Laditka, J. N. (2009). Using hospitalization for ambulatory care sensitive conditions to measure access to primary health care: An application of spatial structural equation modeling. *International Journal of Health Geographics, 8*(1), 51. https://doi.org/10.1186/1476-072X-8-51

Huerta Munoz, U., & Källestål, C. (2012). Geographical accessibility and spatial coverage modeling of the primary health care network in the Western Province of Rwanda. *International Journal of Health Geographics, 11*, 1–11. https://doi.org/10.1186/1476-072X-11-40

Iyer, A., Sen, G., & Östlin, P. (2008). The intersections of gender and class in health status and health care. *Global Public Health, 3*(SUPPL. 1), 13–24. https://doi.org/10.1080/17441690801892174

Jalil, I. A., Rasam, A. R. A., Adnan, N. A., Saraf, N. M., & Idris, A. N. (2018). Geospatial network analysis for healthcare facilities accessibility in semi-urban areas. Proceedings – 2018 IEEE 14th International Colloquium on Signal Processing and Its Application, CSPA 2018, July, 255–260. https://doi.org/10.1109/CSPA.2018.8368722.

Khan, A. A. (1992). An integrated approach to measuring potential spatial access to health care services. *Socioeconomic Planning Sciences, 26*(4), 275–287. https://doi.org/10.1016/0038-0121(92)90004-O

Kim, Y., Byon, Y., & Yeo, H. (2018). Enhancing healthcare accessibility measurements using GIS : A case study in Seoul, Korea. 1–19.

Kirk, J., & Sommer, M. (2006). Menstruation and body awareness: linking girls' health with girls' education. *Tropical Institute (KIT), Special on Gender and Health*, 1–22. http://www.wsscc.org/sites/default/files/publications/kirk-2006-menstruation-kit_paper.pdf%5Cn, http://www.susana.org/_resources/documents/default/2-1200-kirk-2006-menstruation-kit-paper.pdf

Kumar, G., Choudhary, T. S., Srivastava, A., Upadhyay, R. P., Taneja, S., Bahl, R., & Kumar, R. (2019). Utilisation, equity and determinants of full antenatal care in India: Analysis from the National Family Health Survey 4. *BMC Pregnancy and Childbirth, 19*(1), 1–9. https://doi.org/10.1186/s12884-019-2423-8

Langford, M., & Higgs, G. (2006). Measuring potential access to primary healthcare services: The influence of alternative spatial representations of population. *58*(March 2005), 294–306.

Lawal, O., & Anyiam, F. E. (2019). Geo-spatial information science modelling geographic accessibility to primary health care facilities : Combining open data and geospatial analysis modelling geographic accessibility to primary health care facilities : Combining open data and geospatial analysis. *Geo-spatial Information Science, 22*(3), 174–184. https://doi.org/10.1080/10095020.2019.1645508

Luo, W. (2004). Using a GIS-based floating catchment method to assess areas with a shortage of physicians, *10*, 1–11. https://doi.org/10.1016/S1353-8292(02)00067-9

Maleki Rastaghi, M., Barzinpour, F., & Pishvaee, M. S. (2018). A multi-objective hierarchical location-allocation model for the healthcare network design considering a referral system. *International Journal of Engineering, Transactions B: Applications, 31*(2), 365–373. https://doi.org/10.5829/ije.2018.31.02b.22

McGrail, M. R., & Humphreys, J. S. (2014). Measuring spatial accessibility to primary health care services: Utilising dynamic catchment sizes. *Applied Geography, 54*, 182–188. https://doi.org/10.1016/j.apgeog.2014.08.005

McLafferty, S. L. (2003). GIS and health care. *Annual Review of Public Health, 24*(February 2003), 25–42. https://doi.org/10.1146/annurev.publhealth.24.012902.141012

Ministry of Health and Family Welfare, Government of India. (2020). Annual Report 2019–2020. Retrieved from http://www.mohfw.gov.in/

Ministry of Health and Family Welfare, Government of India. (2002). *National Health Policy*, 2002. https://nhm.gov.in/images/pdf/guidelines/nrhm-guidelines/national_nealth_policy_2002.pdf

Ministry of Health and Family Welfare, Government of India. (2013). National Urban Health Mission: Framework for implementation. https://nhm.gov.in/images/pdf/NUHM/Implementation_Framework_NUHM.pdf

Ministry of Health and Family Welfare, Government of India. (2012). *Rural Health Statistics in India*, 2012. https://nhm.gov.in/images/pdf/publication/RHS-2012.pdf

Ministry of Statistics and Programme Implementation, Government of India. (2015). National Accounts Statistics - 2015. Retrieved from https://www.mospi.gov.in/publication/national-accounts-statistics-2015

Mishra, S., Sahu, P. K., Pani, A., & Mehran, B. (2021). Spatial planning framework for development of rural activity centers: Method of location allocation, effect on trip length, and policy implications. *Papers in Applied Geography, 7*(4), 372–391. https://doi.org/10.1080/23754931.2021.1901237

Muggah, E., Graves, E., Bennett, C., & Manuel, D. G. (2013). Ascertainment of chronic diseases using population health data: A comparison of health administrative data and patient self-report. *BMC Public Health, 13*, 16. https://doi.org/10.1186/1471-2458-13-16

Murawski, L., & Church, R. L. (2009). Improving accessibility to rural health services: The maximal covering network improvement problem. *Socioeconomic Planning Sciences, 43*(2), 102–110. https://doi.org/10.1016/j.seps.2008.02.012

National Health Policy (2002). Ministry of Health & Family Welfare, Government of India (2002). *National Health Systems Resource Centre*. https://qps.nhsrcindia.org/sites/default/files/2019-07/National%20Health%20Policy%202002.pdf

Owen, K. K., Obregón, E. J., & Jacobsen, K. H. (2010). A geographic analysis of access to health services in rural Guatemala. *International Health, 2*(2), 143–149. https://doi.org/10.1016/j.inhe.2010.03.002

Patel, V., Parikh, R., Nandraj, S., Balasubramaniam, P., Narayan, K., Paul, V. K., Kumar, A. K. S., Chatterjee, M., & Reddy, K. S. (2015). Assuring health coverage for all in India. *The Lancet, 386*(10011), 2422–2435. https://doi.org/10.1016/S0140-6736(15)00955-1

Penchasky, R., & Thomas, J. (1981). The concept of access. *Medical Care, 19*, 127–140.

Polo, G., Acosta, C. M., Ferreira, F., & Dias, R. A. (2015). Location-allocation and accessibility models for improving the spatial planning of public health services. *PLoS One, 10*(3), 1–14. https://doi.org/10.1371/journal.pone.0119190

Pyrialakou, V. D., Gkritza, K., & Fricker, J. D. (2016). Accessibility, mobility, and realized travel behavior: Assessing transport disadvantage from a policy perspective. *Journal of Transport Geography, 51*, 252–269. https://doi.org/10.1016/j.jtrangeo.2016.02.001

Raghupathi, W., & Raghupathi, V. (2014). Big data analytics in healthcare: Promise and potential. *Health Information Science and Systems, 2*(1), 1–10. https://doi.org/10.1186/2047-2501-2-3

Rahman, S. U., & Smith, D. K. (2000). Use of location-allocation models in health service development planning in developing nations. *European Journal of Operational Research, 123*(3), 437–452. https://doi.org/10.1016/S0377-2217(99)00289-1

Rekha, R. S., Wajid, S., Radhakrishnan, N., & Mathew, S. (2017). Accessibility analysis of health care facility using geospatial techniques. *Transportation Research Procedia, 27*, 1163–1170. https://doi.org/10.1016/j.trpro.2017.12.078

Sedenu et al. (2016). Spatial analysis of the distribution pattern of primary healthcare facilities in spatial analysis of the distribution pattern of primary healthcare facilities in Ile—Ife Metropolis Using Geographic Information System. June. https://doi.org/10.9734/JSRR/2016/22339.

Shah, T. I., Bell, S., & Wilson, K. (2016). Spatial accessibility to health care services: Identifying under-serviced neighborhoods in Canadian urban areas. *PLoS One, 11*(12), 1–22. https://doi.org/10.1371/journal.pone.0168208

Shaw, S., & Sahoo, H. (2020). Accessibility to primary health center in a Tribal District of Gujarat, India: Application of two-step floating catchment area model. *GeoJournal, 85*(2), 505–514. https://doi.org/10.1007/s10708-019-09977-1

SOS Children's Villages International. (2016). International annual report 2015: Our village. https://www.soschildrensvillages.org/getmedia/866c662b-f542-4326-a98b-4c45feff1901/AR-2015-LowRes_FINAL.pdf.

Wang, F., & Luo, W. (2005). Assessing spatial and nonspatial factors for healthcare access: Towards an integrated approach to defining health professional shortage areas, *11*, 131–146. https://doi.org/10.1016/j.healthplace.2004.02.003

Wang, L., Fan, H., & Wang, Y. (2018). Site selection of retail shops based on spatial accessibility and hybrid BP Neural Network. https://doi.org/10.3390/ijgi7060202.

World Health Organization. (1986). Ottawa Charter for Health Promotion: First International Conference on Health Promotion, Ottawa, 21 November 1986. https://www.who.int/teams/health-promotion/enhanced-wellbeing/firstglobal-conference.

Yasenovskiy, V., & Hodgson, J. (2007). Hierarchical location-allocation with spatial choice interaction modelling. *Annals of the Association of American Geographers, 97*, 496–511. https://doi.org/10.1111/j.1467-8306.2007.00560.x

Ye, H., & Kim, H. (2014). Measuring spatial health disparity using a network-based accessibility index method in a GIS environment : A case study of Hillsborough measuring spatial health disparity using a network-based accessibility. *International Journal of Geospatial and Environmental Research, 1*(1), Article 2. http://dc.uwm.edu/ijger/vol1/iss1/2

Zhou, L., & Wu, J. (2012). GIS-based multi-criteria analysis for hospital site selection in Haidian District of Beijing.

Chapter 6
Nature of Health Coping Mechanism

6.1 Introduction

Coping mechanisms are associated with people's livelihoods, whether they belong to socioeconomic groups. However, the magnitude of coping mechanisms differs among the economic, social, and ethnic groups. A coping mechanism is a strategy people use when they face stressful situations such as financial, health, social, and environmental problems. The people's cognitive and behavioral actions reduce, control, and tolerate troublesome environmental situations (Folkman & Lazarus, 1985). Carver (2013) states, "Coping is an effort to prevent or diminish the threat, harm, and loss, or to reduce the distress often associated with those experiences." WHO (1999) defines that "Coping strategies or mechanisms are remedial actions undertaken by people whose survival and livelihood are compromised or threatened." The coping mechanism is deeply influenced and varies by region, community, religion, social group, household, gender, age, season, and historical time. There are many types of coping mechanisms, but Folkman and Lazarus (1985) focused on two basic types of coping mechanisms: problem-focused strategy (PFS) and emotion-focused strategy (EFS). Problem-focused strategies emphasize managing problems rather than emotion, while emotion-focused strategies emphasize using energy to manage situations rather than problems. Healthy coping mechanisms are strategies undertaken by people during adaptation to environmental stress at present and in the future, which include self-help initiatives, practices of personal hygiene, physical exercise, mental pleasure, and spiritual, emotional, and social activities such as asking for help, attempting to find solutions for the issues going on, talking about the problems at hand, etc. The unhealthy coping mechanism can increase vulnerability and risk of the people to environmental stresses such as migration, begging, child labor, violence, prostitution, even selling organs, overeating, drug use, abusing people, crime, etc. and enhance the overburden on the health, economy, and society.

© The Author(s), under exclusive license to Springer Nature Switzerland AG 2025

B. Dutta et al., *Using GIS to Identify Opportunities for Improved Health in Urban Slums*, Global Perspectives on Health Geography,
https://doi.org/10.1007/978-3-031-89617-0_6

The rapid urbanization with migration stream in the urban area has led to slum development with deterioration of the living environment. Therefore, susceptibility to illness is higher in an urban area than its rural counterpart due to the high cost of living; unhygienic living circumstances; competitive and challenging employment opportunities; lack of basic amenities, infrastructure, and facilities; etc., which increases catastrophic cost and disease burden among the urban poor. Moreover, people living in urban slums felt inaccessible and lacked infrastructure for public healthcare services. Such hostile condition makes them more vulnerable to disease progression and calamitous cost burdens. Furthermore, migrants living in urban slum neighborhoods accepted the social structure and resources around them with norms of urban culture and practices (Giddens, 1984). Therefore, slum dwellers intentionally or unintentionally adopt some coping mechanism that may be healthy or unhealthy in their livelihoods, which are different across neighborhoods (Pedersen & Rytter, 2018) and have other health issues or effects (Conde et al., 2018). Being an emerging city, Midnapore town attracted many people to live and work, and these migrated people have adopted different coping mechanisms to maintain their livelihood. This chapter, therefore, seeks to explore the coping strategies of the poor people residing in Midnapore town and to find the problems they encounter while taking coping strategies.

6.2 Foregone Care

Foregone care is the healthcare services needed during illness and health shocks. More explicitly, it is another type of healthcare accessibility for people taken during their health shock in the family, irrespective of the age of family members. For example, family members with chronic diseases have a higher prevalence of foregone care than minor illnesses. Foregone care is the measure of the gap between apparent need and actual accessibility of healthcare services, thereby assessing the performance of the healthcare system of a place. Foregone care may be affected by households' income, available healthcare facilities, accessibility, education, cultural status, beliefs, etc. (Mielck et al., 2009; Bonfrer & Gustafsson-wright, 2017). Figure 6.1 shows the household level data for foregone healthcare, where 41.85% of slum households reported that they had taken foregone care during chronic illness and another 23.26% when they seek. In comparison, 13.65% and 13.01% reported preceding care for maternity care and minor illness, respectively.

However, they reported less frequently foregone care is taken for preventive care, hospitalization, maternity care, and other cases. For foregone care, the urban, primarily poor, prefer government hospitals, primary healthcare centers, private healthcare, and quack doctors (Fig. 6.2). Foregone care is associated with high healthcare expenditure, so poor people either ignore the foregone care or cannot use modern foregone healthcare facilities. However, many other factors may influence the foregone healthcare accessibility, such as costing too much, long waiting time, etc. (Fig. 6.3).

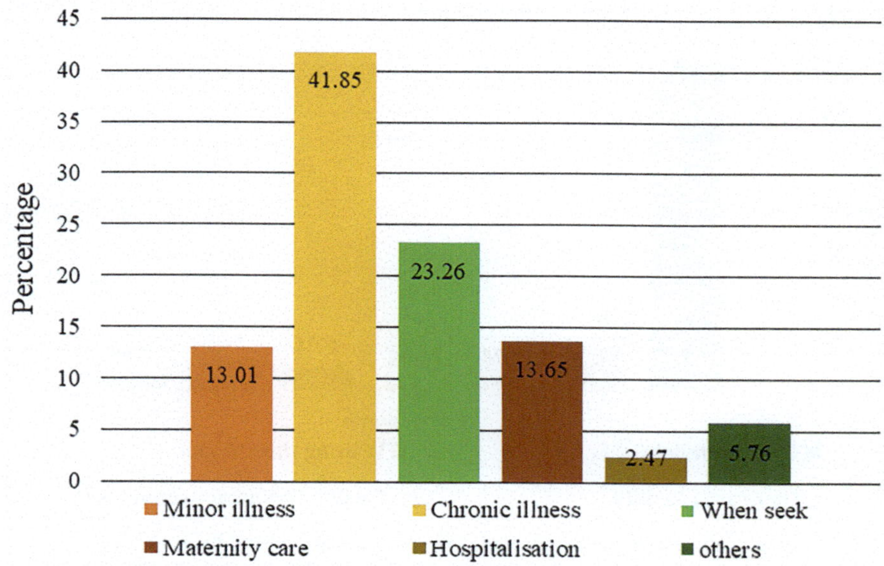

Fig. 6.1 Households visited foregone care (Source: Field survey, 2022)

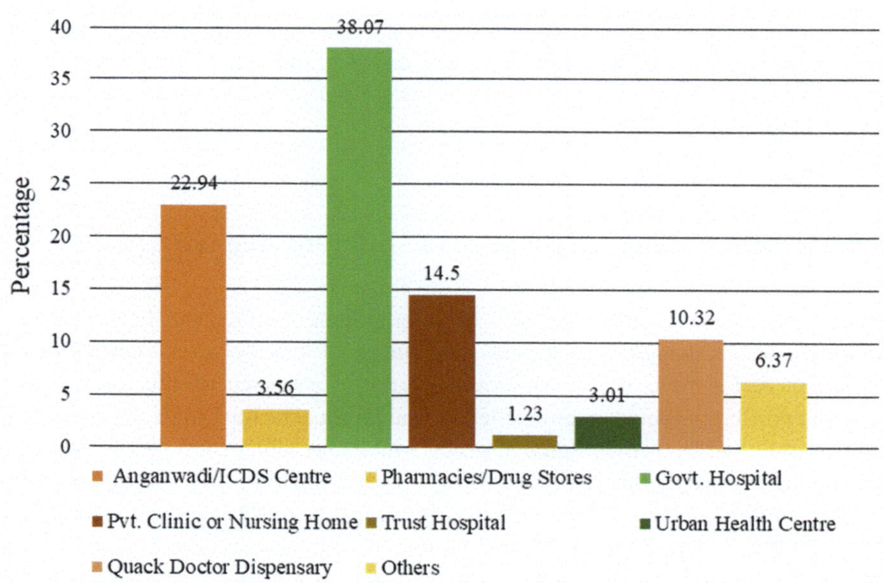

Fig. 6.2 Place of foregone medical care households most visited (Source: Field survey, 2022)

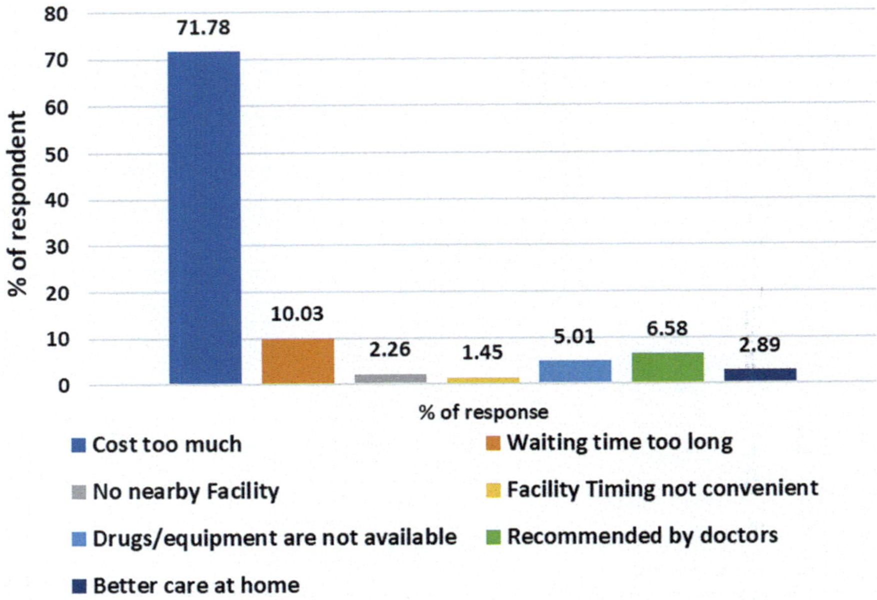

Fig. 6.3 Households most facing obstacles to foregone medical care (Source: Field survey, 2022)

When urban poor were asked about the satisfaction level of the existing foregone care facilities, only 50.44% reported being satisfied and still 20% of households were dissatisfied with the existing foregone care during treatment of their health shocks (Fig. 6.4).

6.3 Health Coping Mechanism of Different Age Groups

Since birth, we are confronted with socio-emotional conflict and health-related problems or challenges through the life stages of psychological development (Folkman & Lazarus, 1985). Moreover, during every stage of life, individuals encounter different socioeconomic, mental, and physical health problems, including cultural, political, and environmental stress, which may affect coping strategies and health outcomes (Folkman & Lazarus, 1985). Therefore, from a psychological point of view, children, adolescents, middle-aged people, and older people encounter different problems within their work and living environment, and they are solved by adopting different types of coping strategies that vary across places. Folkman and Lazarus (1985) opined that coping mechanisms are different at different age groups and vary across places with their daily livelihood; more explicitly, people change their coping strategies with the successive stages of life. For example, adolescents control their psychological problems by working or studying hard while seeking a job, while middle-aged people try to balance family and their earning problems.

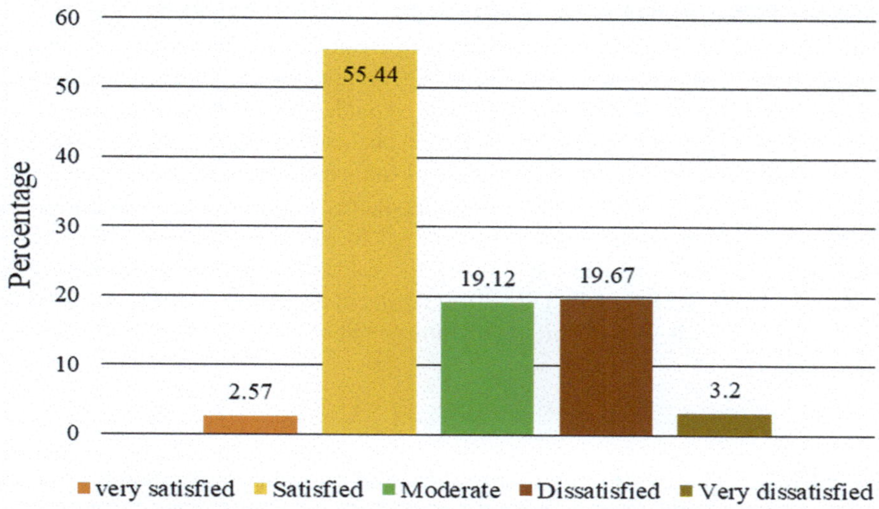

Fig. 6.4 Satisfaction level about healthcare facilities reported by households (Source: Field survey, 2022)

Finally, in old age, people try to recover from chronic illnesses such as diabetes and other aged health problems. Beyond that, at every age, from childhood to old age, one tries to recover from health challenges through adopting different strategies, which may vary across socioeconomic, ethnic, religious, and cultural groups (Aldwin & Revenson, 1987; Baker & Berenbaum, 2007; Almeida, 2005). Therefore, there is a close relationship between age differences and coping strategies (Teachman, 2006) to a broad extent.

In the context of slum neighborhoods, they face challenges from socioeconomic and environmental threats due to living in an unhygienic and dilapidated environment in the urban areas of most developing countries. Most reports and researchers found that slum dwellers are deprived in urban areas and vulnerable to different environmental threats. They encountered different challenges from birth to old age. As slum dwellers are socioeconomically and environmentally vulnerable, they adopt various strategies to cope with their vulnerable situations at different stages of their lives. Thus, the study used a qualitative cum quantitative approach to analyze the coping mechanism of slum dwellers in various age groups.

6.3.1 Data and Method

The Kaiser-Meyer-Olkin (KMO) and Bartlett's test were conducted to test the suitability of principal component analysis (PCA) for the variables, indicating that correlations among variables were appropriate for PCA. Before statistical analysis, all variables were normalized through percentage, average, weightage, or per capita functions. PCA extracted components from the variables and employed Kaiser normalization and varimax rotation methods with eigenvalues higher than 1.0 through the dimension reduction tool of SPSS v26. The components with an eigenvalue greater than 1.0 were identified as the latent variables. The components were then described by signs denoting their influence on health coping strategies (HCS). Positive values (+) indicated higher levels of influence, while negative values (−) indicated lower levels of influence of the components. For each slum neighborhood, the highest loading components were selected to calculate the final HCS index (HCS_{slum}). HCS was calculated by adding the values of the principal component (Cutter et al., 2003; Hummell et al., 2016; Aksha et al., 2019) to arrive at the total HCS score for each spatial unit (156 slum neighborhoods) of Midnapore town, derived as (Eq. 6.1)

$$\mathrm{HCS}_{\mathrm{slum}} = \sum \mathrm{PC}_i \; i = 1,2,3\ldots\ldots n \tag{6.1}$$

where HCS_{slum} indicates the index score of HCS of the slum neighborhoods and PC_i is the principal component. Assume that different components have equal importance on HCS; therefore, no weight was assigned. For each slum neighborhood, the HCS scores were mapped through ArcGIS v10.4 to visualize high and low HCS based on Z-scores. The Local Moran's I test was used to recognize a statistically clustered or random pattern of slum neighborhoods regarding HCS across the study area through estimated autocorrelation. The autocorrelation value close to ±1 denotes the strong positive or negative spatial autocorrelation indicating clustering of high or low values of HCS. The values are near zero, suggesting the non-appearance of spatial patterns of randomness.

6.3.2 Health Coping Mechanism in Childhood (0–10 Years)

Physical, emotional, and behavioral health and well-being vary within the living environment, even in distinctive socioeconomic, cultural, and environmental conditions. It is believed that if the beginning is in good condition, the future will be wealthy. Therefore, childhood is an important stage of life for individuals and a nation. If a child is malnourished, they are stunted (UN, 2019), which subsequently affects their physical, psychological, and behavioral development as well as their future health outcomes (UNICEF, 2020). Another perspective is that problems in terms of socioeconomic and environmental threats may affect children's mental health, poor school regularity and performance, and poor behavioral practices, which will make them vulnerable to society in the future, and they will always be in a vicious cycle of poverty. Therefore, a child is at an important stage of life, fully responsible by their parents with utmost care and nourishment. Primarily, children have three important features such as being psychologically weak with low understanding, not having solution power spontaneously, and always having health-related problems. In this context, parents should have some strategies to maintain their child well in terms of psychological, socialization, and behavioral environment (WHO, 2008). In developing countries like India, urban slum children are more in trouble in terms of nutrition, recreation, health, living space, school, and their psychosocial living environment than their rural counterparts. In childhood, proper nutrition and healthy living make them socially healthy, which is a nightmare or rare feature in urban slum areas. As the urban poor are the vulnerable, deprived group, therefore it is assumed that they have been adopting some strategies during childcare than the affluent group.

The study has undertaken nine broad variables (Table 6.1) in the light of distinct slum livelihood practices in Midnapore town regarding childcare. The PCA has extracted seven components (Table 6.2) among the variables and built up a health coping strategy (HCS) index following the methodological formula for childhood care among the slum neighborhoods. The map (Fig. 6.5) shows that roughly 72% of slum neighborhoods adopt moderate coping strategies and around 15% have not yet adopted active coping strategies. In comparison, only 13% of slum neighborhoods

Table 6.1 Variables for HCS of childcare

Variables		Mean	Std. deviation
Preference for child nutrition	Breastfeeding	29.17	21.937
	Normal home-made food	29.36	20.719
	ICDS supply food	1.73	5.584
	Baby food	4.10	8.935
Taken preschool services for early childcare	Regularly	18.27	16.585
	Occasionally	15.32	17.055
Demand for junk food	Junk food	51.92	23.227
Aware of polio/vaccine	Polio/vaccine	54.68	23.040
Growth monitoring of child	Regularly	19.17	17.486
	Some time	16.86	17.145
	When sick	14.10	16.920
Go to school by	Cycle	20.19	20.176
	Motorcycle	7.12	11.529
	Car	22.88	19.543
Place of recreation	Park	8.85	14.231
	Nearby playground	15.83	15.982
	Nearby roads	9.55	12.357
	Outside town	19.68	19.391
Caregiver during illness or recreation	Mother	28.40	19.854
	Mother and father	18.40	19.493
	Neighbors	5.00	8.908
During illness, go to healthcare centers	Government hospital	55.28	11.245
	Private (quack doctor dispensary)	36.21	12.357
	Others	12.32	8.912

Source: Field survey (2022)

Table 6.2 PCA for HCS of childcare

Variables	Components 1	2	3	4	5	6	7
Breastfeeding	0.328	0.615	0.080	0.322	0.316	−0.233	−0.172
Normal home-made food	0.471	−0.023	0.185	0.012	−0.178	0.605	0.238
ICDS supply food	0.170	−0.066	−0.017	0.019	−0.115	−0.175	0.695
Baby food	−0.007	0.031	−0.122	0.138	0.171	0.175	0.566
Regularly	0.661	−0.169	−0.145	−0.324	0.315	−0.021	0.147
Occasionally	0.030	0.340	0.185	0.685	0.091	−0.132	0.191
Demand junk food	0.734	0.359	0.281	0.068	0.207	0.128	0.030
Aware of polio/vaccine	0.666	0.404	0.178	0.178	0.062	0.330	0.035
Regularly	0.537	0.168	0.200	−0.441	0.285	0.103	0.231
Some time	0.079	−0.188	0.089	0.808	0.095	0.206	0.080
When sick	0.176	0.729	0.209	−0.101	−0.017	0.021	−0.215
Cycle	0.136	0.185	0.641	0.274	0.195	0.110	0.197
Motorcycle	−0.097	0.113	−0.190	0.065	0.024	0.791	−0.091
Car	0.705	0.211	−0.165	0.091	−0.125	−0.307	−0.178
Park	0.279	−0.214	−0.553	0.313	0.297	0.230	−0.223
Nearby playground	0.652	−0.008	0.036	0.101	−0.109	−0.018	0.110
Nearby roads	0.086	−0.051	0.712	0.073	−0.037	−0.127	−0.347
Outside town	0.021	0.746	0.102	−0.033	0.133	0.244	0.345
Mother	0.506	0.103	−0.236	0.022	−0.649	0.231	−0.068
Mother and father	0.172	0.307	0.118	0.166	0.764	0.062	0.034
Neighbors	0.013	0.281	0.761	0.076	0.272	−0.017	−0.132
Government hospital	0.537	0.168	0.200	−0.441	0.285	0.103	0.231
Private (quack doctor dispensary)	0.079	−0.188	0.089	0.808	0.095	0.206	0.080
Others	0.176	0.729	0.209	−0.101	−0.017	0.021	−0.215

Extraction method: principal component analysis
Rotation method: varimax with Kaiser normalization
(a) Rotation converged in 17 iterations

have taken active coping strategies in their childcare. The Local Moran's I index (cluster-outlier analysis) value is 0.05 at p-value <0.005 with a Z-score value of 2.83 (Fig. 6.6), indicating a clustered pattern of slum neighborhoods regarding HCS of their childcare. Therefore, it is directed toward the average health condition among the children living in the slums of Midnapore town, where awareness among the parents needs to be strengthened for the better well-being of their children at the primary stage and in the future. It is also suggested that they look into their socio-economic status and make provisions to improve their existing situation with the help of local authorities, NGOs, and other administrative bodies.

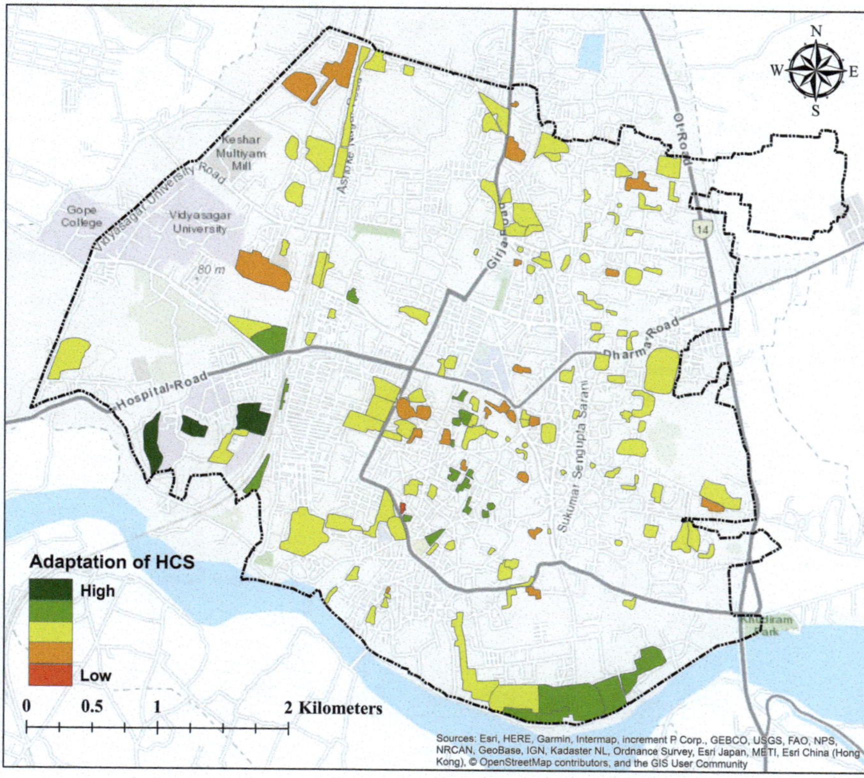

Fig. 6.5 Map shows the health coping mechanism of child care among the slum neighbourhoods

6.3.3 Health Coping Mechanism of Adolescents (11–18 Years)

Physical appearance, voice, psychology, and behavior change during adolescence. It is the phase from which adolescents jump into the responsible mature or adult stage. Thus, it is an essential phase of life. Adolescents often experience changes in physical appearance due to hormonal shifts, which can lead to feelings of embarrassment, increased self-consciousness, irritability, aggression toward family and friends, moodiness, risktaking behavior, curiosity about the opposite sex, and heightened mental and emotional questioning. Therefore, this stage of life faces many challenges like educational pressure (failure or highest position), future of life, responsibility, family and friend relations, sexual desire and love affair problems, homework, menstruation-related problems, and health abuse. These problems are influenced by the family background, socioeconomic and cultural status, friend characteristics, school and living environment, etc.

Slum adolescents are characterized by low socioeconomic families, low levels of literacy, lower school enrollment, engagement in child labor, lower BMI and weight, addiction to tobacco, and related health problems, which are common in developing

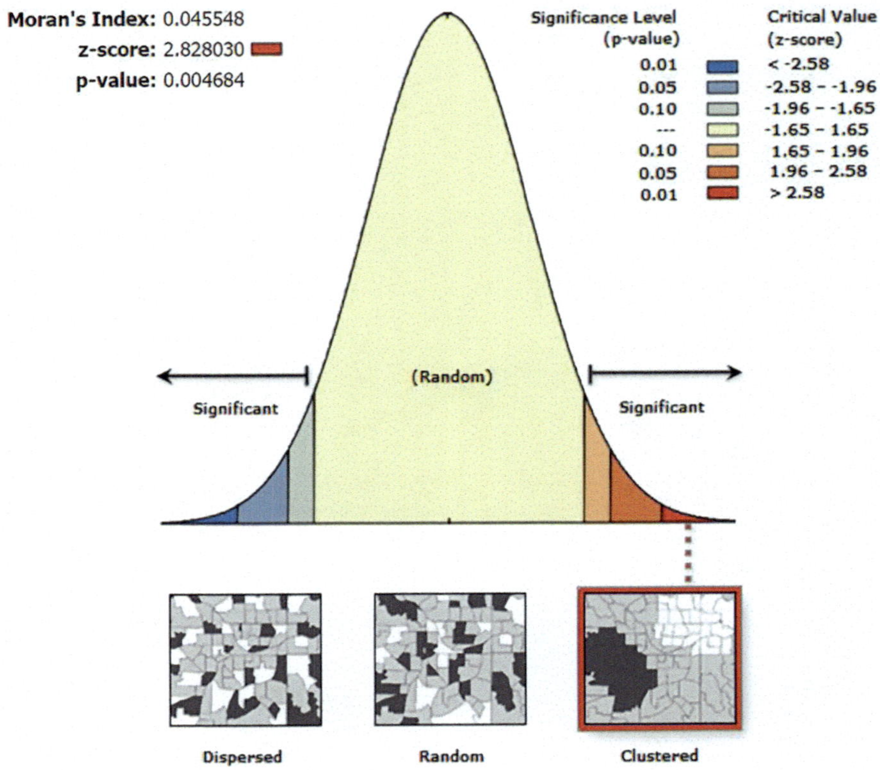

Fig. 6.6 Local Moran's I statistical test of slum neighbourhoods' regarding health coping mechanism of child care

countries like Bangladesh, India, etc. They are always perceived as having more stress than adolescents of affluent families like the responsibility of the family, earning for the family, parental loss, treatment for the self and family members, sexual assaults and love-related issues, and threats to the social position, which make them worse people in the society and enter into the vicious cycle of poverty in future (Izutsu et al., 2006). Due to the stresses of slum adolescents, they fail to achieve educational attainments, physical and mental development, and the potential for productivity in the future. Hence, this phase has to be appropriately handled so adolescents' futures can be bright and out of the poverty cycle and for the country's well-being. Therefore, slum adolescent and their family adopt different strategies to cope with the adverse living situation. The study has thus undertaken 15 variables (Table 6.3) and extracted 6 components by PCA (Table 6.4). A map in ArcGIS represents the index score for adolescents' HCS. The map (Fig. 6.7) shows that adolescents in roughly 50% of slum neighborhoods adopt a moderate level of coping strategies and around 30% have not been adopting active coping strategies, while only 20% have taken active coping strategies. The Local Moran's I index (cluster-outlier analysis) value is −0.02 at p-value <0.60 with a Z-score value of −0.58

Table 6.3 Variables for HCS of adolescents

Variables		Mean	Std. deviation
Go to school by	Motorcycle	5.00	9.739
	Car	16.22	17.090
Aware of health nutrition	Aware of health nutrition	33.72	20.921
Menstruation care	Menstrual safety	30.19	19.125
Relation with siblings or family	Very good	7.50	11.839
	Good	25.13	18.716
	Moderate	11.99	15.130
	Bad	22.37	17.784
Preference for leisure time with	Mother	14.94	16.791
	Mother and father	14.17	15.779
	Neighbors	11.15	14.367
	Relatives	6.28	9.982
Caregiver during illness/stress/depression	Mother	21.86	20.346
	Mother and father	13.97	15.479
	Neighbors	5.83	10.094

Source: Field survey (2022)

Table 6.4 PCA for HCS of adolescents

	Components					
Variables	1	2	3	4	5	6
Motorcycle	−0.014	0.061	−0.018	−0.017	0.085	0.923
Car	0.029	0.692	0.073	0.131	0.289	−0.320
Aware of health nutrition	−0.018	0.680	0.096	0.314	0.058	0.343
Menstrual safety	0.043	0.833	0.086	−0.023	−0.053	0.076
Very good	0.077	0.093	0.056	0.912	0.104	−0.031
Good	0.637	0.519	0.055	−0.173	0.081	−0.015
Moderate	0.180	0.270	−0.008	0.206	0.816	0.142
Bad	0.507	0.390	0.222	0.270	−0.551	0.068
Mother	−0.169	0.400	−0.609	0.311	0.052	0.205
Mother and father	0.222	0.291	0.806	−0.009	0.062	0.033
Neighbors	0.805	0.065	0.023	−0.061	0.272	−0.044
Relatives	0.833	−0.061	0.146	0.227	−0.078	−0.034
Mother	0.003	0.611	−0.619	0.027	0.220	0.094
Mother and father	0.162	0.186	0.777	0.341	−0.042	0.081
Neighbors	0.851	−0.086	0.243	0.028	−0.079	0.045

Extraction method: principal component analysis
Rotation method: varimax with Kaiser normalization
(a) Rotation converged in six iterations

(Fig. 6.8), indicating a random pattern of slum neighborhoods regarding HCS of the adolescents. Therefore, adolescents living in the slums of Midnapore town tend to experience poorer overall health conditions, highlighting the urgent need to raise awareness among both adolescents and their parents to promote better well-being

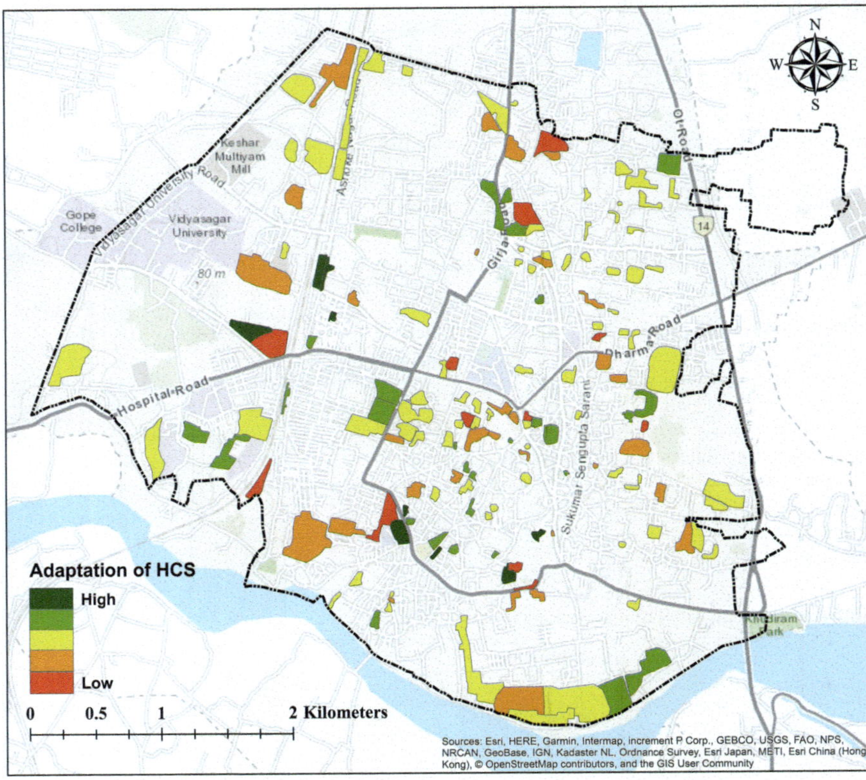

Fig. 6.7 Map shows the adolescents' health coping mechanism among the slum neighbourhoods

now and in the future. It is also suggested that they look into their socioeconomic status and behavioral environment and make provisions to improve their existing situation with the help of local authorities, NGOs, and other administrative bodies.

6.3.4 Health Coping Mechanism of Parents (21–59 Years)

The parental stage of life is a backbone and vital pillar of any family. Family care, in most cases, depends on parental responsibility. Parents have many responsibilities like childcare, earning, providing healthcare to family members, housing maintenance, caring about children's education and aged health, and thinking about the child's future and themselves. Moreover, if family members have a long-term illness or death (child or spouse or parents-in-law), it will be an extra burden on income and mental health. Thus, parental life goes through a lot of stresses and challenges. To cope with this situation, they have taken some strategies for health and well-being. In the context of slum parents, they have faced many challenges and

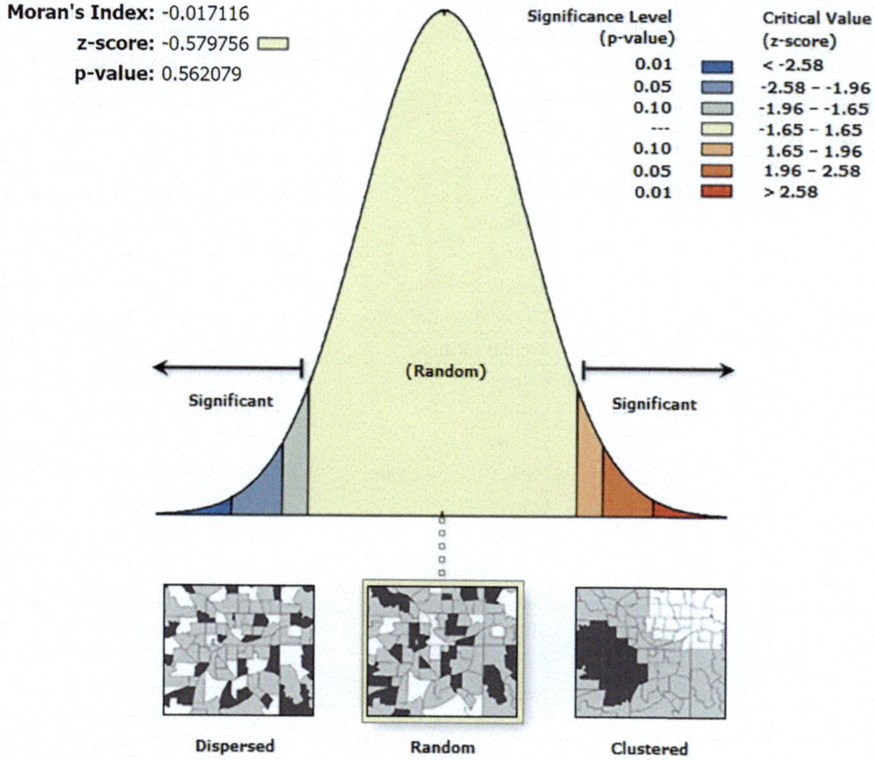

Fig. 6.8 Local Moran's I statistical test of slum neighbourhoods' regarding health coping mechanism of adolescents

stress more than their affluent urban and rural counterparts. The challenges are income and family sustenance; healthcare; child education and other basic amenities cost; unemployment; childcare; extramarital affairs; family violence; death of a spouse, child, or family members; etc. In this situation, parental groups in the slum have developed strategies to cope with and maintain their health and well-being. It is very tough for researchers to identify and quantify the variables related to the coping strategies of parents due to the distinct nature of slum areas, which vary across the place. To understand the coping mechanism of the parental group in the slum of Midnapore town, the study takes into account 24 important variables (Table 6.5), and PCA has extracted 6 components (Table 6.6). A map in ArcGIS represents the parent's index score for the HCS. The map (Fig. 6.9) shows that parents in roughly 45% of slum neighborhoods adopt a moderate level of coping strategies and around 35% have not been adopting active coping strategies, while only 20% have taken active coping strategies. The Local Moran's I index (cluster-outlier analysis) value is 0.03 at *p*-value <0.05 with a *Z*-score value of 1.86 (Fig. 6.10), which signifies a clustered pattern of slum neighborhoods regarding the HCS of the

Table 6.5 Variables for HCS of parental group

Variables		Mean	Std. deviation
Preference of son	Preference of son	53.44	20.566
Perception of family health	Good	77.70	18.018
	Bad	4.75	8.487
Decision about healthcare	Wife	14.43	13.232
	Husband	31.80	15.759
	Jointly	19.67	18.346
	Family members	32.46	20.709
Violence in the family	Physical/sexual abuse	2.30	7.392
	Mental torture	1.31	4.992
	Others	11.97	12.223
Causes of violence in the family	Drinking	3.44	8.344
	Income	1.64	5.532
	Family conflict/care	3.93	8.016
	Others	6.07	9.179
Caregiver during illness/stress/depression	Mother	38.03	18.423
	Father/mother-in-law	10.66	12.365
	Neighbors	13.44	17.405
Preference for leisure time with	Social media	23.11	16.486
	Family	61.80	22.324
	Neighbors	6.23	10.027
	Watching TV	18.36	15.075
Relation with	Relation with family/neighbors	51.31	19.103
	Relation with children	48.69	22.021
	Relation with spouse	53.44	22.795

Source: Field survey (2022)

parents. Therefore, it is directed toward moderate health conditions and well-being among the parents living in the slums of Midnapore town, which need to look into the socioeconomic status and significant problems in their living environment and make provisions to improve their existing situation from the local authorities, NGOs, and other administrative bodies. Further in-depth research is also needed in this context.

6.3.5 *Health Coping Mechanism of the Aged (>60 Years)*

Aged or elder or geriatric (above 60 years) is an end stage of life characterized by some misery due to loneliness, depression, social isolation, loss of mobility, cognitive declination, morbidity, and illness. Moreover, it is common in every geriatric with declining physical, social, and cognitive activities and self-neglected nature.

Table 6.6 PCA for HCS of parental group

Variables	Components								
	1	2	3	4	5	6	7	8	9
Preference son	0.014	−0.289	0.079	0.037	−0.094	0.372	0.458	−0.112	−0.045
Good	0.318	0.185	0.275	0.723	−0.042	0.029	0.049	0.009	0.235
Bad	−0.067	0.185	0.126	−0.815	−0.111	−0.020	−0.006	−0.059	0.055
Wife	−0.458	−0.055	0.135	0.282	0.464	0.254	−0.303	0.132	0.151
Husband	−0.116	0.205	−0.032	0.100	0.047	0.106	0.068	0.833	0.013
Jointly	0.385	−0.011	0.156	0.405	−0.304	−0.166	−0.131	−0.034	−0.308
Family members	0.723	0.084	0.140	0.129	0.049	−0.131	−0.063	−0.327	0.222
Physical/sexual abuse	−0.340	0.340	−0.035	0.147	0.162	−0.076	0.632	−0.282	−0.009
Mental torture	0.253	−0.093	0.152	0.031	0.660	−0.518	−0.119	0.153	−0.083
Others	0.684	0.243	−0.239	−0.255	−0.031	0.355	−0.088	0.261	0.002
Drinking	−0.037	0.266	−0.026	0.123	0.773	0.193	−0.015	−0.261	−0.162
Income	0.275	−0.133	−0.583	−0.467	−0.003	0.042	0.075	−0.015	0.355
Family conflict/care	−0.107	−0.052	−0.051	−0.103	0.027	−0.068	0.798	0.396	−0.095
Others	0.716	0.241	0.237	0.307	−0.167	0.091	−0.179	0.091	−0.108
Mother	0.046	0.053	0.042	0.039	0.017	0.174	−0.115	−0.003	0.815
Father/mother-in-law	0.849	−0.030	−0.089	0.119	−0.044	0.002	−0.058	−0.097	0.040
Neighbors	0.470	0.033	0.262	0.257	0.084	0.454	−0.149	0.044	−0.333
Social media	−0.051	0.002	−0.040	0.024	−0.053	−0.782	0.022	−0.146	−0.239
Family	0.343	0.079	0.800	0.056	−0.215	−0.010	−0.091	0.109	0.207
Neighbors	−0.192	0.156	0.011	−0.099	0.633	−0.059	0.173	0.170	0.193
Watching TV	0.093	−0.028	−0.869	0.048	−0.232	−0.088	−0.066	0.142	0.035
Relation with family/neighbors	0.001	0.936	0.031	0.020	0.031	−0.026	0.075	0.090	−0.005
Relation with children	0.060	0.863	0.042	−0.186	0.032	0.019	−0.116	0.091	0.004
Relation with spouse	0.234	0.854	0.092	0.123	0.224	0.011	0.031	0.018	0.076

Extraction method: principal component analysis
Rotation method: varimax with Kaiser normalization
(a) Rotation converged in 15 iterations

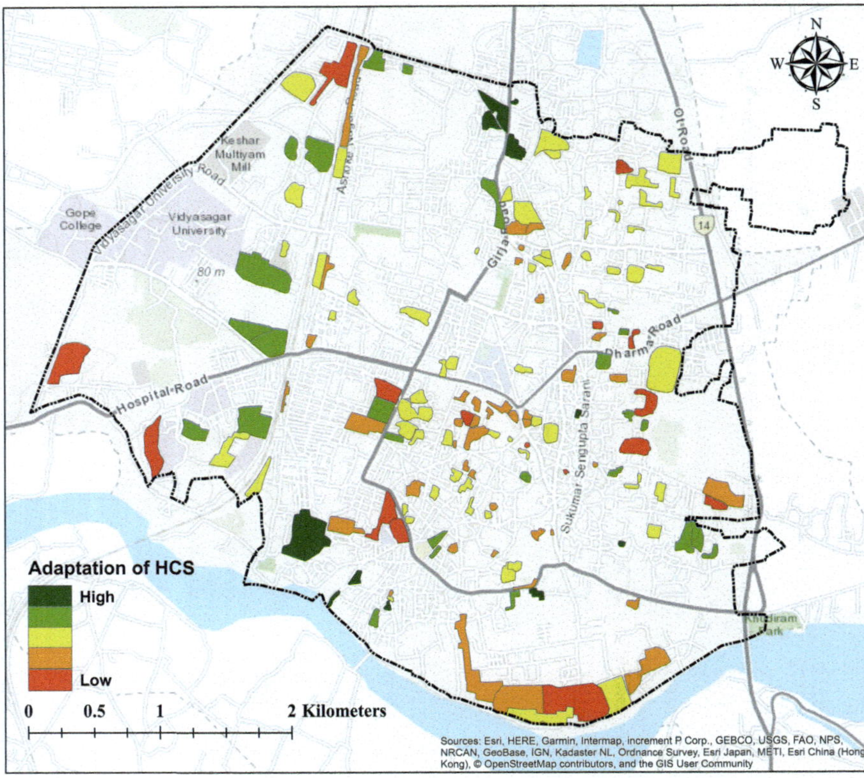

Fig. 6.9 Map shows the parental health coping mechanism among the slum neighbourhoods

This stage of life is always associated with depression due to inaccessibility to healthcare services, death of a spouse and family members, bereavement, poor health, loss of income, dependency on caregivers, loss of mobility, etc. This depression can lead to mental health disorders, a decline in quality of life, and overall well-being, and may contribute to the onset of chronic illnesses, increased risk of non-suicidal mortality, and even death. Therefore, coping strategies are limited to taking action in this stage of life. However, socioeconomic status, type of family, nature of employment, monetary dependency, living environment, cultural status and belief, healthcare facilities and services, caregiving, loneliness, government policy, etc. are associated with geriatric people's health well-being and coping strategies.

In India, the elderly (above 60 years) population is increasing steadily; for example, geriatric people were 12.1 million in 1901, which increased to 104 million (8.6% of the total population) in 2011 and is expected to become 173 million by 2026. WHO reported that mental health problems are the leading cause of morbidity among geriatric people in India. It is also evidenced from the different research that physical and mental well-being is worse among the elderly in a slum than the urban affluent and rural counterparts. Generally, a slum is characterized by high monetary

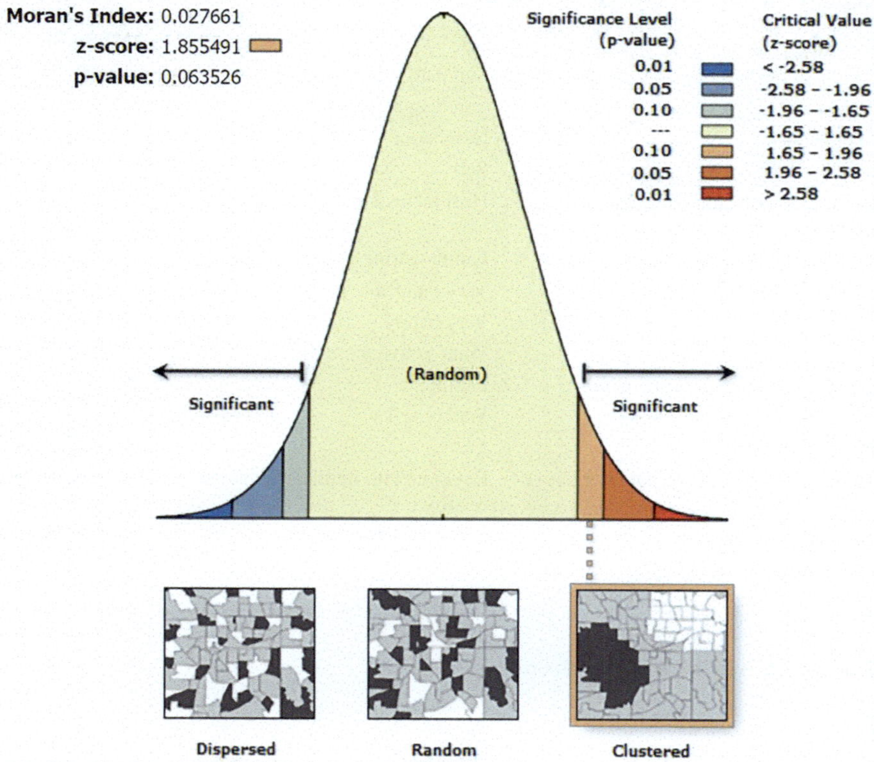

Fig. 6.10 Local Moran's I statistical test of slum neighbourhoods' regarding health coping mechanism of parental

dependency on the young, chronic illness, less living space (in terms of the number of living rooms), poverty, and an unhygienic environment. Therefore, one can imagine the physical and mental well-being of the geriatric people living in a slum. As living in poor socioeconomic status, the aged have been taking coping strategies somehow. Thus, the study was undertaken to understand that group's living conditions and well-being. To understand the coping mechanism of the aged group in the slum of Midnapore town, the study takes into account eight broad variables (Table 6.7), and PCA has extracted five components (Table 6.8). A map in ArcGIS represents the index score for aged HCS. The map (Fig. 6.11) shows that roughly 40% of slum neighborhoods' aged group adopts a moderate level of coping strategies, while around 22% have not been adopting active coping strategies, and only 38% have taken active coping strategies. The Local Moran's I index (cluster-outlier analysis) value is −0.02 at *p*-value <0.40 with a *Z*-score value of −0.76 (Fig. 6.12), implying a random pattern of slum neighborhoods regarding HCS of the aged. Therefore, it is directed toward lower health conditions and well-being among the aged living in the slum.

Table 6.7 Variables for HCS of aged group

Variables		Mean	SD
Caregiver during illness/stress/depression	Son/daughter	11.67	14.540
	Son/daughter-in-law	12.95	15.708
	Neighbors	5.32	9.327
	Self	2.95	7.114
Taken regular health checkups and medicine	Health checkup and medicine	11.41	14.567
Satisfaction about family caring	Family caring	26.92	19.398
Feel about life	Very painful	4.36	9.979
	Very happy	10.13	14.096
	Want to live more time	5.45	9.459
Preference for leisure time with	Family	25.90	19.003
	Watching TV	5.71	11.535
Religious activity	Yes	20.96	17.881
Relation with family/neighbors/relatives	Relation with family/neighbors/relatives	19.74	18.208
Relation with children	Relation with children	21.60	17.358

Source: Field survey (2022)

Table 6.8 PCA for HCS of aged group

	Components				
Variables	1	2	3	4	5
Son/daughter	−0.046	0.137	0.305	0.807	−0.136
Son/daughter-in-law	0.813	0.091	0.202	−0.146	−0.209
Neighbors	−0.354	0.729	0.036	−0.338	0.015
Self	0.062	−0.130	0.086	0.060	0.906
Health checkup and medicine	0.722	−0.076	−0.067	0.277	0.071
Family caring	0.439	0.636	0.293	0.189	−0.114
Very painful	0.571	−0.082	0.244	−0.034	0.130
Very happy	0.173	−0.030	0.060	0.738	0.177
Want to live more time	−0.014	0.756	−0.083	0.045	−0.194
Family	0.446	0.706	0.093	0.318	0.127
Watching TV	0.220	−0.253	0.663	0.178	−0.190
Yes	0.643	0.185	0.176	0.163	0.085
Relation with family/neighbors/relatives	0.301	0.070	0.826	0.225	0.143
Relation with children	0.030	0.458	0.769	0.033	0.258

Extraction method: principal component analysis
Rotation method: varimax with Kaiser normalization
(a) Rotation converged in seven iterations

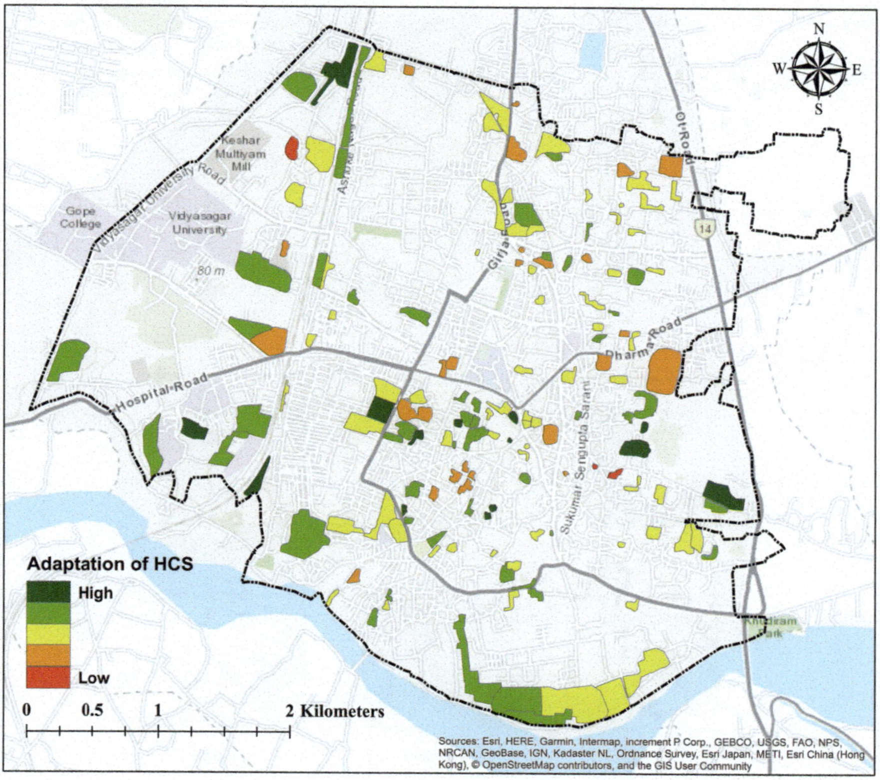

Fig. 6.11 Map shows the aged health coping mechanism among the slum neighbourhoods

Midnapore town needs to look into the aged health status and monitoring by camp, aged healthcare services, and significant problems in their living environment and make provisions to improve their existing situation from the local authorities, NGOs, and other administrative bodies. The Government of India and West Bengal have recently launched "Ayushman Bharat" and "Swasthya Sathi" for national and state health protection. Nonetheless, further in-depth research is also needed in this context.

6.4 Overall Health Coping Mechanism (WYHCS and HSCS)

People in the world face different environmental, socioeconomic, natural, and health shocks in their life span despite their economic status. However, in developing countries, health shock and the "medical poverty trap" (Whitehead et al., 2001) are more common among the poor due to socioeconomic and environmental circumstances. Urban poor encounter shocks associated with their daily livelihood and

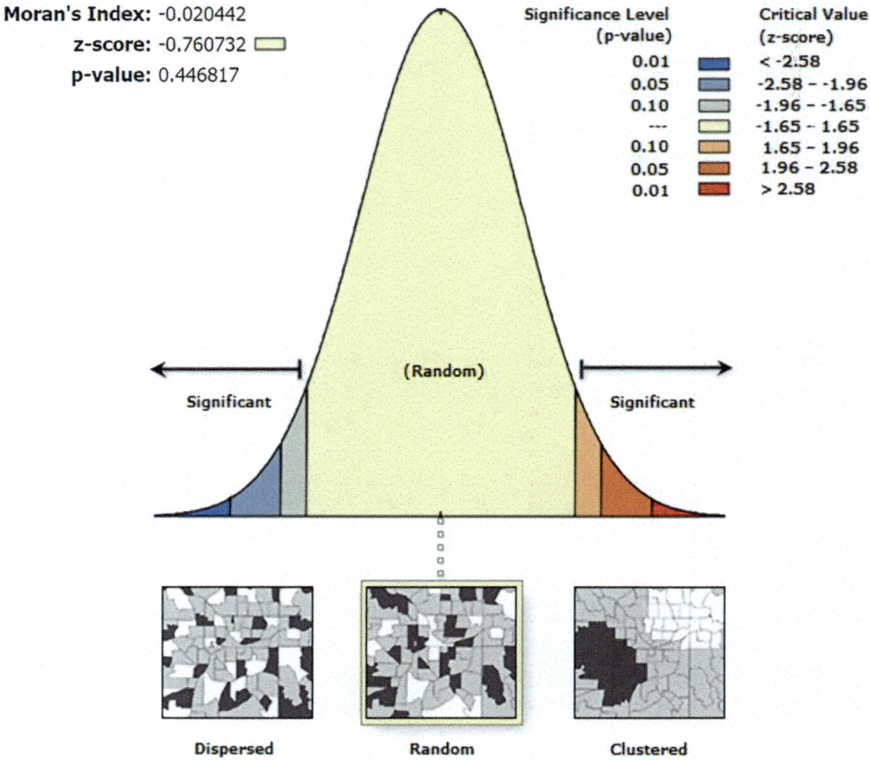

Fig. 6.12 Local Moran's I statistical test of slum neighbourhoods' regarding health coping mechanism of aged

health shocks throughout the year. Health shocks throughout the year are largely driven by factors that are idiosyncratic in nature (Dhanaraj, 2016), including lack of basic amenities, unemployment, seasonal weather changes, limited access to healthcare, illness within the family, and inadequate housing infrastructure, etc., are the leading causes of health shocks throughout the year. For example, during seasonal changes and the monsoon periods, children and the aged are more likely to have colds and coughs, and due to indoor pollution, women face respiratory problems. To date, different studies have examined the cause of health shocks in the urban poor. All agree that the reasons behind health shocks among the urban poor are poverty, lack of personal hygiene awareness, future planning, etc. Therefore, they are adopting coping strategies throughout the year to overcome the health shocks, such as selling fixed assets, mortgaging household assets, cutting down costs on other livelihood purposes, stopping children's education, diversifying employment and credit loans, resorting to acquaintances and government facilities, and many more. Nevertheless, sometimes adaptive strategies are limited among the poor and have some adverse effects on family welfare and living standards, thus falling them again into the poverty trap. Moreover, throughout the year, health shocks increase the extra economic burden of healthcare costs on family income, reducing the quality of life and increasing the "medical poverty trap." The study, thus, tries to explore the coping strategies among the urban poor throughout the year and during the health shocks, as well as the role of livelihood assets in taking preferred health shock coping strategies by introducing a framework of the whole-year coping strategies and health shock coping strategies approach (WYHCS and HSCS).

6.4.1 Data and Method

Worldwide health shocks and coping strategies are associated with the livelihood system and area's multidimensional concept. Most of the research assesses risk perceptions and coping strategies in the different frameworks rather than the livelihood system. This study introduces a framework of the whole-year coping strategies and health shock coping strategies approach (WYHCS and HSCS) to assess the perceptions of health risk and the risk management strategies of slum households in Midnapore town. Comprehensively, simple random sampling was employed to survey 800 households with well-structured questionnaires from 156 notified slum neighborhoods. The households (ranging from five to ten) in each slum neighborhood were selected according to their population composition. Household heads or other senior members were targeted from the selected households for the interview. The average age of the household heads or other senior members at the time of the survey was 48 years. The mean household size is 4.90 persons, of whom 3.54 are literate and 1.43 are employed persons with an approximate monthly income of Rs. 2526.72 (Table 6.9). The data collected on health risk perception (illness of their own or their families) in connection with different age groups and their coping strategies to health shock throughout the year was based on in-depth interviews. The data collected from 800 households was converted slum-wise (156) using the average method. PCA was used to develop the WYHCS and HSCS index to understand better the coping strategies for eliminating methodological errors. The PCA was accomplished in SPSS v26, which extracts the components with their cardinality based on the influences of the components. The Pearson correlation coefficient and

Table 6.9 Characteristics of sample slum households

Characteristics	Average
Households' health shock experiences (yearly)	2.45
Monthly per capita income (rupees)	2526.72
Household size (persons)	4.90
Age of household head (years)	48.02
Literate in a household (persons)	3.54
Employed in a household (persons)	1.43

Source: Field survey (2022)

mapping have compared the results with the GIS algorithm. Cluster-outlier analysis based on Anselin Local Moran's I statistics tool in ArcGIS v10.4 was used to identify spatial clusters or grouping whether slum neighborhoods in the context of WYHCS and HSCS are in the spatial pattern (clustered or random) or other forms based on the Z-score of each neighborhood. Furthermore, a multinomial logistic regression model (MNLR) is used to find the role of livelihood assets in coping strategies among slum households. Similarly, OLS was used to find the degree of relationship between health shocks and coping strategies. Descriptive statistics were also estimated (Table 6.10) based on collected SPSS v26 software (IBM) data. The methodological framework is shown in Fig. 6.13, guided by the complete study.

6.4.2 *Result and Discussion*

6.4.2.1 Results of Principal Component Analysis and Construction of WYHCI and HSCI

The Kaiser-Meyer-Olkin (KMO) and Bartlett's tests were conducted to determine the variables' suitability for principal component analysis (PCA). The value of the KMO is 0.762 and Bartlett's test of sphericity χ^2 (270) = 192.06, $p < 0.01$, indicating

Table 6.10 Descriptive statistics of the variables

Asset types	Variables		Description	Mean	SD
Human	Employment capability of household		Percentage of employment capability of the household	40.21	9.86
	Educational level		Percentage of the literate person in the household	81.60	10.40
Physical	Household fixed asset		Percentage of household having minimum assets (production and appliance items)	19.42	12.05
	Housing condition	House type	Percentage of households made with concrete	21.97	19.70
		Number of rooms	Number of rooms household have	2.273	0.5184
Financial	Income		Monthly cash income of a household	12,520.66	4140.29
	Credit loan		Households taking loans at credit	49.78	21.02
Social	Number of acquaintances		Several acquaintances households have	0.526	0.24
	Closeness with acquaintances		Correlation with acquaintances	0.517	0.25
	Government facility		Percentage of households taking government facilities	11.46	6.56

Source: Field survey (2022)

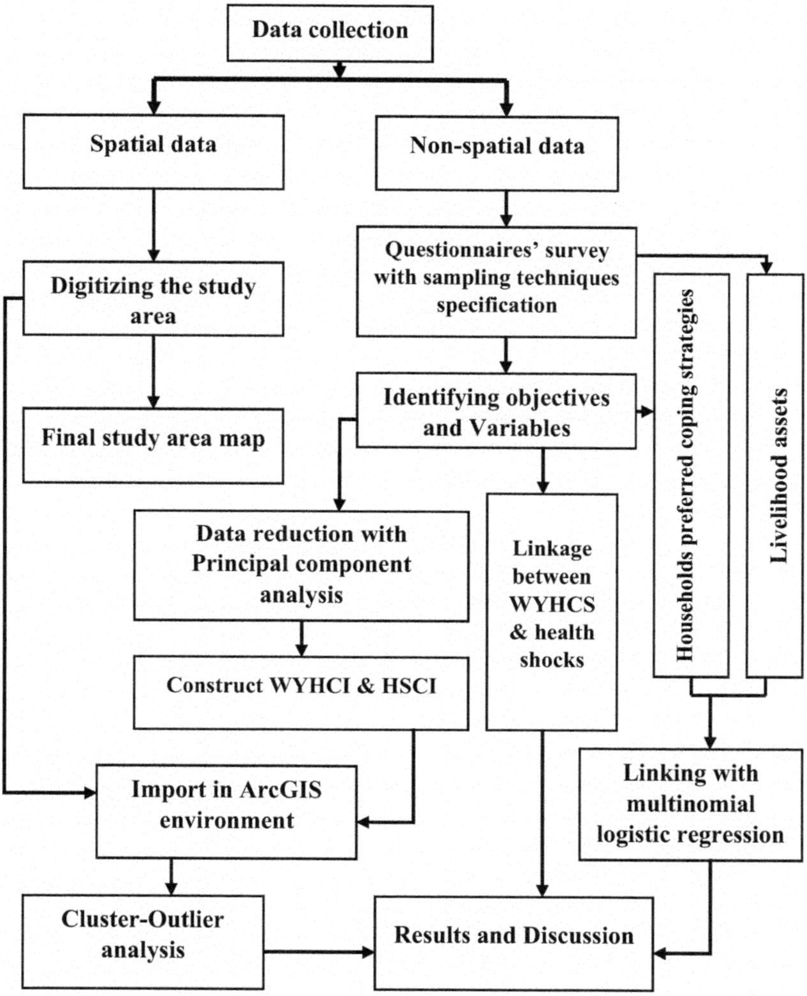

Fig. 6.13 Flowchart of the methodology for WYHCSI and HSCSI

that correlations among the variables were appropriate for PCA. PCA extracted components from the variables and employed Kaiser normalization and varimax rotation methods with eigenvalues higher than 1.0 through the dimension reduction tool of SPSS v26. Varimax rotation with Kaiser normalization converged in 8–12 iterations, explaining 65.66% and 69.23% variance for the components together, as summarized in Tables 6.11 and 6.12. Extracted components with an eigenvalue greater than 1.0 were identified as the latent variables and assigned cardinality (±) (Aksha et al., 2019; Maria et al., 2016). Cardinality described each component's influence on coping strategies (Maria et al., 2016). Positive values (+) indicated higher levels of influence, while negative values (−) indicated lower levels of influence of the components (Tables 6.11 and 6.12). The ambiguous components with

Table 6.11 Rotated component matrix for practicing coping strategies for the whole year

Component	Cardinality	Variables	Loading value	Explained variance (%)
PC1	(+)	Cleaning of dress	0.734	13.359
		Keep nails short	0.673	
		Timely food intake	0.756	
		Take a shower with warm water	−0.440	
PC2	(+)	Finish all the tasks earlier in the morning	0.725	9.211
		Drainage management	0.586	
		Do not get out of the house until the sun comes out	0.493	
		Use oil/body lotion cream to prevent skin disease	0.827	
PC3	(+)	Washing hands before meals and after using the toilet	−0.569	7.294
		Healthy eating habits	0.768	
		Do not get out at noon	−0.556	
		Take extra rest at home	0.767	
PC4	(+)	Drinking water treatment	0.838	7.181
		Maintain sewerage system	−0.336	
		Drink boiled water	0.889	
PC5	(+)	Drink much water comparatively	0.893	5.892
		Use warm clothes	0.673	
PC6	(+)	Access of toilet	0.695	5.289
		Take oral saline/coconut water/fruit juice	−0.682	
PC7	(+)	Use of menstruation care	0.450	4.869
		Physical activity for health	0.569	
		Drink hot water/tea/coffee	−0.511	
PC8	(+)	Brush at morning	0.739	4.444
PC9	(+)	Try not to get in the rain	0.880	4.355
		Use mosquito net	−0.438	
PC10	(+)	Take bath regularly	0.729	3.765
		Keep household clean	0.769	

Extraction method: principal component analysis
Rotation method: varimax with Kaiser normalization
(a) Rotation converged in 12 iterations

cardinality positive and negative (±) values were excluded due to their negligible effect on health coping strategies. For each slum neighborhood, selecting variables in each loading component was averaged. Finally, the coping strategies' index (WYHCSI and HSCSI) was calculated by adding the values of the principal component to arrive at the total WYHCSI and HSCSI scores (Eqs. 6.2 and 6.3).

Table 6.12 Rotated component matrix for adapting coping strategies in health shocks

Component	Cardinality	Variables	Loading value	Explained variance (%)
PC1	(+)	Seek religious or spiritual help	0.521	16.072
		Stop child education	0.542	
		Wife working as a maidservant	−0.802	
		Moved or ceased living with relatives	0.733	
PC2	(+)	Understand the situation	0.879	12.443
		Credit loan	0.900	
PC3	(+)	Sell fixed assets	0.437	9.883
		Resort to acquaintances	−0.652	
		Government facility	0.779	
PC4	(+)	Cutting down costs on others' livelihood purposes	0.486	8.600
		Use saving	−0.876	
PC5	(+)	Diversifying of employment	0.841	8.036
		Buy health insurance	0.600	
PC6	(+)	Mortgaged household assets	0.836	7.459
PC7	(+)	Working as child labor	−0.909	6.739

Extraction method: principal component analysis
Rotation method: varimax with Kaiser normalization
(a) Rotation converged in eight iterations

$$\begin{aligned}\text{WYHCSI}_{\text{Slum}} &= \text{PC1} + \text{PC2} + \text{PC3} + \text{PC4} + \text{PC5} + \text{PC6} \\ &\quad + \text{PC7} + \text{PC8} + \text{PC9} + \text{PC10}\end{aligned} \tag{6.2}$$

$$\text{HSCSI}_{\text{Slum}} = \text{PC1} + \text{PC2} + \text{PC3} + \text{PC4} + \text{PC5} + \text{PC6} + \text{PC7} \tag{6.3}$$

We assume different components have equal importance on overall coping strategies, and therefore, we did not assign any weight to the components. To identify the spatial clusters of the magnitude of the practices of coping strategies, WYHCSI and HSCSI scores of the slum neighborhoods were mapped in ArcGIS through cluster-outlier analysis spatial statistic tool.

WYHCSI and HSCSI scores have been grouped into five quintiles based on estimated SD values from high (>1.5 SD) to low (<−1.5 SD). Afterward, the total WYHCSI and HSCSI scores have been mapped through ArcMap (Figs. 6.14 and 6.15). The map of the total WYHCSI and HSCSI (Figs. 6.14 and 6.15) exhibit marked differences among the slum neighborhoods in the context of coping strategies. Figure 6.14 reveals that only 29.56% of slum neighborhoods have taken active coping strategies. Almost 47.21% adopted moderate coping strategies, while 23.23% have not adopted active coping strategies regarding their lifestyle to overcome the hostile situation in their daily livelihood throughout the year (WYHCS), distributed more clustered than randomly across the study area. In contrast, Fig. 6.15 reveals that only 24.55% of slum neighborhoods have taken active coping

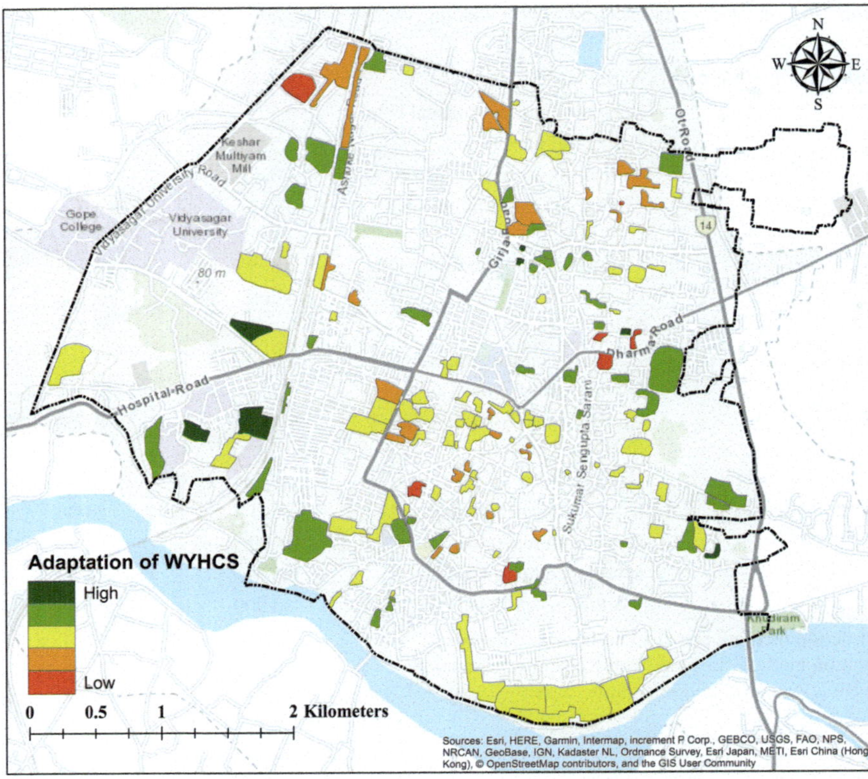

Fig. 6.14 Spatial distribution of slums based on WYHCSI score

strategies, and almost 25.74% adopted moderate level of coping strategies, while 49.71% have not been adopted active coping strategies during the health shocks (HSCS) in the family distributed more clustered than randomly across the study area. The results from Figs. 6.14 and 6.15 disclose the remarkable difference between WYHCS and HSCS among the slum neighborhoods characterized by the low standard quality of life. The reason behind the phenomenon is that most of the slum neighborhoods are under vulnerable socioeconomic status, which involuntarily falls them into high health risks such as lack of housing accommodation, lack of access to drinking water, lack of proper food and nutrition, as well as lack of healthcare facilities. Moreover, high family size and fewer literate and employed persons may contribute to their health security, living standard, and overall quality of life. More explicitly, low income per month (<Rs. 5000) is intrinsically associated with health risks and health shocks. Due to low income levels, slum dwellers face difficulties in their livelihood beyond meeting health expenditures. The observation through field survey reveals that the occupation status of the dwellers is volatile and vulnerable as most males are rickshaw or toto pullers. At the same time, a female is engaged in insecure servant work with a low salary. The loss of jobs and

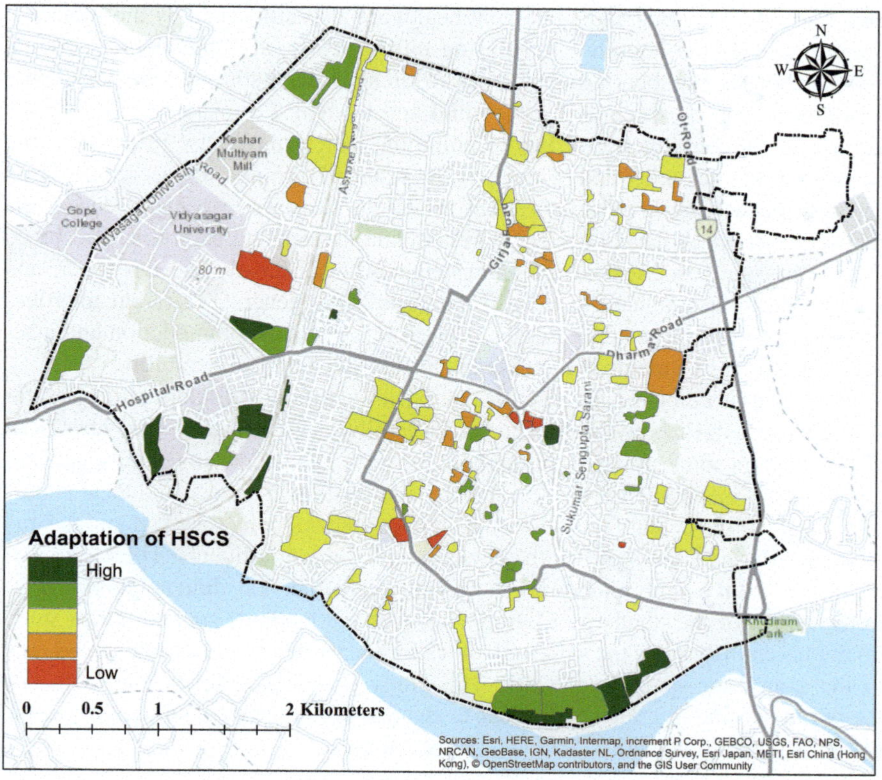

Fig. 6.15 Spatial distribution of slums based on HSCSI score

lack of self-employed workers push households toward taking an informal loan or mortgaged household assets, which may influence their daily livelihood and social vulnerability and lead them into a poverty trap and vicious cycle of health shocks. As a consequence, to meet the expenditure of health shock, they again become susceptible to vulnerability and a "medical poverty trap."

A low level of literacy has high social vulnerability and vice versa in the context of knowledge, perception, and management skills. Education may help to change the family's superstitious nature, strengthening the family's economic planning and culture, thereby reducing health shock and economic and social vulnerability. Moreover, an educated or literate person can decide what ought to be or what is wrong in the family, thereby taking strategies to reduce the family stress and the burden of health shocks. Women's literacy can help to reduce gynecological problems during the menstruation period, express their problems to the doctor as a most educated family does, and reduce the mortality rate. The study investigated the context of personal hygiene and found that most women (23%) use cloth during their menstruation period rather than sanitary napkins in the present era. Low education may also affect healthcare practices for children and the aged, reducing morbidity and mortality. Maternal and adolescent health are also influenced by the socioeconomic status of the households. Housing infrastructure, accommodation, spacing between houses, number of rooms, separate kitchens, and bathrooms are considered a health indicator, especially for the child and women. For example, no separate kitchen is the consequence of indoor pollution, which may cause respiratory diseases among the child and women. Low-standard rooms and accommodation severely impact the breed of different infectious diseases, especially during the monsoon period. A flawed sewerage system and no separate bathroom also breed dengue mosquitoes. Most of the slum neighborhoods mostly use tap water, and without purifiers, stagnant water and dampness may generate different microbes (zoonotic) and cause related diseases. Therefore, overall accommodation and inaccessibility to basic amenities increase the health shock and vulnerability.

From the aforementioned circumstances, it is difficult to adopt strategies for slum dwellers during any health shocks the family faces. However, they have been trying to adopt some strategies throughout the year. The underlying causes and situation made it inevitable for the researchers, administrative bodies, and local people to understand the actual facts about the slum people and why they cannot take any strategies during illness or health shocks than the urban affluents. Therefore, it is clear that "slum neighborhoods are more vulnerable than non-slum neighborhoods" in the milieu of the standard of living. However, the methods applied in this study do not necessarily yield the same outcomes for the variables and do not spread over in the same time-space framework. This study is also a complex one, with key themes in our findings. However, it can evaluate whether the most vulnerable people of Midnapore municipality live in the worst slums.

6.4.2.2 Results of Cluster-Outlier Analysis

The cluster-outlier analysis detects statistically significant groupings based on a fixed distance band with a Euclidean distance of 928.36 meters in the ArcGIS environment. The deployment of Anselin Local Moran's I test is to find out the local clusters' coexistence, such as high (HH) and low (LL) cluster and high (HL) and low (LH) outlier association (Figs. 6.16 and 6.17) about WYHCSI and HSCSI among the slum neighborhoods. In the case of high (HH) and low (LL) personal hygiene clusters, the adjacent polygon's value is similar. Figure 6.16 reveals that active coping strategies (HH) throughout the year are found in only six slum neighborhoods, and very low or least chosen coping strategies (LL) are found in only one or two slum neighborhoods. Therefore, the similarity of WYHCS is not uniform, which indicates that dissimilarity is found among the slum neighborhoods. A maximum number of slum neighborhoods are in the insignificant cluster category, which signifies that slum neighborhoods are characterized by an average level of coping strategies throughout the year.

Figure 6.17 reveals that active coping strategies (HH) during health shocks are found in only 12 slum neighborhoods, and very low or least chosen coping strategies (LL) are found in 7 slum neighborhoods. Therefore, the similarity of HSCS is

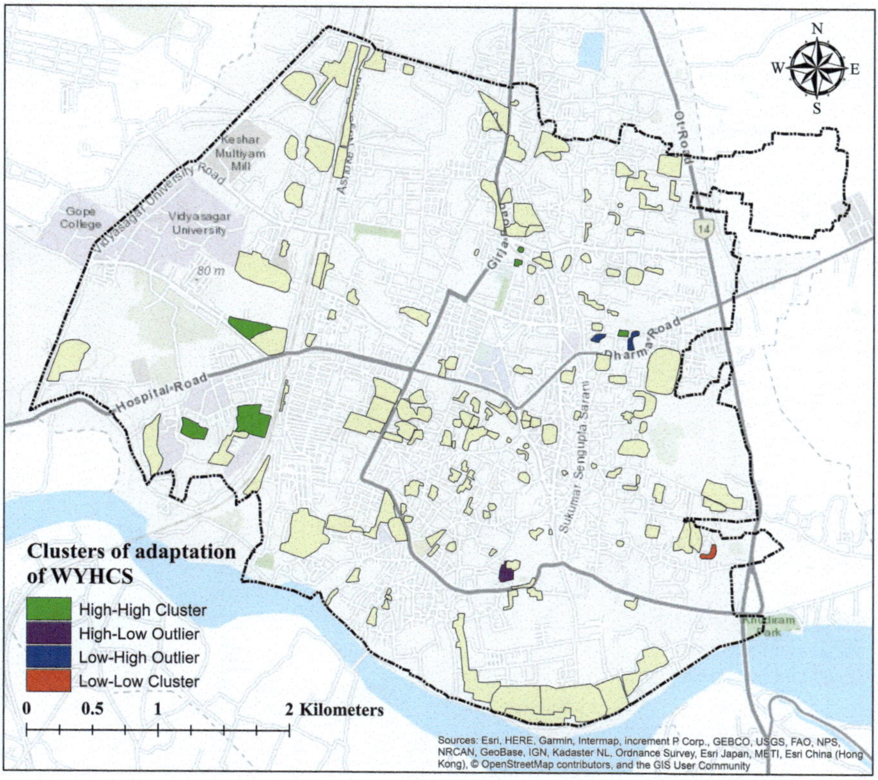

Fig. 6.16 Spatial clustering of slums following WYHCS

not uniform, indicating dissimilarities among the slum neighborhoods. A maximum number of slum neighborhoods are in the insignificant cluster category, which signifies that slum neighborhoods are characterized by the average level of coping strategies during health shocks in the family. For the outlier, the value of the adjacent polygon is dissimilar (Andrews et al., 2020; Martín et al., 2020). As a whole, the cluster-outlier analysis represents the similar and dissimilar slum clusters in the context of the index score of both coping strategies. A similar cluster is limited, which signifies a marked difference among the slum neighborhoods in the context of adopting coping strategies throughout the year and during health shocks.

6.5 Health Risk Perception

The study incorporated the mixed method of surveying and collected information about the perception of the sources of health risk and their prevalence time. Health risks are always associated with low-income people, which is quantitatively

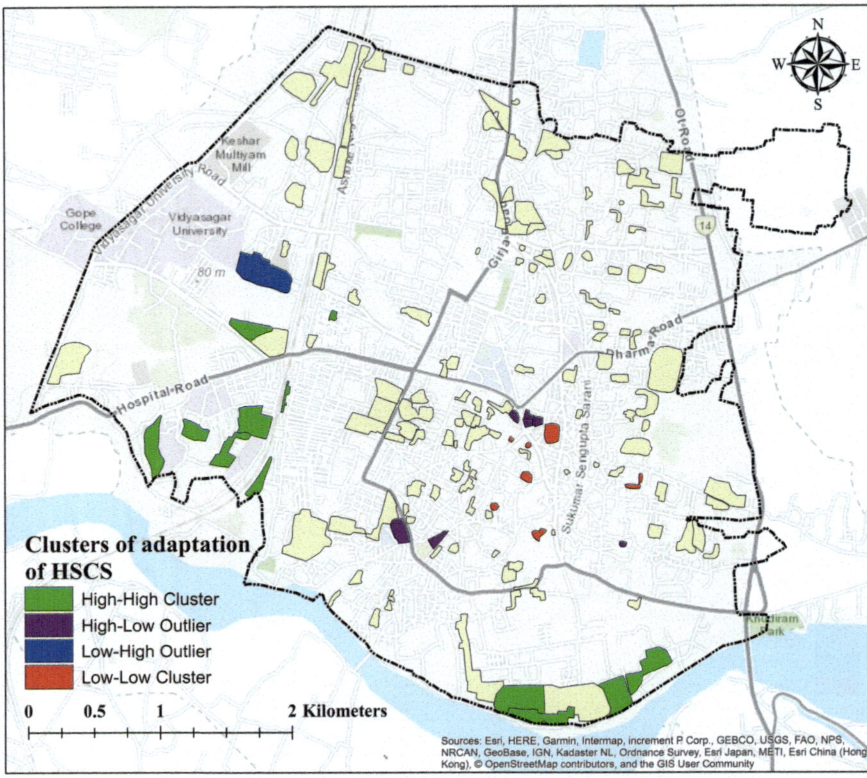

Fig. 6.17 Spatial clustering of slums following HSCS

estimated based on the probable variables in the study (in Chap. 3) and finding the slum neighborhoods with high health risks. Here, the study incorporates only the perception of the sources of health risk among the slum households in terms of their living circumstances throughout the year. Slum households across the study area confronted a wide-ranging health shock from different sources, as reported in Fig. 6.18. Slum people are more likely to report that the highest share of households (39.73%) is faced with housing conditions followed by pollution (indoor and outside) within their living premises (28.99). Beyond the two sources of health risk they sensed in their livelihood, other sources of health risk are hereditary, bad health practices like smoking and alcohol consumption, changes in lifestyle, and high levels of stress for maintaining daily livelihoods and health shocks.

On the other hand, households were also asked about the time of prevalence of the health risks they had encountered. Slum dwellers reported that the seasonal prevalence rate of health risk is high (52.45%), followed by yearly (21.27%) and monthly (17.4%) health risk (Fig. 6.19). During monsoon hot weather, different microbes breed that severely affects the children as they are more susceptible to pathogens and live in a dirty environment. Most of the research found that

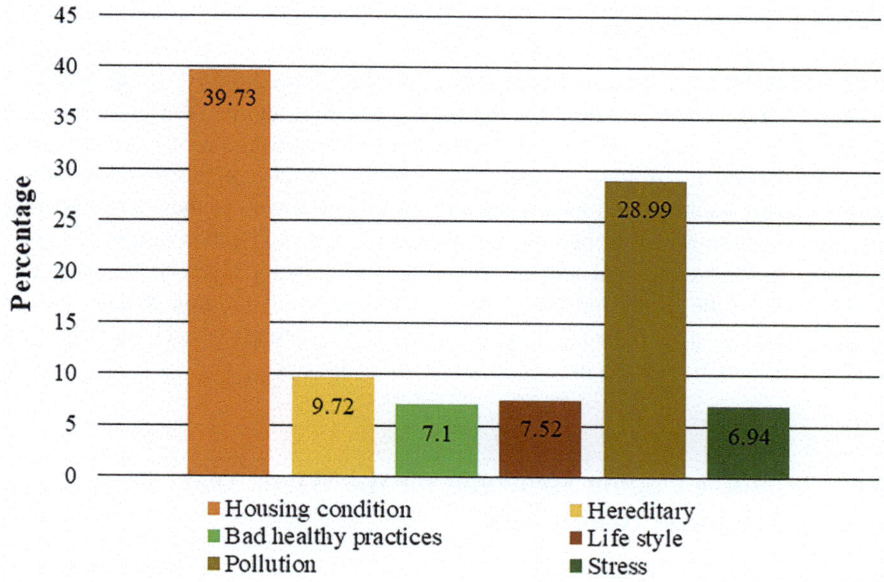

Fig. 6.18 Proportion of households facing sources of health risk (Source: Field survey, 2022)

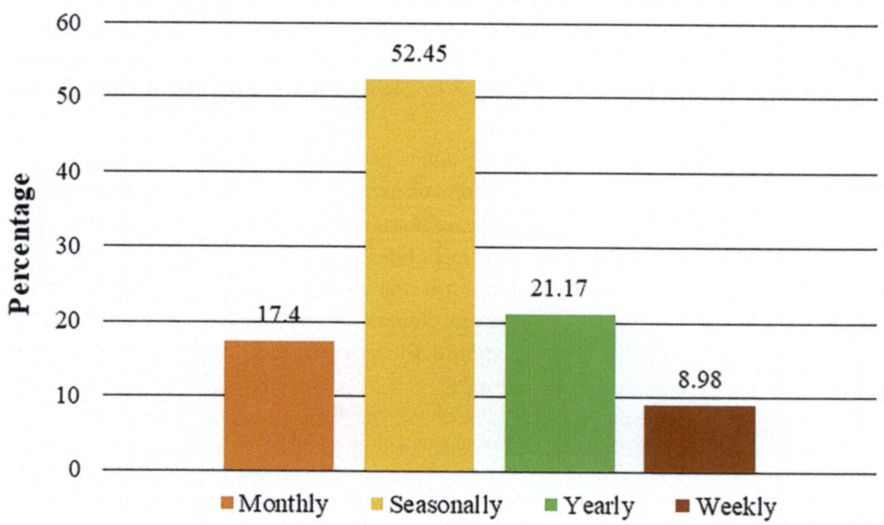

Fig. 6.19 Interval of sickness experienced by households (Source: Field survey, 2022)

environmental conditions and accommodation of the households as well as the family indoor environment are directly linked with health risks as they would face a range of difficulties during the monsoon and summer season due to urban flash floods and hot temperatures (Aiello et al., 2008; Dewi & Kurniawan, 2019). During monsoon season with excessive rainfall, urban flash floods cause dampness in the room and stagnant water, which may cause breeding grounds for different infectious microbes (such as dengue mosquito), as most slum households developed across urban sewerage channels. Besides, during the hot, dry summer season, excessive temperature may cause sweating and dehydration because the roof is mostly covered by asbestos, tin, or tiles. Furthermore, members of the slum households bathed in open places as they don't have a separate bathroom as well as they used community toilets. Moreover, lower room density and no separate kitchen may cause congestion, overcrowding, and indoor pollution, respectively, severely affecting the health status of the family members, especially women and children. A flawed sewerage system and no separate bathroom can inevitably spread diseases, and mosquitos (such as dengue) have a dreadful impact on health risks.

6.6 Adaptive Coping Strategies for the Whole Year to Minimize Health Risk

Risk management strategies, the so-called coping strategies, are common in every household, whether socioeconomically rich or poor. However, the perception of risk management strategies is most relevant, and it varies across the community level. Risk management strategies of slum dwellers throughout the year and during health shocks are displayed in Tables 6.13 and 6.14. The survey reveals that slum dwellers will likely adopt different coping strategies when encountering multiple livelihood

Table 6.13 Adaptive coping strategies for health shocks

Coping strategies	Mean	SD
Sell fixed assets	8.85	10.593
Mortgaged household assets	1.28	5.414
Understand the situation	48.08	26.178
Seek religious or spiritual help	6.73	9.716
Cutting down costs on others' livelihood purposes	13.33	12.918
Stop children education	12.05	12.787
Working as a child labor	2.24	6.679
Wife working as a maidservant	16.47	24.118
Moved or ceased living with relatives	11.22	12.357
Diversifying of employment	40.03	9.97
Credit loan	44.23	25.094
Buy health insurances	3.59	8.343
Use savings	10.64	13.663
Resort to acquaintances	50.71	24.445
Resort to a government facility	11.41	6.643

Source: Field survey (2022)

Table 6.14 Practicing of coping strategies for the whole year to minimize health risk

Types		Coping strategies	Mean	SD
Awareness in personal hygiene and healthy lifestyle practices		Brush at morning	58.85	17.596
		Take bath regularly	82.76	9.809
		Cleaning of dress regularly	34.94	14.480
		Washing hands before meals and after using the toilet	87.44	13.578
		Keep nail short	43.33	15.338
		Timely food intake	24.10	12.693
		Menstruation care	28.33	19.305
		Access of toilet	87.56	16.279
		Physical activity for health	2.18	7.029
		Healthy eating habits	15.83	16.809
Awareness in households' hygiene practices		Keep household clean	30.45	14.999
		Drinking water treatment	18.27	16.575
		Maintain sewerage system	11.41	18.012
Awareness in seasonal coping strategies	Summer season	Do not get out during noon	26.22	19.455
		Drink much water comparatively	58.27	20.483
		Finish all the tasks earlier in the morning	12.69	15.876
		Take extra rest at home	7.05	11.258
		Take oral saline/coconut water/fruit juice	1.67	5.546
	Rainy season	Try not to get wet in the rain	20.58	17.236
		Prevent mosquitos and insects	50.96	21.123
		Drink boiled water	21.54	17.786
		Drainage management	6.92	13.754
	Winter season	Drink hot water/tea/coffee	52.82	19.170
		Do not get out of the house until the sun comes out	5.77	10.099
		Take shower with warm water	9.68	11.828
		Use warm cloths	37.37	19.113
		Use oil/body lotion/cream to prevent skin disease	7.56	15.796

Source: Field survey (2022)

shocks. To safeguard the sustainability of livelihoods, most slum dwellers are adopting mixed risk management strategies to reduce their health risks and livelihood problems. From the sample households, based on the percentage of all responses, it is found that "resort to acquaintances" (50.71%) and "understand the situation" (48.08) are the most frequently chosen strategy, and "diversification of employment" (40.03%) and "credit loan" (44.23%) are two other common strategies adopted during health shocks. Other significant strategies are "use savings," "cutting down cost on other livelihood purposes," "working as a maidservant," "stop children's education," and "resort to government facility," which are valid to some extent for the slum dwellers, while the least chosen strategies by the proportion of slum dwellers are "promote child labor," "sell or mortgaged fixed assets," "buy health insurance," etc. (Table 6.13).

The significant statistics of coping strategies adopted by slum dwellers throughout the year in the study area were also collected and expressed in percentages, as shown in Table 6.14. The study investigated that, on average, more than 75% of slum dwellers regularly brush in the morning, take baths daily, wash their hands before meal and after using the toilet, and skip their meals less frequently. Nevertheless, in the same way, more than 50% of slum dwellers do not have food in a timely manner, keep their nails short, and clean their dresses daily, and women are less aware of their menstruation care. It is clear from the observation that slum dwellers have a piece of knowledge of coping strategies; instead, they do not know about standard living practices. The study also investigated the household's strategies during different seasons, such as summer, rainy, and winter. The most significant strategies that slum dwellers are adopting reveal that during the summer season, they "drink much water comparatively" (58.27%), and during the rainy season, they use a mosquito net and oil to "prevent mosquitos and insects" (50.96%), and during the winter season, they "drink hot water/tea/coffee" frequently (52.82%), while other strategies are not found so far as significant (Table 6.14). Overall, they have less frequent awareness about their health and promote healthy livelihood. This finding is also stimulating because slum neighborhoods within similar geospatial environments exhibit a marked difference in adaptive coping strategies and exposure to different health risk levels. However, the social, economic, and built environment attributes of the slum neighborhoods are different in terms of coping strategies, which in turn affect health risk levels.

6.7 Linkage Between Coping Strategies and Health Shocks

Health shocks are closely associated and interlinked with coping strategies. People try to overcome health shocks by adopting different strategies. Urban poor are a socioeconomically backward community in the urban area and face different health shocks in their daily livelihood at different ages. The study investigated different adaptive strategies for health shocks of the urban poor (Table 6.13) throughout the year in Midnapore town. Furthermore, the study employed a simple Pearson correlation to estimate the relation between WYHCSI and health shock incidents, which displays (Table 6.15) the strong negative correlation (−0.862) at a high significant level ($p < 0.01$). Moreover, the OLS regression (Tables 6.16 and 6.17) also shows

Table 6.15 Pearson correlation between health shocks and WYHCSI

		Health shock incident	WYHCSI value
Health shock incident	Pearson correlation	1	−0.862**
WYHCSI value	Pearson correlation	−0.862**	1

**Correlation is significant at the 0.01 level (two-tailed)

Table 6.16 OLS regression between health shocks and WYHCSI

Coefficients[a]

Model		Unstandardized coefficients		Standardized coefficients			95.0% confidence interval for B	
		B	Std. error	Beta	*t*	Sig.	Lower bound	Upper bound
1	(Constant)	32.923	0.997		33.028	0.000	30.954	34.892
	WYHCSI value	−0.571	0.027	−0.862	−21.149	0.000	−0.625	−0.518

[a]Dependent variable: health shock incident

Table 6.17 Model summary of OLS between WYHCSI and health shocks

Model summary[a]

Model	*R*	*R*-squared	Adjusted *R*-squared	Std. error of the estimate
1	0.862[b]	0.744	0.742	3.275

[a]Dependent variable: health shock incident
[b]Predictors: (constant), WYHCSI value

that the beta value (−0.862) is at a highly significant level. The adjusted R^2 value (0.744) with SE (3.275) signifies that those households who practice coping strategies throughout the year are less likely to face health shock and vice versa. In contrast, the result also signifies that low-level practices of coping strategies contribute to the high intensity of health shock incidents among the poor in the study area.

The general thought is to believe that health security is coherently related to the quality of life of the households. The ordinary least squares (OLS) regression was applied to estimate the relationship between health shocks and coping strategy indicators. Similarly, this method was used to explore the impact of the variables of coping strategies on health shocks as shown in Tables 6.18 and 6.19. Based on beta value, the result exhibits (Table 6.18) that "mortgaged household assets" and "wife working as a maidservant" are the significant ($p < 0.006$) coping strategies of the urban poor, which have an impact during their health shock incident in the family. Other coping strategies that are less significant ($p < 0.20$) but slum dwellers have been taking while managing health shocks are "stop child education," "move or cease living with relatives," "credit loan," and "use savings." However, selling fixed household assets, cutting down costs on other livelihood purposes, working as child labor, diversifying employment, and resorting to acquaintances do not have significant coping strategies and do not have a significant impact on the reduction of health shocks. The OLS regression model also shows that the *R*-value (0.45) and the adjusted R^2 value (0.141) with SE (5.978) signify that the model is validated at a highly significant level (Table 6.19).

Table 6.18 OLS regression between health shocks and the adaptive coping strategies

Coefficients[a]						
Model		Unstandardized coefficients		Standardized coefficients		
		B	Std. error	Beta	*t*	Sig.
1	(Constant)	8.402	2.840		2.958	0.004
	Sell fixed assets	−0.029	0.050	−0.047	−0.574	0.567
	Mortgaged household assets	0.258	0.092	0.217	2.793	0.006
	Cutting down cost on other livelihood purposes	−0.094	0.042	−0.189	−2.262	0.025
	Stop child education	0.112	0.044	0.223	2.578	0.011
	Working as child labor	−0.032	0.076	−0.033	−0.417	0.678
	Wife working as maidservant	0.071	0.024	0.264	2.932	0.004
	Moved or ceased living with relatives	0.056	0.045	0.108	1.269	0.206
	Diversifying of employment	0.023	0.050	0.035	0.453	0.651
	Credit loan	0.026	0.020	0.100	1.261	0.209
	Use saving	0.091	0.039	0.193	2.328	0.021
	Resort to acquaintances	−0.015	0.020	−0.055	−0.710	0.479

[a]Dependent variable: health shock incident

Table 6.19 Model summary of OLS regression between health shocks and the adaptive coping strategies

Model	*R*	*R*-squared	Adjusted *R*-squared	Std. error of the estimate
1	0.450[a]	0.202	0.141	5.978

[a]Predictors: (constant), resort to acquaintances, mortgaged household assets, wife working as a maidservant, diversifying of employment, credit loan, use savings, working as child labor, sell fixed assets, cutting down cost on other livelihood purposes, moved or ceased living in with relatives, stop child education

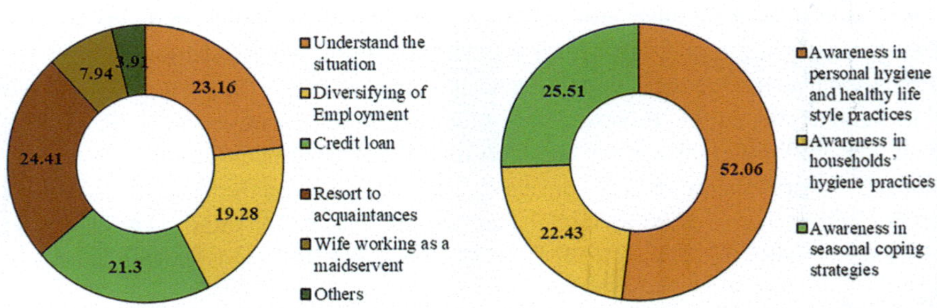

Fig. 6.20 Households preferred coping strategies in health shocks (HSCS) and throughout the year (WYHCS) (Source: Field survey, 2022)

6.8 Households' Preferred Health Coping Strategies

The study has also estimated the most significant health coping strategies and their role in the reduction of health shocks by statistical analysis. However, as a whole, based on the percentage of respondents from sample slum households after the door-to-door survey, the most revealing preferred health coping strategies are summarized in Fig. 6.20. During the survey, slum dwellers are asked what are their preferred coping strategies pursued throughout the year and during health shocks to enhance their sustainability of livelihood as well as the perception of strategies taken to cope with their existing situation. All the responses were recorded carefully and summarized into a broad category by percentage. Concerning their health coping strategies, slum dwellers reported that "resort to acquaintances" (24.41%) and "understand the situation" (23.16%) have been perceived as two admirable tools as well as "credit loan" (21.30%) and "diversification of employment" (19.28%) are two other popular health coping strategies when encountered health shocks. For the proportion of slum dwellers responding, "awareness in personal hygiene and healthy lifestyle practices" (52.06%) is the most often preferred strategy throughout the year, while "awareness in household hygiene practices" (25.51%) and "awareness in seasonal coping" (22.43) are the two other essential risk management strategies adopted among the slum dwellers throughout the year. Therefore, regarding the preferred coping strategies, "resort to acquaintances" (24.41%) and "awareness in personal hygiene and healthy lifestyle practices" (52.06%) are regarded as the most useful for managing livelihood and health risks compared to the others. However, livelihood assets are the driving predictor or independent variables that may affect the practices of WYHCS and HSCS in slum neighborhoods.

6.9 Role of Livelihood Assets in Choosing Preferred Health Shock Coping Strategies (HSCS)

Livelihood assets have been driving components of coping strategies. Based on the percentage of dependent and independent variables derived from the survey, Table 6.20 shows the role of assets in adaptive coping strategies estimated from the multinomial logistic regression (MNLR) model. The analysis reveals that "credit loan" is the only significant component affecting slums' adaptive coping strategies of "credit loan." The regression coefficient (*B*) for this component (0.171) is significantly negative at the 0.01 level, and the exponentiation of the coefficients (Exp(*B*)) is >1. This means households taking more "credit loans" are likely to adopt more adaptive coping strategies than those taking "understand the situation." While employment status, educational level, and credit loan are shown to be three components that influence slums' adaptive coping strategy of "diversifying employment". The regression coefficient of "employment status" and "credit loan" is significantly positive at 0.01 level with Exp(*B*) >1, whereas "educational level" is significantly negative with Exp(*B*) <1 at 0.05 significant level. This result indicates that the lower

Table 6.20 Role of assets in health shock coping strategies, parameter estimated from the multinomial logistic regression (MNLR) model

Dependent variable	Independent variable	*B*	S.E.	Wald	Exp (*B*)
Credit loan ($n = 33$)	Intercept	−9.330	5.809	2.580	–
	Employment	0.000	0.058	0.000	1.000
	Educational level	0.009	0.048	0.035	1.009
	Household fixed asset	0.009	0.034	0.079	1.010
	House type	0.002	0.023	0.012	1.002
	Number of rooms	−0.309	0.848	0.133	0.734
	Household monthly income	0.000*	0.000	2.988	1.000
	Credit loan	0.171***	0.034	25.902	1.187
	Number of acquaintances	−1.998	2.477	0.651	0.136
	Closeness with acquaintances	−1.475	1.587	0.864	0.229
	Government facility	0.037	0.065	0.317	1.038
Diversifying of employment ($n = 30$)	Intercept	−17.381**	8.536	4.146	–
	Employment	0.521***	0.127	16.896	1.684
	Educational level	−0.140**	0.064	4.836	0.869
	Household fixed asset	0.060	0.049	1.528	1.062
	House type	−0.022	0.030	0.525	0.979
	Number of rooms	−1.824	1.439	1.607	0.161
	Household monthly income	0.000	0.000	1.106	1.000
	Credit loan	0.121***	0.037	10.917	1.128
	Number of acquaintances	3.212	2.552	1.584	24.818
	Closeness with acquaintances	1.596	2.231	0.512	4.932
	Government facility	−0.176	0.112	2.458	0.838
Resort to acquaintances ($n = 38$)	Intercept	−10.335*	5.440	3.609	–
	Employment	−0.012	0.052	0.053	0.988
	Educational level	0.021	0.044	0.220	1.021
	Household fixed asset	0.022	0.032	0.478	1.022
	House type	0.012	0.021	0.312	1.012
	Number of rooms	−0.883	0.926	0.911	0.413
	Household monthly income	0.000	0.000	0.533	1.000
	Credit loan	0.115***	0.029	15.825	1.122
	Number of acquaintances	8.887***	2.237	15.783	7238.956
	Closeness with acquaintances	−0.606	1.499	0.163	0.545
	Government facility	0.007	0.056	0.018	1.008

*Significant at the 0.10 level
**Significant at the 0.05 level
***Significant at the 0.01 level
The reference category is understand the situation
Cox and Snell R-squared is 0.820, Nagelkerke R-squared is 0.875, and McFadden is 0.620
Regarding goodness-of-fit, the Pearson chi-square is 513.542, df is 378, and sig. = 0.000

the level of education households have, the more reluctant they are not to diversify their employment while maintaining their livelihood strategies compared to understanding the situation. In contrast, households having diversifying employment status and credit loan are more likely to practices of adaptive coping strategies as to maintaining their sustainability of livelihood and uncertainty. Additionally, "credit loan" and a number of acquaintances are shown to be two components that influence slums' adaptive coping strategies of "resort to acquaintances" in which the regression coefficient of both variables is significantly positive at the 0.01 significant level with Exp(*B*) >1. This result indicates that households with credit loans and an adequate number of acquaintances are more reluctant to adopt adaptive coping strategies for health shocks and unavoidable circumstances than to understand the situation. Household monthly income is a significant predictor at a 0.10 significant level but does not affect adaptive coping strategies as the value of Exp(*B*) is absolute 1.0.

6.10 Conclusion

A coping mechanism is a strategy or method associated with daily human life and manages every aspect of the problem in livelihood. In this chapter, we discussed the healthy coping mechanisms of slum neighborhoods of Midnapore municipality. In terms of health coping mechanisms, the study examines how slum neighborhoods manage various issues that are directly related to health. Most slum dwellers have experienced seasonal health shocks due to detrimental housing conditions and pollution (indoors and outside). Urban poor have adopted different coping strategies during the year and health shocks. Among the coping strategies, the most common are finding jobs for all family members, resorting to acquaintances, taking credit loans, being aware of personal hygiene and quality of life, etc. Whole-year coping strategies (WYHCS) among the slum neighborhoods reveal moderate to high in comparison to health shock coping strategies. The health shock coping strategies (HSCS) among the slum neighborhoods exhibit relatively moderate to low. The model (OLS and Pearson correlation) estimates a strong negative correlation between health shock and WYHCS, i.e., with increasing WYHCS, health shock will be reduced. The employed model estimates that credit loans, employment status, educational level, and several acquaintances are primarily affecting the health shock coping strategies. Health coping strategies (HCS) among the different age groups indicate that childhood, adolescents, and the working group of slum dwellers among the neighborhoods adopted moderate to low levels of health coping strategies, while the aged group of slum dwellers comparatively adopted moderate to high level of health coping mechanism. This feature indicates the lack of awareness, knowledge, and practices of health coping strategies among the expected age group.

References

Aiello, A. E., Larson, E. L., & Sedlak, R. (2008). Personal health bringing good hygiene home. *American Journal of Infection Control, 36*(10S), 152–165. https://doi.org/10.1016/j.ajic.2008.09.009

Aksha, S. K., Juran, L., Resler, L. M., & Zhang, Y. (2019). An analysis of social vulnerability to natural hazards in Nepal using a modified social vulnerability index. *International Journal of Disaster Risk Science, 10*(1), 103–116. https://doi.org/10.1007/s13753-018-0192-7

Almeida, D. M. (2005). Resilience and vulnerability to daily stressors assessed via diary methods. *Current Directions in Psychological Science, 14*(2), 64–68. https://doi.org/10.1111/j.0963-7214.2005.00336.x

Andrews, M. R., Tamura, K., Claudel, S. E., Xu, S., Ceasar, J. N., Collins, B. S., Langerman, S., Mitchell, V. M., Baumer, Y., Powell-wiley, T. M., Andrews, M. R., Tamura, K., Claudel, S. E., Xu, S., Ceasar, J. N., Collins, B. S., Langerman, S., Mitchell, V. M., & Baumer, Y. (2020). *Geospatial analysis of neighborhood deprivation index (NDI) for the United States by county. 5647*. https://doi.org/10.1080/17445647.2020.1750066.

Baker, J. P., & Berenbaum, H. (2007). Emotional approach and problem-focused coping: A comparison of potentially adaptive strategies. *Cognition and Emotion, 21*(1), 95–118. https://doi.org/10.1080/02699930600562276

Bonfrer, I., & Gustafsson-wright, E. (2017). *Health shocks, coping strategies, and foregone healthcare among agricultural households in Kenya. 1692*. https://doi.org/10.1080/17441692.2015.1130847.

Carver, C. S. (2013). COPE inventory [measurement instrument database for the social sciences]. Retrieved from http://www.midss.org/content/cope-inventory

Conde, P., Gutiérrez, M., Sandín, M., Díez, J., et al. (2018). Changing neighborhoods and residents' health perceptions: The heart-healthy hoods qualitative study. *International Journal of Environmental Research and Public Health, 15*(8), 1617.

Cutter, S. L., Carolina, S., Boruff, B. J., Carolina, S., Shirley, W. L., Carolina, S. (2003). Cutter et al 2003. In Soc. Sci (Vol. 84, pp. 242–261).

Dewi, Y. S., & Kurniawan, D. (2019). The influence of personal hygiene and healthy lifestyle on environmental sanitation, *6*, 148–153.

Dhanaraj, S. (2016). Effects of parental health shocks on children's schooling: Evidence from Andhra Pradesh, India. *International Journal of Educational Development, 49*, 115–125. https://doi.org/10.1016/j.ijedudev.2016.03.003

Folkman, S., & Lazarus, R. S. (1985). Study of emotion and coping during three stages of a collage examination. *Personality and Social Psychology, 48*(1), 150–170. https://webs.wofford.edu/steinmetzkr/Teaching/Psy150/Lecture PDFs/Coping.pdf

Giddens, A. (1984). *The constitution of society*. University of California Press.

Hummell, B. M. L., Cutter, S. L., & Emrich, C. T. (2016). Social vulnerability to natural hazards in Brazil. International Journal of Disaster Risk Science, 7(2), 111–122. https://doi.org/10.1007/s13753-016-0090-9

Izutsu, T., Tsutsumi, A., Islam, A. M., Kato, S., Wakai, S., & Kurita, H. (2006). Mental health, quality of life, and nutritional status of adolescents in Dhaka, Bangladesh: Comparison between an urban slum and a non-slum area. *Social Science and Medicine, 63*(6), 1477–1488. https://doi.org/10.1016/j.socscimed.2006.04.013

Maria, B., Hummell, D. L., Cutter, S. L., & Emrich, C. T. (2016). Social vulnerability to natural hazards in Brazil. *International Journal of Disaster Risk Science*. https://doi.org/10.1007/s13753-016-0090-9

Martín, et al. (2020). Hot spot analysis versus cluster and outlier analysis : An enquiry into the grouping of rural accommodation in Extremadura (Spain). *International Journal of Geo-information, 8*, 176. https://doi.org/10.3390/ijgi8040176

Mielck, A., Kiess, R., Knesebeck, O. V. D., Stirbu, I., & Kunst, A. E. (2009). Association between forgone care and household income among the elderly in five Western European countries analyses based on survey data from the SHARE-study. *BMC Health Services Research.*

Pedersen, M. H., & Rytter, M. (2018). Rituals of migration: An introduction. *Journal of Ethnic and Migration Studies, 44*(16), 2603–2616.

Revenson, T. A. (1987). Does coping help? A reexamination of the relation between coping and mental health aid win normative aging study Boston Veterans Administration Outpatient Clinic. *Journal of Personality and Social Psychology, 53*(2), 337–348.

Teachman, B. A. (2006). Aging and negative affect: The rise and fall and rise of anxiety and depression symptoms. *Psychology and Aging, 21*(1), 201–207. https://doi.org/10.1037/0882-7974.21.1.201

United Nations. (2019). World population prospects 2019: Highlights. Department of Economic and Social Affairs, Population Division. https://doi.org/10.18356/13bf5476-en

United Nations Children's Fund. (2020). The state of children's health 2020. https://www.unicef.org/health.

Whitehead, M., Dahlgren, G., & Evans, T. (2001). Can low-income countries escape the medical poverty trap. *The Lancet, 358*(September), 833–836. https://pubmed.ncbi.nlm.nih.gov/11564510/

World Health Organization. (1999). World health report 1999: Making a difference. World Health Organization.

World Health Organization. (2008). The World Health Report 2008: Primary health care – Now more than ever. World Health Organization. https://iris.who.int/handle/10665/43912]

Chapter 7
Conclusion and Implications

7.1 Is Urban Poor Really Poor in Health?

Urbanization is faster in developing countries, with overpopulation within limited space and challenges of livelihood sustainability, health, and well-being. The different reports revealed that in India, the urban population is increasing at an alarming rate in small cities rather than mega or metropolitan cities. However, the nature of urbanization in India is population growth rather than development. The continuous rural-to-urban migration flux enriches the city's population. Still, substantial infrastructural development is not evident simultaneously, resulting in inequitable distribution of basic amenities and inaccessibility to essential services. Moreover, poor migrant people occupy the most vacant and unprivileged land in urban areas with developing slum neighborhoods along with dilapidated living environments. It is also remarkable that the slum population is increasing faster with pseudo-urbanization (Ray, 2017) and is the most deprived group in urban areas. To keep in mind the following phenomena, the present study explored the livelihood system and coping strategies of slum dwellers in a small city characterized by distinct socio-physical morphology. Midnapore is a fast-growing city; its growth rate of built-up area was high between 2001 and 2017, about 35.39% (Dinda et al., 2018). Naturally, slum neighborhoods also developed in the administrative boundary; for instance, the decadal growth rate of the slum population between 2001 and 2011 was 35.54% (Census of India, 2011) within the same limited space. Generally, continuous rural-to-urban migration in recent years is accountable for the development of slums and squatter settlements that occupy most of the vacant land illegally. Unplanned urban development is also responsible for detrimental living and health conditions. In this concern, the present study refers to the spatial statistical and geospatial analytical framework based on field-based primary data collection from slum neighborhoods of Midnapore municipality. This study has explored living conditions, socioeconomic status, lifestyle behavior and health awareness,

© The Author(s), under exclusive license to Springer Nature Switzerland AG 2025

B. Dutta et al., *Using GIS to Identify Opportunities for Improved Health in Urban Slums*, Global Perspectives on Health Geography,
https://doi.org/10.1007/978-3-031-89617-0_7

healthcare infrastructure and provision, utilization and accessibility of healthcare services, and survival strategies among the slum neighborhoods in the study area.

The case study material generally shows that sample households are multidimensionally poor, with the least earning members engaged in informal sectors, which increases the burden of family expenditure. The high child dependency ratio, large family size, illiteracy, early marriage, fewer savings, etc. contribute to their socio-economic backwardness and the incidence of poverty. The unhygienic living and premises in underprivileged conditions with a lack of personal hygiene made them socially vulnerable and distressed quality of life, which diminished their social well-being and health security and subsequently increased health risks and health shock incidence. Also, less nutritious food intake, high levels of drugs (bidis, gutka, pan masala, etc.), and alcohol consumption increase the burden of family expenditure and health outcomes. Waterlogging conditions, lack of awareness of personal health and their living environment, etc. increase the disease's prevalence. The study finds a strong correlation between personal hygiene and health shock incidents. Therefore, colds and fevers are more common among slum dwellers, and disease prevalence is most common among children and adolescents. Mass health awareness and personal hygiene are deficient in slum neighborhoods, for example, women are less aware of their menstruation care.

Health infrastructure and the provision of the healthcare system in this context are most effective in controlling disease progression, morbidity, and mortality. However, the finding is that the community healthcare center is absent in the study area, which is incompatible with India's healthcare system norms. PHC and SHC are not sufficient for the study area, as per NUHM. Moreover, there is a severe lack of health workers who can conduct continuous health awareness camps and provide healthcare equipment. For example, a proportion of slum dwellers were satisfied with primary care and public facilities due to a lack of knowledge about healthcare centers and less frequent service provision; however, there was a relatively higher level of satisfaction with the services by quack doctors and government hospitals. It is also remarkable that the provision of healthcare and disease prevention equipment is not satisfactory, as reported by the slum dwellers. Therefore, the utilization and accessibility of healthcare services are incubators for the slum dwellers in the study area. Slum women mostly preferred government hospitals for their antenatal and delivery care. However, surprisingly, they preferred both public clinics and government hospitals for their postnatal care and sterilization due to the low cost of standard facilities. During minor health problems, they mostly preferred quack doctors, but for major issues, they used government healthcare facilities. Child vaccination and early childhood care have been taken at government hospitals. At the same time, only the polio vaccine is more common in the study area. Half of the total slum households did not have health insurance cards such as "Swasthya Sathi" or "Ayushman Bharat" until the end of 2019, meaning there was a severe lack of awareness, information, and services. Thus, proper allocation of healthcare centers and services is needed for the urban slum dwellers.

In these circumstances, the urban poor have been adopting different strategies to cope with the vulnerable situation in their daily livelihoods. The major livelihood survival strategies they adopted during the health shocks and throughout the year are finding jobs for all family members, resorting to acquaintances, taking credit loans, being aware regarding personal hygiene and quality of life, etc., because of experience with health shocks seasonally due to detrimental housing condition and pollution (indoor and outside). However, coping strategies throughout the year are more prevalent than the health shock coping strategies among slum dwellers, and there is a strong negative correlation between health shock and whole-year coping strategies. However, an employed model estimated that credit loans, employment status, educational level, and number of acquaintances are the important controlling factors affecting health shock coping strategies. Health coping strategies among slum people vary across age groups. The study explored childhood, adolescents, and the working group of slum dwellers in the neighborhoods who adopted moderate to low levels of health coping strategies, while the aged group of slum dwellers comparatively adopted moderate to high levels of health coping mechanisms. Still, 40% of the households have not yet adopted any strategies, and more than half of households have faced challenges in adopting health coping strategies due to the high expenses of healthcare services as well as less accessibility to those services; therefore, households are visited foregone medical healthcare during chronic illness preferably at government hospital with a moderate level of satisfaction. Hence, the possibilities for improving the environment through individual action are limited.

This recent outbreak of slum neighborhoods brought much requisite attention to researchers who can help to provide science-based structural mitigation. The present study refers to the geospatial analytical framework that has been able to distinguish between geospatial techniques and the application of spatial statistical tools for assessing livelihood strategies of slum neighborhoods of the city. The findings of this study suggest that the environmental and socioeconomic vulnerability of the slum neighborhoods is of high risk for their health conditions, and results of the prevalence of fever, gastro, diarrhea, skin, urine infection, kidney diseases, etc. are expected. The results demonstrate that the social and economic disparities among slum neighborhoods indicate the population's survival strategies and health shock outcomes. The study also finds that significant social differences exist among the slum neighborhoods, exclusively those living in detrimental environmental settings and with diverse religions. The framework of this study reflects healthcare planning for Midnapore municipality of West Bengal. Using a GIS-based approach, the adopted methodology evaluates existing healthcare facilities and improves equitable healthcare facilities for future demand. Geographic barriers should be reduced to discover new innovative technologies such as telehealth and information and communication technology (ICT). This study supports measuring geographic accessibility, infrastructure development, road network, and environmental development and recommends a potential optimal location for developing new HCF in the deprived health services area. The study is based on primary survey data, which can be periodically reviewed to monitor changes and measure the impacts of urban programs and policies. Moreover, the following study is the indication of awareness to the public as well as policymakers, government, researchers, and nongovernmental organizations for resource allocation, potential impacts, recovery, and risk reduction strategies, thereby reducing the health risk of households in the slum neighborhoods of any small and large cities in the light of integrated health management planning. However, this study was limited by the absence of attribute data and statistics, for example, the number of beds, physicians, and nurses per healthcare center, quality of services with available facilities, people's satisfaction level, and other criteria to be painstaking for future research. Therefore, more refinement of the data structure with the deployment of more spatially significant variables in compliance with advanced geospatial techniques (Health GIS) may be appended high exactitude to prototype research as well as provide meaningful insights in addressing the nature of health coping mechanisms among the community over a spatial frame.

7.2 Policy Implications and Recommendations

The descriptive analysis presented in the chapters reflects the urban programs and policies implemented in Midnapore municipality, which have significantly shaped the distribution and development of notified slum neighborhoods and their living circumstances. After analyzing the overall livelihood system of the slum neighborhoods, the study explored multiple problems associated with health coping mechanisms that impose the implementation of short-term and long-term solutions and actions for slum upgrading and equality in urban morphology. Therefore, required changes should be reflected in policies and laws, the government's attitude, the general public and service providers, NGOs, different organizations, and the community. Based on the empirical findings and to meet the smart city norm, the following recommendations may be implemented:

7.2.1 Affordable Housing through Architectural Development

The provision of affordable housing is the primary concern among the slums in India. Unplanned housing with material conditions played an important role in congestion, indoor pollution, and mental and physical health problems, subsequently affecting their income generation. Still, 20% of houses in the study area are characterized by semi-pucca. Therefore, the following recommendation should be made in favor of the slum neighborhood in the study area:

- Strengthening the "Indira Awas Yojana."
- Collective housing may be provisioned in congested areas, where the spacing between buildings should be less than 1 m.
- Housing construction should be made according to the family size; for example, a large family size requires two- and three-floor buildings with the help of architecture planning (Fig. 7.1), thereby managing the space problems or limitations.

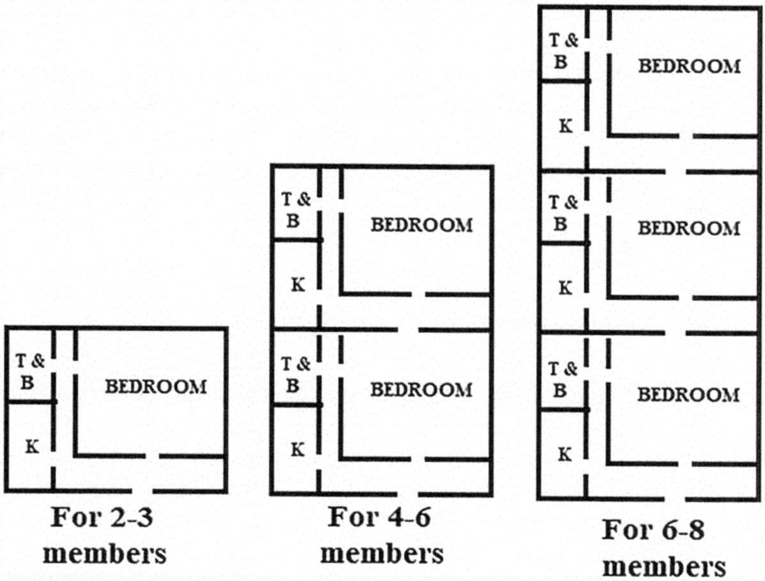

Fig. 7.1 Proposed architectural design of affordable houses for the slum neighbourhoods (T & B = Toilet and bathroom, K = Kitchen)

- Relocate the household living across the drainage to vacant land with housing construction within or on the periphery of the town by proper planning.

7.2.2 Policy for Migrated and Illegal Residence on Vacant Land

As usual, slum neighborhoods are in congested areas that may subsequently become more congested due to migration from rural areas. Therefore, a slum is the worst for the standard of living, which may be overcome by overlooking migration policies and living in urban areas on vacant lands. A strict policy should be implemented so that illegal migration can be restricted, and, in this context, GIS DBMS can be very helpful in managing the space continuum.

7.2.3 Tenure Security and Property Rights

Most of the slum dwellers do not have land ownership or property rights. They are living on the government's vacant land, lacking basic services. Therefore, it is recommended that land ownership and property rights should be given prime priority so that slum dwellers can access government facilities with basic services as well as loan facilities that can improve their livelihood system and well-being. Property rights are also recommended to the owner; they cannot sell anyone further. For example, sometimes slum dwellers sell their half property or land for money and live in a limited space. Although the government of West Bengal has been initiated to provide tenure security to the urban poor, that should be strengthened with proper administration.

7.2.4 Housing Improvement/Renovation

The decrepit or old houses in the slum area may increase injuries, dampness, and a breeding ground for microbes, which can increase disease prevalence. Therefore, a particular policy should be imposed to renovate the old houses at an affordable cost or sometimes free of cost for the better health and well-being of the families living in the slums of Midnapore municipality. It is also recommended that new houses be constructed for the decrepit dwellings.

7.2.5 Increasing the Field of Employment and the Job Market for the Vulnerable Households

A "smart card" should be allotted to poor and vulnerable households, which can help the authorities find jobless households and put them into the job market.

- Provision may be designed, such as informal sector to formal sector transformation in the premises of their living place; in this context, local authorities and architecture department sought to be most effective.
- Job market-related information should be provided to the household through mobile services, social media, and every slum neighborhood through the digital board.
- Promote SHG (self-help group), a small business with subsidies, and a place for business activity.
- Low-grade municipal or government services should be preserved for the slum people as enacted by law of provision by local authorities.
- Special skill development training should be provided for the slum young to better develop in the job market.
- Special provisions should be reflected in the policies of local authorities for the most vulnerable poor people living in slum neighborhoods.

7.2.6 *Expanding Microfinance and Credit Systems (Loan Facilities)*

Microfinance and credit systems are the most effective poverty reduction tools with inclusive growth for poor people. Local government or authorities should provide proper loan facilities to the urban poor associated with SHGs, small-scale businesses, small-scale industries, building renovation, etc. and collaborate with the national and local bank, Samabay Samity (collective groups), during the needs of finance in the emergency or sometimes in health shocks. This provision should be reflected in the community development programs.

7.2.7 *Participatory Development of Infrastructures*

Lack of infrastructure such as drainage, toilets with bathrooms, roads, community hall, building maintenance and renovation, electricity, sewerage system, etc. are the common features among the slums in India. According to UN-Habitat (Fig. 7.2), Infrastructure development through the participation of slum communities is one of the effective upgrading processes in these circumstances. This upgrading process would generate job opportunities for the slum dwellers and minimize the developmental cost.

Another recommendation is the involvement of only slum dwellers in the maintenance (such as solid waste management, cleanness of drainage, repairing of roads,

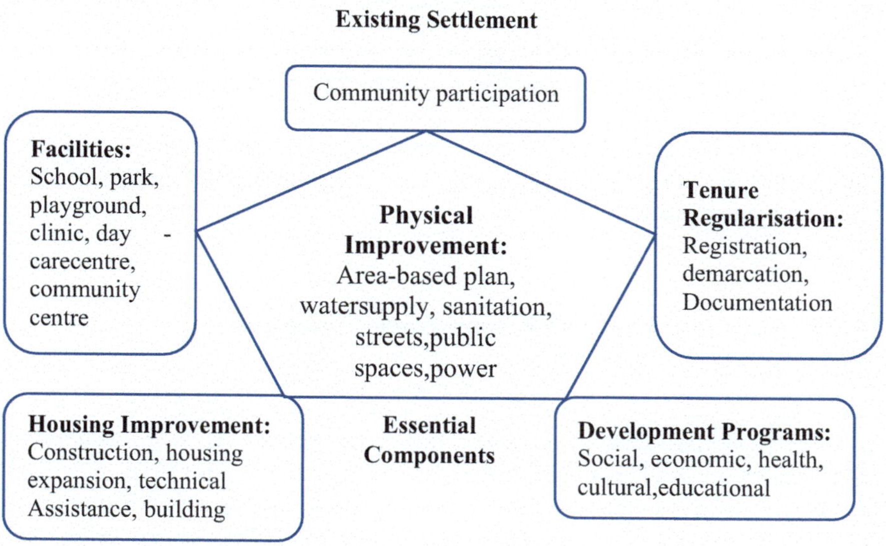

Fig. 7.2 Framework of participatory slum upgrading (adopted from UN-Habitat, 2012)

awareness camps, health-related facilities, etc.) of living premises and community development. This provision makes them aware of their community, which is guided and funded by local authorities, NGOs, researchers, etc.

7.2.8 Solid Waste Management and Personal Hygiene

Solid waste management is not so satisfactory among the slums in India. The study reveals that solid waste remains scattered and collected infrequently. This situation may be caused by a lack of awareness among the dwellers and local authorities. This condition may lead to the production of different microbes, which may severely affect children, adolescents, and people of a certain age, and the result is the disease prevalence in the study area. Therefore, more attention should be given to waste management through a participatory approach as more slum dwellers may be engaged in this context.

7.2.9 Women's Empowerment

The position of women in the slum area is worse than in the non-slum urban area. Empowering women in the slum area can change the family and social structure. Although the government of West Bengal has initiated welfare programs such as "Lakshmir Bhandar," nonetheless, the work participation rate is low among the women living in the slum. Therefore, a definite percentage of the seats should be reserved for women, which must be reflected in the policy of the local authorities.

- Strengthening the SHGs with the participation of women.
- Job (formal or informal) provisions should favor widows and vulnerable households.
- Women should also be engaged in community development programs and government representatives for healthcare facilities such as ASHA, household information collectors, etc.

7.2.10 Promote Free and Compulsory Education up to the 12th Class

Education is the backbone of a society. It reduces superstition, helps to develop mental health, and makes it suitable for participation in the job market. Free and compulsory education up to the 12th standard is strictly provisioned for slum children with laws and regulations. This may help to think about better well-being, maintain personal hygiene and aspiration for a job, develop the ability to think positively, make the right decision, and change cultural traits and societal structure. For vulnerable households, dresses, books, pens, etc. should be provided by the local authorities, NGOs, and urban affluent. Special schools for poor children in urban areas should be provided if needed.

7.2.11 Savings with Pension Schemes

Savings can be used during actual needs and emergencies or health shocks. Savings with pension schemes may be adequate for the urban poor as they can save a small amount of income rather than extravagance with future securities at the old stage. In this regard, pension schemes with life insurance should be provisioned for every urban poor through a small amount of deposits. Regular savings would be good habits for the urban poor, benefiting their needs in an emergency.

7.2.12 Healthcare Facility Center with Affordable Healthcare Services

Primary healthcare facility centers are essential to the healthcare system, as they prevent disease progression and provide affordable services to people of different age groups. The study's findings reveal that primary healthcare centers are deficient in some parts of the town, and due to the lack of health centers, they travel more than 1 km to the nearest healthcare center. Moreover, existing healthcare centers provide services 1–2 days a week, sometimes a month, with a working duration of 4–5 h and a lack of doctors and nurses. Therefore, it is recommended to establish a new healthcare center as proposed by the study and strengthen healthcare personnel in the existing healthcare center with 30 days × 12 h services to provide affordable healthcare services and reduce the disease progression as well as healthcare costs among the urban poor in the study area. Frequent health checkups of women and children may also prevent mortality and morbidity rates among poor households.

7.2.13 Promote the Use of Modern Contraception and Limit Population Growth

The study finds a severe lack of contraception care and family planning among married couples, which may account for the population growth in the slum areas of Midnapore town. In this context, local authorities should promote the use of the methods of modern contraception and birth control by providing contraceptives to poor married couples along with awareness about family, which can subsequently limit the steady population growth, thereby reducing the congestion and room density in the slum areas.

7.2.14 Provide Insurance Facilities

Public and private entrepreneurs provided compensation through insurance for property damage, healthcare costs, life loss, and return benefits, with savings for the stakeholders. For the urban poor, the following insurance facility should be recommended:

- Provision of property insurance scheme for building damage and renovation with low-cost EMI.
- Strengthening the health insurance scheme, for example, “Swasthya Sathi” or “Ayushman Bharat,” provided by the government of West Bengal and India, respectively.
- Life insurance for accidental or health shock-related deaths with minimum cost or deposits.
- A student credit card should be provided for higher education; for example, the government of West Bengal has launched a “student credit card,” which has to be strengthened as quickly as possible.

7.2.15 Proper Distribution System

A lack of knowledge, education, awareness, and accessibility to resources and services characterizes a slum. Therefore, a proper distribution system should be implemented by the local authorities and NGOs in the urban poor areas, such as services related to government facilities, health insurance and registration of other facilities, maintenance of the existing facilities, disease prevention equipment, and other digital facilities at home. In this context, “Duare Sarkar” (government services at home) is the most effective initiation or launch by the government of West Bengal, which solves the major problems of the households within a short duration as well as aims to access the services properly at home and information regarding basic facilities for the poor people. Therefore, it is recommended that the local government should encourage and promote “Duare Sarkar”-related programs frequently to minimize the deprivation of essential services and government facilities among the urban poor.

7.2.16 *Frequent Health Awareness Programs and Monitoring*

Health awareness programs are most beneficial in rural and urban slum areas, intending to provide awareness regarding health, health behavior, personal hygiene, education, employment, disease risk, superstition, behavioral change, etc., organized and conducted by local authorities, NGOs, institutions, and other organizations. As the slum is the hot spot of superstition, disease prevalence, unemployment, low education, low health status and well-being, malnutrition, etc., there need to be frequent awareness camps in the urban poor area regarding:

- Awareness of health checkups and maintaining personal hygiene.
- Awareness of the use of modern menstruation care, such as the use of a sanitary napkin.
- Awareness of family planning and malnutrition.
- Awareness in education attainment and job seeking.
- Awareness of changes in health behavior, mentality, and superstitious nature.
- Awareness in the early marriage of the girl.
- Awareness camps should be for solid waste management and the cleanness of the living environment.
- A counseling camp should be arranged for the adolescent to monitor the physical, mental, and behavioral changes.
- A postering campaign (display board) (Fig. 7.3) should be provisioned in each slum neighborhood to contain information regarding healthcare centers, distance

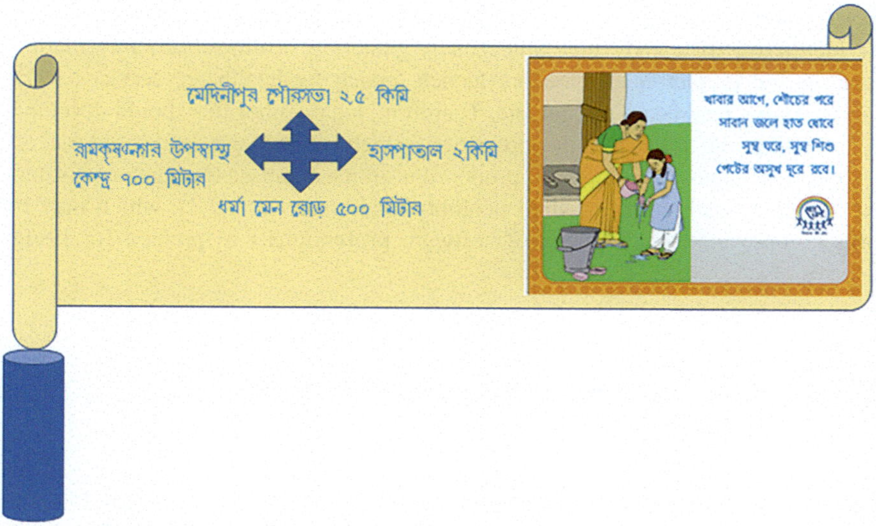

Fig. 7.3 Display board and poster campaign for information on healthcare facility centre's and others

to the nearest hospitals, government offices, and major roads, with personal hygiene-related graphics or pictures.

7.2.17 Mapping Layout with GIS Database

Mapping of slum neighborhoods with their essential services such as roads, drainage, healthcare centers, schools, etc., with collaboration with Google, may enrich slum-related information as well as identify and maintain those services with the utmost care by the local authorities. Mapping every aspect of slum neighborhoods is also helpful to the community development authorities and local bodies for future planning intervention and the researchers for further in-depth analysis such as proximity analysis, air quality analysis, etc. The study tried to map the limited services of the slum neighborhoods; nonetheless, the study recommended mapping and creating a database regarding every housing unit, source of water, location of a community hall, religious place, school, etc., which can be helpful to the local bodies to restrict the further proliferation of slum household in the slum neighborhoods and restrict the illegal construction.

7.2.18 Health GIS

Geographic information systems (GIS) are geospatial DBMS tools that can manage, gather, analyze, and visualize data through the map. Health GIS is the utilization of GIS tools for better healthcare. Health GIS may be an effective tool in the study area for identifying risk and vulnerable zones, gathering health-related information, tracking and predicting disease prevalence, managing healthcare services while improving the healthcare system, etc. Therefore, local authorities should develop a Health GIS portal with a mobile application, which can help the slum people access healthcare services. The study has proposed a simple database management system with mapping techniques of slum neighborhoods (see Sect. 7.3), which may be further improved with the help of software professionals to prepare for health security.

7.2.19 Slum Community Development Authority Board (SCDAB)

A slum community development authority board should be formulated in the municipal authority under which all planning frameworks would be implemented, especially for slum upgrading.

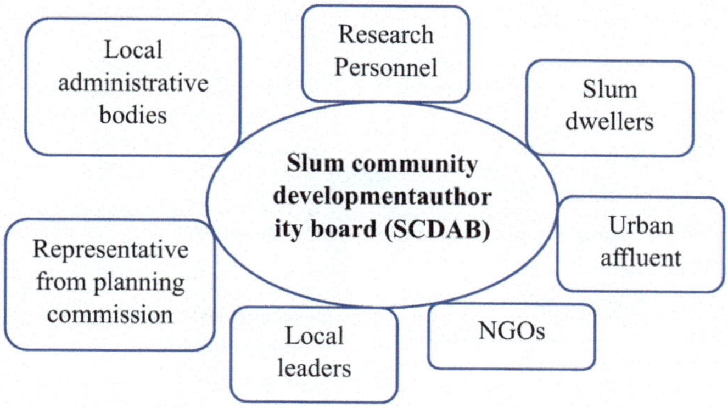

Fig. 7.4 Proposed structure of slum community development authority board (SCDAB)

The structure of the board (Fig. 7.4) would be a participatory approach by slum dwellers from each slum unit, local bodies, the urban affluent of each ward, and people's representative (political local leader or councilor). The SCDAB will discuss the existing problem and provide an affordable solution (Fig. 7.5), which the local authority will implement under the norm of a smart city.

7.2.20 Approaches for slum Upgrading

A slum is different from place to place and has distinct socioeconomic characteristics. Therefore, the planning intervention would be unique and situational for each slum worldwide. The study proposes an upgrading framework (Fig. 7.6) based on distinct problems and available resources in the socioeconomic and demographic setting. The study also designs a "SMART" approach (Fig. 7.7) that would help the local authorities, NGOs, and other administrative bodies in policymaking and intervention to upgrade the slum neighborhoods in Midnapore municipality.

Other recommendations are:

1. Implement government policies strictly without political bias, as slums are the hot spot of small-scale crime, thereby controlling their attitudes and behaviors.
2. Build local government partnerships with the World Bank and other agencies.
3. Infrastructural development by using participatory saving of the community.
4. Taxation with an insurance scheme for the essential services.

The above recommendations should be made on area-based planning or ward-wise or slum neighborhood level through intense observation by architecture, planning commission, and local authority over community participation, demand, and actual needs. At the same time, the surrounding rural areas should parallel speed up developmental programs (Mitra, 1992) to restrict migration flux in the poor slum neighborhoods and free them from the pool of poverty.

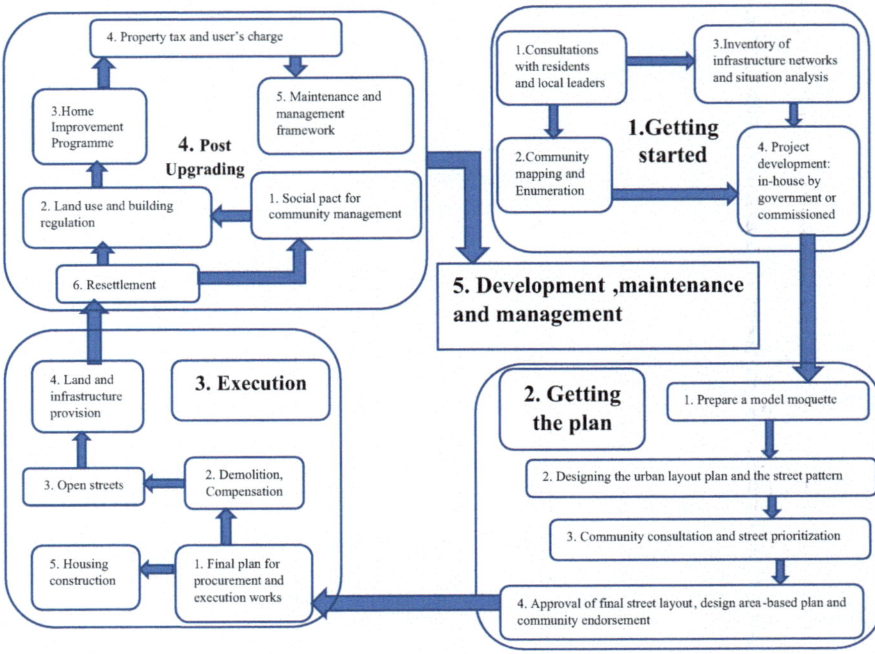

Fig. 7.5 The slum upgrading design and implementation cycle (adopted from UN-Habitat, 2012)

7.3 Managing Urban Health Using GIS

Geographic information systems (GIS) have been used in health geography to analyze spatial data of public health, which investigates the health outcomes and provision of healthcare facilities. Now, GIS enables the mapping, monitoring, and modeling of spatial data related to diseases, healthcare facilities, detection of disease outbreaks, and health security, as well as predicting risk, disease prevalence, and places for new healthcare infrastructure and targeting interventions and policies. Research on public health has been increasing daily, including daily livelihoods, health promotion activities, and community coping strategies using geospatial and statistical techniques. Public health is a growing concern in the world, and in this context, healthcare infrastructure, accessibility, services, planning, etc. are essential tasks to manage the disease progression and prevalence. To manage the healthcare system, Health GIS or geo-health with geographic information systems (GIS) technology has been the most spectacular tool widely used in developed countries since the 1990s. However, developing countries have now begun to use this spatial analytical tool in the public health sector. Health GIS is the utilization of GIS tools for better healthcare. A geographic information system (GIS) is a geospatial DBMS tool that can manage, gather, analyze, and visualize data through the map.

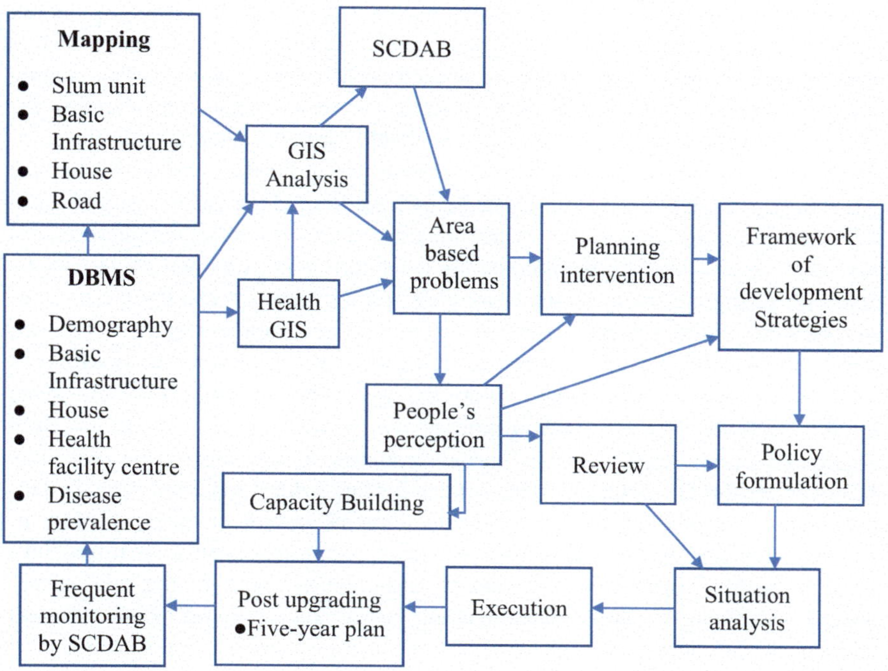

Fig. 7.6 Proposed slum upgrading framework for the study area

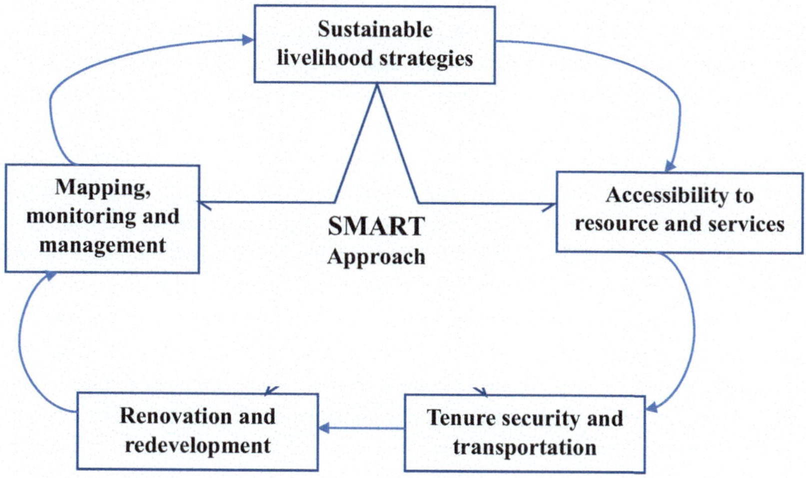

Fig. 7.7 SMART approach for slum upgrading for the study area

Health GIS integrates GIS technology across spatial scales to link population, health, and environmental data. It enables correlation analysis and spatial mapping to identify health risks and vulnerability factors. Moreover, Health GIS mapping finds disease prevention and control strategies to some extent, and thereby, the implications of those strategies are made available for better healthcare at the community level. However, the lack of GIS technology and infrastructure, inadequate skill, and unavailability of health-related data have limited the use of Health GIS in developing countries. Spatial data is essential for administrative bodies and researchers to analyze geo-health at the community level during different periods in compliance with changing health-related variables and the input of changing data over time. GIS can find the desired result of a particular geographical space. Moreover, geo-health analysis in collaboration with GIS technology answers the question of "3W" (what, where, and when) from the geographical point of view since Hippocrates in the fifth and fourth centuries and Dr. John Snow in the 1850s. Both geographers emphasized locational disease patterns on the map for the first time. Moreover, Health GIS focuses on the selection of new developmental sites (healthcare centers), the prediction or pattern of phenomena, and the planning of interventions and strategies for the healthcare system and community health. Therefore, GIS in public health informatics is a growing technological tool for the relationship between locational diseases and influencing factors and mapping disease clusters and location-based queries.

Thus, Health GIS plays a significant role in the need-based approach to answering the "3W," i.e., What are the problems, where and when to intervene, and provision of better healthcare, enhanced accessibility to the services, made solution of cost-effective delivery modes (ESRI), and provide data availability of the research community for further predictive analysis. Furthermore, Health GIS helps to allocate a healthcare center, health personnel, and equipment to the government health sector, as well as the availability details and timing of physicians, care, and services through the application in the Android mobile phone by the community. Therefore, Health GIS has become an essential tool to the local bodies, state, and central government for identifying the trends of the health status of the communities, tracking the spread of infections and chronic diseases, and improving the service quality and healthcare system and need-based solutions through the software and mobile application like as "Aarogya Setu." Health GIS can visualize the health status associated with the demography with statistical analysis and tests incorporating various socioeconomic demographic and environmental data. There are now many DBMSs with GIS integrating software or applications with minimum cost available (such as Map Pro, ArcView, etc.), and administrative authorities should select one based on the local context and available database.

Health GIS may be an effective tool in the study area for identifying risk and vulnerable zones, gathering health-related information, tracking and predicting disease prevalence, managing healthcare services while improving the healthcare system, etc. Therefore, local authorities should develop a Health GIS portal with a mobile application, which can help the slum people access healthcare services. The study has proposed a simple database management system with mapping techniques of slum neighborhoods, which may be further improved with the help of software professionals to prepare for health security. The study has analyzed different health and healthcare aspects based on GIS and proposed a simple user-based system or KIOSK that integrates the GIS spatial technology to visualize the overall health and well-being and healthcare services. A database is created and integrated with a geographic information system (GIS) to analyze health-related aspects of slum neighborhoods of Midnapore municipality. The proposed designing strategies are employed for local policymaking, data acquisition, and examining health-related information. A user-based or KIOSK system analyzes users' data across the municipality. Selective variables are taken into the Microsoft Visual Basic Application (VBA) software database management system to construct the health security index. A user (administrative or public) gets their index score on selected suitable variables for slum neighborhoods of Midnapore municipality through simple data processing. The suggested demo process (Figs. 7.8, 7.9 and 7.10) visualized health security and many other health-influencing parameters through a basic visual application. The kiosk or user-editable database page of health security shows the selected health security variables as a component of the health security index. Firstly, the input percentage data of selected variables is averaged and normalized. The normalized score is the outcome health security index score of health security linked with the GIS platform, which represents spatial disparity mapping for slum neighborhoods in health security.

a

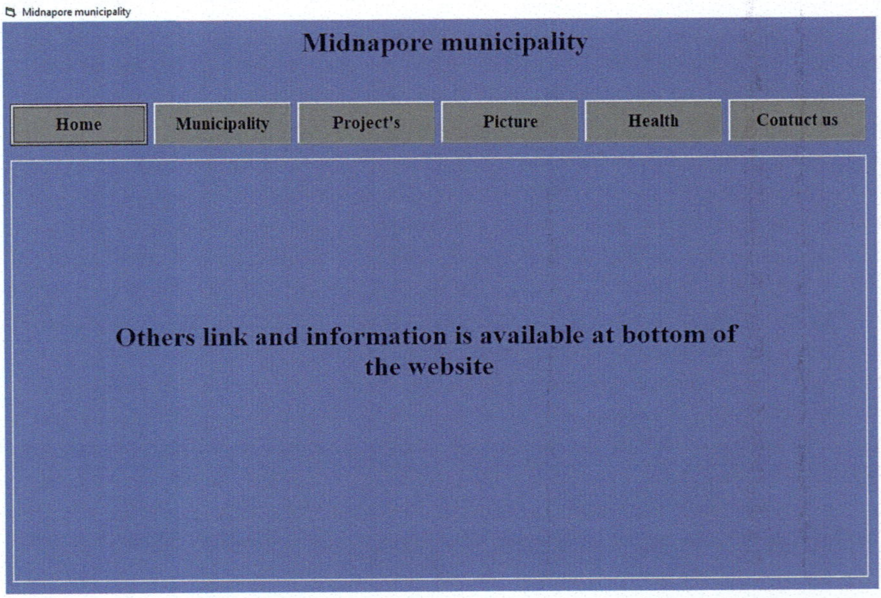

b

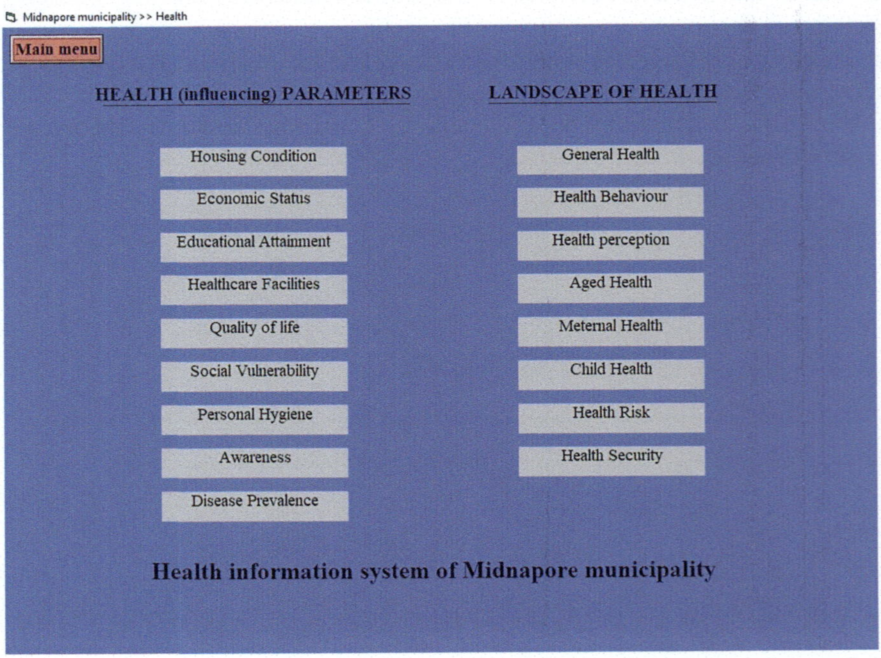

Fig. 7.8 (**a**) Health related information page (Prepared through GIS and Visual Basics Software). (**b**) Health related information page (Prepared through GIS and Visual Basics Software)

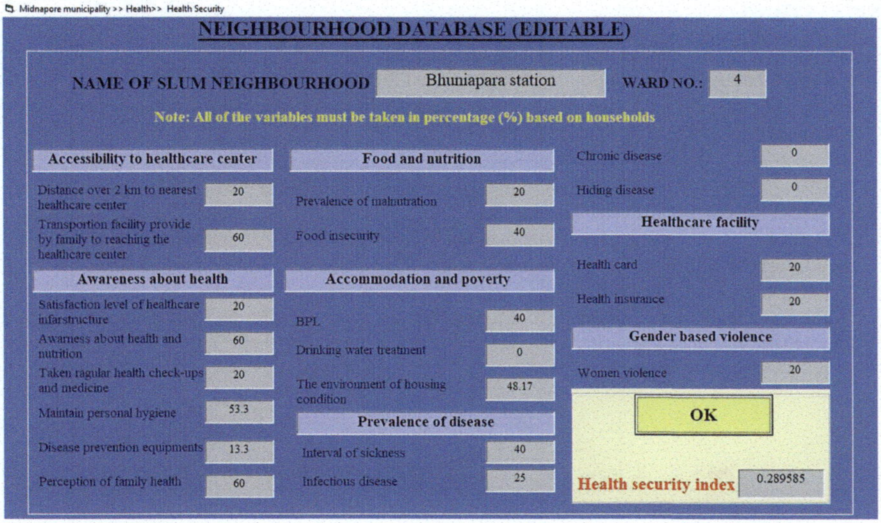

Fig. 7.9 Details information of Bhuniapara station in a single page of KIOSK (Preparedthrough GIS and Visual Basics Software)

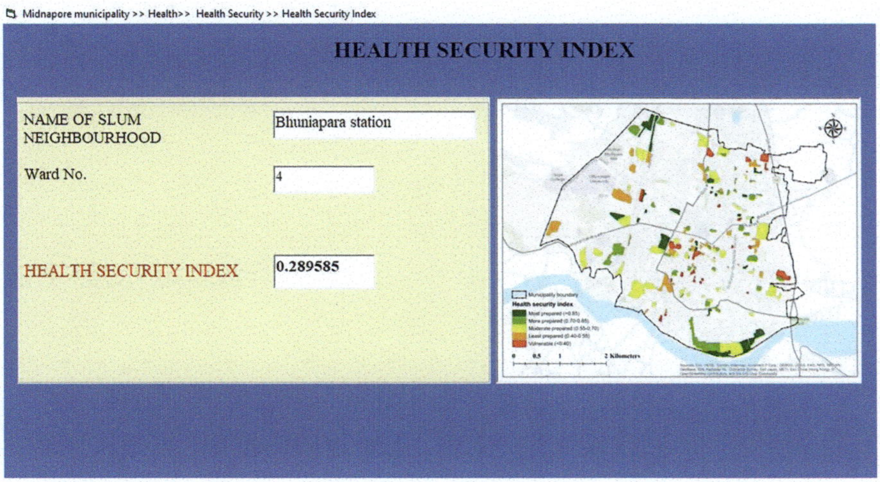

Fig. 7.10 Health security index of Bhuniapara station with selected health variables (Prepared through GIS and Visual Basics Software)

References

Census of India. http://censusindia.gov.in/; (2001 and 2011).

Dinda, S., Chaterjee, N. D., Das, K., & Ghosh, S. (2018). *Integration of GIS and statistical approach in the mapping of urban sprawl and predicting future growth in Midnapore town.* Springer.

Mitra (1992). The 1991 national survey on status of rural water supply and sanitation for DPHE/ UNICEF, Dhaka. Final Report, Bangladesh.

Ray, B. (2017). Quality of life in selected slums of Kolkata: A step forward in the era of pseudo-urbanization. *Local Environment, 22*(3), 365–387. https://doi.org/10.1080/13549839.2016.1205571

United Nations Human Settlements Programme (UN-Habitat). (2012). State of the world's cities 2012/2013: Prosperity of cities. UN-Habitat.

Index

© The Editor(s) (if applicable) and The Author(s), under exclusive license to Springer Nature Switzerland AG 2025
B. Dutta et al., *Using GIS to Identify Opportunities for Improved Health in Urban Slums*, Global Perspectives on Health Geography,
https://doi.org/10.1007/978-3-031-89617-0